distributions in statistics

distributions-2

A WILEY-INTERSCIENCE PUBLICATION

JOHN WILEY & SONS, New York · London · Sydney · Toronto

PRINTED IN THE U.S.A.
ISBN 0 471 44627 0

10 9 8 7 6 5 4 3 2

General Preface

The purpose of this compendium is to give an account of the properties and uses of statistical distributions at the present time. The idea was originally suggested by one author (Samuel Kotz) while he was visiting the University of North Carolina at Chapel Hill as a Research Associate in 1962–1963. He felt that a collection of the more important facts about the commonly used distributions in statistical theory and practice would be useful to a wide variety of scientific workers.

While preparing the book, it became apparent that "important" and "commonly" needed to be rather broadly interpreted because of the differing needs and interests of possible users. We have, therefore, included a considerable amount of detailed information, often introducing less frequently used distributions related to those other which are better-known. However, we have tried to adhere to our original intention of excluding theoretical minutiae of no apparent practical importance. We do not claim our treatment is exhaustive, but we hope we have included references to most practical items of information.

This is not an introductory textbook on general statistical theory; it is, rather, a book for persons who wish to apply statistical methods, and who have some knowledge of standard statistical techniques.

For this reason no systematic development of statistical theory is provided beyond what is necessary to make the text intelligible. We hope that the book will also prove useful as a textbook and a source book for graduate courses and seminars in statistics, and possibly for discussion or seminar groups in industrial research laboratories.

Some arbitrary choices have been unavoidable. One such choice deserves special explanation: methods of estimation for parameters occurring in the specification of distributions have been included, but tests of significance have been excluded. We felt that the form of distribution is usually well-established in situations associated with estimation, but not in cases using tests of significance. At this point we would like to emphasize the usefulness of a rough classification of distributions as "modelling" or "sampling" distributions. These classes are neither mutually exclusive nor exhaustive. However, it is true that some distributions (e.g., Wishart or correlation coefficient) are almost always encountered only as a consequence of application of sampling theory to well-established population distributions. Other distributions (e.g., Weibull or logistic) are mostly used in the construction of models of variations in real populations.

We have tried to arrange the order of discussion to produce a text which develops as naturally as possible. Although some inversions have appeared desirable in the discussion of some distributions, there are broad similarities in all the treatments and the same kind of information is provided in each case. The "historical remarks" found in the discussions are not intended to be full accounts, but are merely to help assess the part played by the relevant distribution in the development of applied statistical methods.

We have included at the end of each chapter a fairly comprehensive bibliography including papers and books mentioned in the text. This inevitably produces some repetition of titles, but is, in our opinion, important to a reader who would like a more extended treatment of a particular topic.

References are indicated by numbers in square brackets in the text e.g., [4]. References and equations are numbered separately in each chapter.

The production of a work such as the present necessarily entails the assembly of substantial amounts of information, much of which is only available from published papers and even less readily accessible sources. We would like to thank Miss Judy Allen, Miss Lorna Lansitie, Mr. K. L. Weldon, Mr. H. Spring, Mr. B. G. A. Kelly, Mr. B. Zemmel, Mr. P. Cehn, Mr. W. N. Selander and Dr. T. Sugiyama for much help in this respect.

We are particularly indebted to Professor Herman Chernoff, who read our manuscript, for valuable comments and suggestions on matters of basic importance; and to Dr. D. W. Boyd and Mr. E. E. Pickett for their assistance in compilation of the preliminary drafts of the chapters on non-central χ^2 distribution, quadratic forms, and Pareto and inverse Gaussian distributions.

We would also like to record our appreciation of Miss D. Coles, Miss V. Lewis, Mrs. G. Bem, Miss C. M. Sawyer and Miss S. Lavender for their typing work and for other technical assistance.

The assistance of our wives, especially in proofreading, cannot be overestimated.

The financial support of the U. S. Air Force Office of Scientific Research and the stimulating assistance of Mrs. Rowena Swanson of the Information Sciences Division are sincerely appreciated.

The hospitality of the Department of Statistics at Chapel Hill during the summer of 1966 and Cornell University during the summer of 1968 aided in the solution of organizational difficulties connected with this project. The help of the Librarians at the University of Toronto, the University of North Carolina at Chapel Hill, Temple University, and Cornell University is also much appreciated.

In the course of the preparation of these volumes, we were greatly assisted by existing bibliographical literature on Statistical Distributions. Among them are F. A. Haight's *Index to the Distributions of Mathematical Statistics*, J. A. Greenwood and H. O. Hartley's *Guide to Tables in Mathematical Statistics*, G. P. Patil and S. W. Joshi's *Bibliography of Classical and Contagious Discrete Distributions*, W. Buckland's *Statistical Assessment of the Life Characteristic* and L. S. Deming's *Selected Bibliographies of Statistical Literature 1930 to 1957.*

Last but not least, we would like to thank the authors from many parts of the globe, too numerous to be mentioned individually, who so generously supplied us with the reprints of their publications and thus facilitated the compilation of these volumes.

Norman L. Johnson; Samuel Kotz

Preface

The preface to "Continuous Univariate Distributions–1" applies also to present volume. We would like, here, to note that the reader may find Chapter 29 rather more difficult to read than the other chapters. However, the results are reasonably simple to use. We would like, once again, to note that Chapter 33 contains some information on Mills' ratio, which may be regarded as an extension of Chapter 13.

Chapters 26–32 contain distributions which arise in the theory of random samples from normal populations, being distributions of statistics calculated from such samples. These distributions can, however, be used independently; it is of interest to attain an appreciation of their various properties.

Contents

22

Logistic Distribution

1. Definition and Introduction

This distribution is most simply defined in terms of its cumulative distribution function:

(1)
$$F_X(x) = [1 + \exp\{-(x - \alpha)/\beta\}]^{-1}$$
$$= \tfrac{1}{2}[1 + \tanh\{\tfrac{1}{2}(x - \alpha)/\beta\}] \qquad \text{with } \beta > 0.$$

It can be seen that (1) defines a proper cumulative distribution function with

$$\lim_{x\to-\infty} F_X(x) = 0; \qquad \lim_{x\to\infty} F_X(x) = 1.$$

The corresponding probability density function is

(2)
$$p_X(x) = \beta^{-1}[\exp\{-(x - \alpha)/\beta\}][1 + \exp\{-(x - \alpha)/\beta\}]^{-2}$$
$$= (4\beta)^{-1}\operatorname{sech}^2\{\tfrac{1}{2}(x - \alpha)/\beta\}.$$

The distribution is sometimes called the *sech-square(d) distribution.*

The function on the right-hand side of (1) has been used extensively to represent growth functions (with x representing 'time'). We will be primarily concerned with its use as a distribution function (which can, of course, include situations in which the random variable represents 'time'). It is worth noting that methods developed for fitting the logistic as a growth curve (e.g. Erkelens [13]) can also be applied to fit the *cumulative* logistic distribution.

2. Genesis

The use of the logistic function as a growth curve can be based on the differential equation

(3) $$\frac{dF}{dx} = c[F(x) - A][B - F(x)]$$

where c, A and B are constants with $c > 0$, $B > A$. In verbal form, (3) can be interpreted as: rate of growth = [excess over initial (asymptotic) value A] × [deficiency compared with final (asymptotic) value B].

Solution of (3) leads to

(4) $$F(x) = \frac{BDe^{x/c} + A}{De^{x/c} + 1},$$

where D is a constant. As $x \to -\infty$, $F(x) \to A$; as $x \to \infty$, $F(x) \to B$ (if $D \neq 0$). The function $F(x)$ represents 'growth' from a lower asymptote A to an upper asymptote B. In order to make $F(x)$ a proper cumulative distribution function, we put $A = 0$, $B = 1$; equation (4) then becomes

(5) $$F(x) = \frac{De^{x/c}}{1 + De^{x/c}} = [1 + D^{-1}e^{-x/c}]^{-1}$$

which is of form (1) with $c = \beta$; $D = e^{-\alpha/\beta}$.

Equation (3) has been used as a model of autocatalysis. This is the name applied to a chemical reaction in which a catalyst M transforms a compound G into two compounds J and K and J itself acts as a catalyst for the same reaction. If M_0, G_0 = original concentrations of M, G respectively and y = common value of concentrations of J and K at time t, then the law of mass action, in this case, is

(6) $$dy/dt = c_1M_0(G_0 - y) + c_2y(G_0 - y)$$

(c_1 and c_2 are 'catalytic constants' for the actions of E, J respectively).

The right-hand side of (6) can be rearranged to read

(6)′ $$c_2\left(y + \frac{c_1}{c_2}M_0\right)\left[\left(G_0 + \frac{c_1}{c_2}M_0\right) - \left(y + \frac{c_1}{c_2}M_0\right)\right]$$

which is the same form as (3) with F, x replaced by $\left(y + \frac{c_1}{c_2}M_0\right)$, t respectively and with $c = c_2$, $A = 0$; $B = G_0 + \frac{c_1}{c_2}M_0$.

The logistic distribution arises in a purely statistical manner as the limiting distribution (as $n \to \infty$) of the standardized mid-range (average of largest and smallest sample values) of random samples of size n. This result was given by Gumbel [20]. Gumbel and Keeney [22] (see also Gumbel and Pickands [23]) showed that a logistic distribution is obtained as the limiting distribution of an appropriate multiple of the 'extremal quotient', that is (largest value)/(smallest value). (See Chapter 21, Section 8.)

Talacko [47] has shown that the logistic is the limiting distribution (as $r \to \infty$)

of the standardized variable corresponding to $\sum_{j=1}^{r} j^{-1}X_j$ where the X_j's are independent random variables each having a type 1 extreme value distribution (see Chapter 21, Equation (1)).

Dubey [11] has shown that the logistic distribution can be obtained as a mixture of extreme value distributions

$$\Pr[X \leq x \mid \eta] = 1 - \exp[-\eta\beta \exp\{-(x - \alpha)/\beta\}] \quad (x > 0; \eta, \beta > 0)$$

(obtained by putting $\theta = \alpha + \beta \log(\eta\beta)$ in Equation (1) of Chapter 21), with η having an exponential distribution with density function

$$p_\eta(y) = \beta e^{-\beta y} \qquad (y > 0).$$

For then

$$\begin{aligned}\Pr[X \leq x] &= 1 - \beta \int_0^\infty \exp[-\beta y\{1 + \exp(-(x - \alpha)/\beta\}]\, dy \\ &= [1 + \exp\{-(x - \alpha)/\beta\}]^{-1}\end{aligned}$$

which is identical with (1).

3. Historical Remarks

An early reference to the use of the logistic as a growth curve is in a paper by Verhulst [51]. The use of the curve for economic demographic purposes has been very popular from the end of the nineteenth century onwards. Some interesting examples are to be found in Fisk [18], Pearl [35], and Perks [36].

4. Generating Functions and Moments

Making the transformation $Y = (X - \alpha)/\beta$, we obtain, from (2), the probability density function of Y:

$$p_Y(y) = e^{-y}(1 + e^{-y})^{-2} = \tfrac{1}{4} \operatorname{sech}^2 \tfrac{1}{2}y \tag{7}$$

The cumulative distribution function of Y is

$$F_Y(y) = (1 + e^{-y})^{-1}. \tag{8}$$

Equations (7) and (8) are standard forms for the logistic distribution. (They are not the *only* standard forms. Equations (12) and (13), page 5, expressing the distribution in terms of mean and standard deviation, can also be regarded as 'standard'.)

The moment generating function of the random variable Y with probability density function (7) is

$$(9) \qquad E(e^{\theta Y}) = M_Y(\theta) = \int_{-\infty}^{\infty} e^{-(1-\theta)y}(1 + e^{-y})^{-2}\,dy$$
$$= \int_0^1 \xi^{-\theta}(1 - \xi)^{\theta}\,d\xi \qquad \text{(with } \xi = (e^y + 1)^{-1})$$
$$= B(1 - \theta, 1 + \theta)$$
$$= \pi\theta \operatorname{cosec} \pi\theta.$$

The characteristic function, $E(e^{itY})$, is $\pi t \operatorname{cosech} \pi t$. The moments of Y may be determined from (9), or by direct integration from (7). Using the latter method (with $r > 0$)

$$(10) \qquad E(|Y|^r) = 2\int_0^{\infty} y^r e^{-y}(1 + e^{-y})^{-2}\,dy$$
$$= 2\int_0^{\infty} y^r \sum_{j=1}^{\infty} (-1)^{j-1} j e^{-jy}\,dy$$
$$= 2\Gamma(r + 1) \sum_{j=1}^{\infty} (-1)^{j-1} j^{-r} \qquad (\text{for } r > 0)$$
$$= 2\Gamma(r + 1)(1 - 2^{-(r-1)})\zeta(r) \qquad (\text{for } r > 1),$$

where $\zeta(r) = \sum_{j=1}^{\infty} j^{-r}$ is the Riemann zeta function (see Chapter 1).

The cumulants are (for r even) $\kappa_r(Y) = 6(2^r - 1)B_r$ where B_r is the rth Bernoulli number. If r is odd, $\kappa_r(Y) = 0$.

The distribution of Y is symmetrical about $y = 0$. Putting $r = 2, 4$ in (10)

$$\operatorname{var}(Y) = E(Y^2) = 2 \cdot 2(1 - 2^{-1})(\pi^2/6) = \pi^2/3$$
$$\mu_4(Y) = 2 \cdot 24(1 - 2^{-3})(\pi^4/90) = 7\pi^4/15.$$

The first two moment-ratios of the distribution are

$$\sqrt{\beta_1} = \alpha_3 = 0; \quad \beta_2 = \alpha_4 = 4.2.$$

The mean deviation is $2\sum_{j=1}^{\infty} (-1)^{j-1} j^{-1} = 2 \log_e 2$. Hence, for the logistic distribution

$$\frac{\text{mean deviation}}{\text{standard deviation}} = \frac{2\sqrt{3}\log_e 2}{\pi} = 0.764.$$

Returning to the original form of the distribution (1), and remembering that $X = \alpha + \beta Y$, we see that

$$(11) \qquad E(X) = \alpha; \qquad \operatorname{var}(X) = \beta^2\pi^2/3.$$

The moment-ratios (and the ratio of mean deviation to standard deviation) are, of course, the same for X as for Y. The cumulative distribution function of X can be expressed in terms of $E(X) = \xi$ and $\text{var}(X) = \sigma^2$ in the standard form:

$$F_X(x) = [1 + \exp\{-\pi(x - \xi)/(\sigma\sqrt{3})\}]^{-1}. \tag{12}$$

The corresponding probability density function is

$$p_X(x) = \frac{\pi}{\sigma\sqrt{3}}[\exp\{-\pi(x - \xi)/(\sigma\sqrt{3})] \cdot [1 + \exp\{-\pi(x - \xi)/(\sigma\sqrt{3})\}]^{-2}. \tag{13}$$

The information generating function ($(u - 1)$-th frequency moment) corresponding to the probability density function (7) is

$$\begin{aligned} T_Y(u) &= \int_{-\infty}^{\infty} e^{-uy}(1 - e^{-y})^{-2u}\, dy \\ &= \int_{-\infty}^{\infty} [e^y/(1 + e^y)]^u(1 + e^y)^{-u}\, dy \\ &= \int_0^1 \xi^{u-1}(1 - \xi)^{u-1}\, d\xi \\ &= B(u,u) \\ &= [\Gamma(u)]^2/\Gamma(2u). \end{aligned} \tag{14}$$

The entropy is:

$$\begin{aligned} -T_Y'(1) &= -2\Gamma(1)\Gamma'(1)/\Gamma(2) + 2[\Gamma(1)]^2\Gamma'(2)/\Gamma(2) \\ &= 2(\psi(2) - \psi(1)) \\ &= 2. \end{aligned}$$

5. Properties

Gumbel [21] noted the properties

$$p_Y(y) = F_Y(y)[1 - F_Y(y)] \tag{15}$$

$$y = \log_e [F_Y(y)/\{1 - F_Y(y)\}] \tag{16}$$

for $p_Y(y)$, $F_Y(y)$ defined as in (7) and (8).

The simple explicit relationships between y, $p_Y(y)$ and $F_Y(y)$ render much of the analysis of the logistic distribution attractively simple. The further fact that the logistic distribution has a shape similar to that of the normal distribution makes it profitable, on suitable occasions, to replace the normal by the logistic to simplify the analysis without too great discrepancies in the theory. Such substitution must be done with care, and understanding of the similarities between the two distributions.

If the cumulative distribution functions $G_1(x) = \frac{1}{\sqrt{2\pi}} \int_{-\infty}^{x} e^{-\frac{1}{2}u^2}\, du$ and $G_2(x) = [1 + \exp(-\pi x/\sqrt{3})]^{-1}$ of the standardized normal and logistic distributions are compared, the differences $G_2(x) - G_1(x)$ vary in the way shown in Figure 1. (Since both $G_1(x)$ and $G_2(x)$ are symmetric about $x = 0$, only the values for $x \geq 0$ are shown.) It can be seen that the maximum value of $G_1(x) - G_2(x)$ is about 0.0228, attained when $x = 0.7$. This maximum may be reduced to a value less than 0.01 by changing the scale of x in G_1 and using $G_1(16x/15)$ as an approximation to $G_2(x)$. This, also, is exhibited graphically in Figure 1.

Birnbaum and Dudman [8] devoted considerable attention to comparison of distributions of order statistics from normal and logistic distributions. Plackett [38] compared coefficients in best linear unbiased estimators (using order statistics) and Gupta and Shah [25] compared the distribution of range in samples of sizes 2 and 3 from the two distributions.

It should be noted that although there is a close similarity in shape between the normal and logistic distributions, the value of β_2 for logistic is 4.2, considerably different from the value ($\beta_2 = 3$) for the normal distribution. The difference may be attributed largely to the relatively longer 'tails' of the logistic distribution. These can have a considerable effect on the fourth central moment, but a much smaller relative effect on the cumulative distribution function. (We may also note that whereas the standard normal curve has points of inflexion at $x = \pm 1$, those of the logistic are $x = \pm(\sqrt{3}/\pi) \log_e (2 + \sqrt{3}) = \pm 0.53$.)

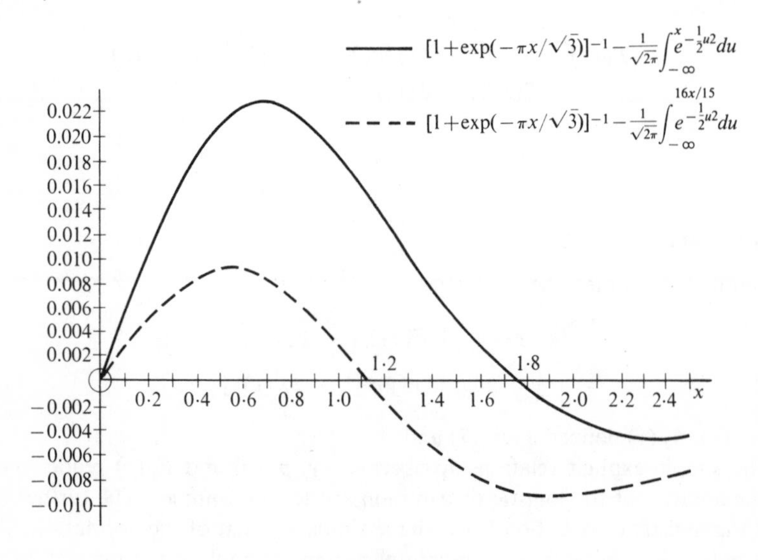

FIGURE 1

Comparison of Logistic and Normal Cumulative Distribution Functions

6. Estimation

The maximum likelihood estimators, $\hat{\xi}$, $\hat{\sigma}$ of the parameters ξ, σ in (13) based on a mutually independent set of random variables $X_1, X_2, \ldots, X_n$, each having this distribution, satisfy the equations

$$n^{-1} \sum_{i=1}^{n} [1 + \exp\{\pi(X_i - \hat{\xi})/\hat{\sigma}\sqrt{3})\}]^{-1} = \tfrac{1}{2} \tag{17.1}$$

$$n^{-1} \sum_{i=1}^{n} \left(\frac{X_i - \hat{\xi}}{\hat{\sigma}}\right) \frac{1 - \exp\{\pi(X_i - \hat{\xi})/(\hat{\sigma}\sqrt{3})\}}{1 + \exp\{\pi(X_i - \hat{\xi})/(\hat{\sigma}\sqrt{3})\}} = \frac{\sqrt{3}}{\pi}. \tag{17.2}$$

For large n

$$n \operatorname{var}(\hat{\xi}) \doteqdot (9/\pi^2)\sigma^2 \doteqdot 0.91189\sigma^2 \tag{18.1}$$

$$n \operatorname{var}(\hat{\sigma}) \doteqdot \{9/(3 + \pi^2)\}\sigma^2 \doteqdot 0.69932\sigma^2. \tag{18.2}$$

Equations (17) must be solved by trial and error.

Taking advantage of the similarity in shape between the logistic and normal distributions, initial values of $\hat{\xi}$ and $\hat{\sigma}$ might be taken as the maximum likelihood estimators

$$\overline{X} = n^{-1} \sum_{i=1}^{n} X_i, \qquad \text{and} \qquad \sqrt{n^{-1} \sum_{i=1}^{n} (X_i - \overline{X})^2},$$

respectively, appropriate to the latter distribution. Improvements could then be made, using (17.1) and (17.2), by applying, for example, the Newton-Raphson method.

Approximate linearization of the maximum likelihood equations (17.1) and (17.2) has been effected by Plackett [38]. He gave coefficients in estimators $\hat{\xi}'$, $\hat{\sigma}'$, which are quite similar to those of the best linear unbiased estimators for parameters of normal distributions, even when the sample size is no greater than 10. Another method of approximate linearization has been proposed by Tiku [50]. Fisk [17] has described the maximum likelihood method of estimation based on grouped or truncated data.

The parameters ξ, σ may also be estimated by the sample mean and standard deviation, m_1', $\sqrt{m_2}$ respectively. The asymptotic efficiency of m_1' is 91.2%; that of $\sqrt{m_2}$ is 87.4%. Gupta *et al.* [24] show that the actual efficiency of m_1', as an estimator of ξ, and of $\sqrt{m_2}$, as an estimator of σ, is greater than the asymptotic efficiency when the sample size is small. These estimators are, however, less efficient (about 10% less for m_1', considerably more so for $\sqrt{m_2}$) than the appropriate best linear unbiased estimators (see Section 8).

A number of methods of fitting parameters of logistic curves were developed in connection with their use as growth curves. Descriptions of such methods can be found in Erkelens [13], Oliver [33], Pearl [35], Rasor [39], Silverstone [45], and Will [52]. Many of these are of a heuristic nature, and are not based directly on probabilistic considerations, but are nevertheless useful in obtaining quick estimators of the parameters. (The fitting of the distribution is simpler than the fitting as a growth curve, since in the former case there is no need to fit values for A and B, see eq. (3).)

From (12), the expected value of

$$f_x = \frac{\text{number of } X\text{'s} \leq x}{n}$$

is $[1 + \exp\{-\pi(x - \xi)/(\sigma\sqrt{3})\}]^{-1}$. One method of fitting consists of plotting $\log[f_x/(1 - f_x)]$ against x and fitting (often by eye) a straight line

$$\log[f_x/(1 - f_x)] = \hat{a} + \hat{b}x \tag{19.1}$$

to the data so obtained. Comparison of coefficients in (19.1) and

$$\log[E(f_x)/\{1 - E(f_x)\}] = \frac{\pi}{\sigma\sqrt{3}}(-\xi + x) \tag{19.2}$$

leads to the estimators

$$\hat{\sigma} = \pi/(\hat{b}\sqrt{3}); \quad \hat{\xi} = -\hat{a}/\hat{b}.$$

Although various refinements (e.g. in the fitting of the line and in reducing bias) can be introduced, this is quite effective as a quick method. Similar methods are not as effective when a growth curve is being fitted, and values of the upper, and possibly also the lower, asymptotes have to be estimated (Oliver [33]).

When only quantal response data (i.e. proportions of observations exceeding certain specified values) are available, special methods must be used (see Section 10).

7. Order Statistics

The distributions of order statistics for random samples from a logistic distribution have a simple explicit form, deriving from the simple explicit form of the cumulative distribution function of the logistic distribution.

If $Y_1' \leq Y_2' \leq \cdots \leq Y_n'$ be the order statistics corresponding to a random sample from the distribution with probability density function (7), then

$$p_{Y_j'}(y) = \frac{\Gamma(n+1)}{\Gamma(j)\Gamma(n-j+1)} e^{-(n-j+1)y}(1 + e^{-y})^{-(n+1)} \qquad (y > 0). \tag{20}$$

The probability integral transformation

$$Y_j'' = (1 + e^{-Y_j'})^{-1} \qquad (j = 1, 2, \ldots, n)$$

produces, of course, a set of ordered rectangular variables. The percentage points of the distribution of Y_j' follow directly from those of Y_j'', since

$$\begin{aligned} \Pr[Y_j' \leq y] &= \Pr[Y_j'' \leq (1 + e^{-y})^{-1}] \\ &= I_{(1+e^{-y})^{-1}}(j, n - j + 1) \qquad (y > 0). \end{aligned} \tag{21}$$

Gupta and Shah [25] have given a table (for $n = 1(1)10$ and, for selected order statistics, $n = 11(1)25$, and percentage points 0.5, 0.75, 0.9, 0.975, and 0.99) to 4 decimal places, by which the use of Incomplete Beta function tables can be avoided. They also give the distribution of the range $(Y'_n - Y'_1) = W$. The probability density function is

$$(22) \qquad p_W(w) = \frac{\sqrt{\pi}\,\Gamma(n)}{2\sqrt{2}\,\Gamma(n+\frac{1}{2})}[1 + \cosh \tfrac{1}{2}w]^{-(n-\frac{1}{2})} \times F(\tfrac{1}{2},\tfrac{1}{2};n+\tfrac{1}{2};(1-\cosh\tfrac{1}{2}w)) \qquad (w > 0)$$

where $F(a,b;c;x) = 1 + \dfrac{ab}{c}\dfrac{x}{1!} + \dfrac{a(a+1)b(b+1)}{c(c+1)}\dfrac{x^2}{2!} + \cdots$ is the hypergeometric function.

Shah [43] obtained the joint distribution of W and the mid-range ($M = \frac{1}{2}(Y'_1 + Y'_n)$). The joint probability density function is

$$(23) \qquad p_{M,W}(m,w) = \frac{n(n-1)[\sinh \frac{1}{2}w]^{n-2}}{4[\cosh m + \sinh \frac{1}{2}w]^n} \qquad (w > 0).$$

From (20) the moment generating function of Y'_j is

$$(24) \qquad E(e^{\theta Y'_j}) = \frac{\Gamma(j+\theta)\Gamma(n-j-\theta+1)}{\Gamma(j)\Gamma(n-j+1)}.$$

The cumulant generating function is

$$\log \Gamma(j+\theta) + \log \Gamma(n-j-\theta+1) - \log \Gamma(j) - \log \Gamma(n-j+1),$$

whence the rth cumulant is

$$(25) \qquad \kappa_r(Y'_j) = \psi^{(r-1)}(j) + (-1)^r \psi^{(r-1)}(n-j+1).$$

(Note that $\kappa_r(Y'_j) = (-1)^r \kappa_r(Y'_{n-j+1})$ as is to be expected from the symmetry of the logistic distribution.)

Gupta and Shah [25] give explicit formulas for the first four moments of order statistics in samples of size $n = 1(1)10$. They also give a general formula for product moments (of any order) of two order statistics. Birnbaum and Dudman [8] give tables of numerical values of the mean and standard deviation to 5 decimal places of each order statistic, for samples of sizes $n = 1(1)10$, 15, 20, 25; (and some selected order statistics for $n = 100$) as part of their comparison of the logistic and normal distributions, already referred to in Section 5. Gupta *et al.* [24] give variances and covariances to 8 decimal places for all order statistics in samples of sizes $n = 10(5)25$.

8. Estimation Using Order Statistics

Despite the simple form of the joint probability density functions of the order statistics corresponding to a random sample from a logistic distribution, the variance-covariance matrix does not have a simple analytic form, as in the cases of the exponential and uniform distributions. As a consequence, we are very nearly in the same position in regard to construction of best linear unbiased estimators of the parameters ξ and σ, using order statistics, as for the normal distribution. It is, again, necessary to rely on numerical calculation, using tables (such as those referred to in Section 7) rather than neat analytical formulas.

Gupta *et al.* [24] show that the efficiency (relative to the Cramér-Rao lower bounds (18)) of the best linear unbiased estimator of ξ increases from about 95% for $n = 5$ to about 98% for $n = 25$; for σ the increase is from about 80% to about 90%.

There are explicit *approximate* formulas available (Gupta and Gnanadesikan [26]) for best linear unbiased estimators based on k selected order statistics $X'_{n_1}, X'_{n_2}, \ldots, X'_{n_k}$ (with $1 \leq n_1 < n_2 < \cdots < n_k \leq n$) from n independent random variables $X_1, X_2, \ldots, X_n$ each having probability density function (13). These formulas should give useful results when n is large, while n_1/n and n_k/n are not "too near" to 0 or 1 respectively. This method is based on large-sample approximations to the expectations, variances and covariances of order statistics (see, e.g., Ogawa [32]).

The first of these approximate formulas to be described here can be used to estimate ξ, σ being known. It is

$$\xi^* = \frac{\displaystyle\sum_{i=1}^{k+1}\left(1 - \frac{n_i}{n} - \frac{n_{i-1}}{n}\right) \times \left[\frac{n_i}{n}\left(1 - \frac{n_i}{n}\right)X'_{n_i} - \frac{n_{i-1}}{n}\left(1 - \frac{n_{i-1}}{n}\right)X'_{n_{i-1}}\right] - K_3\sigma}{K_1} \tag{26}$$

where

$$K_1 = \sum_{i=1}^{k+1}\left(\frac{n_i}{n} - \frac{n_{i-1}}{n}\right)\left(1 - \frac{n_i}{n} - \frac{n_{i-1}}{n}\right)^2$$

$$K_3 = \sum_{i=1}^{k+1}\left(1 - \frac{n_i}{n} - \frac{n_{i-1}}{n}\right) \times \left[\frac{n_i}{n}\left(1 - \frac{n_i}{n}\right)\log_e\left\{\frac{n_i}{n}\Big/\left(1 - \frac{n_i}{n}\right)\right\} - \frac{n_{i-1}}{n}\left(1 - \frac{n_{i-1}}{n}\right) \times \log_e\left\{\frac{n_{i-1}}{n}\Big/\left(1 - \frac{n_{i-1}}{n}\right)\right\}\right],$$

(X'_{n_0} and $X'_{n_{k+1}}$ are each defined to be zero). Note that the coefficients in ξ^* depend only on the ratios $\dfrac{n_i}{n}$ ·

The variance of ξ^* is approximately $(\sigma^2 n^{-1})K_1^{-1}$. For given k, this last quantity is minimized if $n_1, n_2, \ldots, n_k$ are chosen to maximize K_1. This is achieved by taking $n_i = ni/(k + 1)$. (In practice, of course, the nearest suitable integer value would be taken.) With these values of the n_i's we have the estimator

$$\xi^{**} = \frac{6}{k(k+1)(k+2)} \sum_{i=1}^{k} i(k+1-i)X'_{n_i} \tag{27}$$

with $\operatorname{var}(\xi^{**}) = (9/\pi^2)(\sigma^2/n)[k^{-1}(k+2)^{-1}(k+1)^2]$.

The Cramér-Rao lower bound for an unbiased estimator of ξ is $(9/\pi^2)(\sigma^2/n)$. The relative efficiency of ξ^{**} is approximately $k^{-1}(k+2)^{-1}(k+1)^2$. This increases with k, from a minimum of 75% when $k = 1$, up to 100% as k increases. (When $k = 1$, $\xi^{**} = \text{median}\,(x_1,x_2,\ldots,x_n)$.) It may be noted that (27) is the estimator obtained by Blom's [9] method. It is also the estimator obtained by multiplying Jung's [30] estimator by a constant to make it unbiased. The formula for estimating σ, ξ being known, is

$$\sigma^* = \frac{Y - \xi K_3}{K_2} \tag{28}$$

where

$$Y = \sum_{i=1}^{k+1} \left[\frac{n_i}{n}\left(1 - \frac{n_i}{n}\right) \log_e \left\{\frac{n_i}{n} \Big/ \left(1 - \frac{n_i}{n}\right)\right\} - \frac{n_{i-1}}{n}\left(1 - \frac{n_{i-1}}{n}\right) \times \log_e \left\{\frac{n_{i-1}}{n} \Big/ \left(1 - \frac{n_{i-1}}{n}\right)\right\}\right]$$
$$\times \left\{\frac{n_i}{n}\left(1 - \frac{n_i}{n}\right) X'_{n_i} - \frac{n_{i-1}}{n}\left(1 - \frac{n_{i-1}}{n}\right) X'_{n_{i-1}}\right\} \Big/ \left[\frac{n_i}{n} - \frac{n_{i-1}}{n}\right]$$

and

$$K_2 = \sum_{i=1}^{k+1} \left[\frac{\dfrac{n_i}{n}\left(1 - \dfrac{n_i}{n}\right) \log_e \left\{\dfrac{n_i}{n}\left(1 - \dfrac{n_i}{n}\right)\right\}}{\dfrac{n_i}{n} - \dfrac{n_{i-1}}{n}} - \frac{\dfrac{n_{i-1}}{n}\left(1 - \dfrac{n_{i-1}}{n}\right) \log_e \left\{\dfrac{n_{i-1}}{n}\left(1 - \dfrac{n_{i-1}}{n}\right)\right\}}{\dfrac{n_i}{n} - \dfrac{n_{i-1}}{n}} \right]^2$$

with approximate variance $(\sigma^2/n)K_2^{-1}$.

Gupta and Gnanadesikan [26] give a detailed comparison of the estimators of σ (ξ not being known) obtained by Blom's [9], [10] and Jung's [30] methods. They conclude that these estimators have high efficiencies.

(Table 1, taken from Gupta and Gnanadesikan [26] — preliminary version — gives the coefficients a_i in Jung's formula $\sum a_i(X'_{n-i+1} - X'_i)$ modified to make it an unbiased estimator of σ.)

The general problem of maximizing K_2, and so minimizing $\operatorname{var}(\sigma^*)$, is rather complex. However, if only two order statistics are to be used then $n_1 \doteqdot 0.103n$, and $n_2 \doteqdot 0.897n$ gives approximately the minimum value $1.0227\sigma^2/n$ for $\operatorname{var}(\sigma^*)$. The Cramér-Rao lower bound for the variance of an unbiased estimator is $9(3+\pi^2)^{-1}\sigma^2/n$, and so the estimator

$$\sigma^* = 0.4192[X'_{n_1} + X'_{n_2}]$$

TABLE 1

Coefficients of the $(n - i + 1)$th Order Statistic X'_{n-i+1} in the Linear Estimator of σ (by Jung's Method) Modified to Make it Unbiased.

n \ i	1	2	3	4	5	6	7	8	9	10	11	12	13	Variance σ^2
5	.3538	.2038	0											.1706
6	.2907	.2024	.0715											.1372
8	.2125	.1767	.1147	.0396										.0985
10	.1663	.1503	.1170	.0737	.0251									.0769
15	.1062	.1048	.0955	.0813	.0636	.0436	.0222	0						.0496
20	.0774	.0787	.0758	.0700	.0622	.0528	.0422	.0307	.0187	.0063				.0366
25	.0605	.0625	.0618	.0592	.0553	.0504	.0445	.0381	.0310	.0236	.0159	.0080	0	.2903

Computed by using the same approximate covariance matrix as used in Blom's method.

(with n_1 and n_2 as given above) has approximate efficiency 68.38%.

(It may be noted that to get an improved estimator of this form, *four* quantities are needed.)

If neither ξ nor σ are known, the approximate best linear estimators, obtained by similar methods, are

$$\text{(29)} \qquad \begin{aligned} \hat{\xi}^* &= \Delta^{-1}(K_2X - K_3Y); \\ \hat{\sigma}^* &= \Delta^{-1}(-K_3X + K_1Y) \end{aligned}$$

where

$$\Delta = K_1K_2 - K_3^2$$

and

$$X = \sum_{i=1}^{k+1} \left(1 - \frac{n_i}{n} - \frac{n_{i-1}}{n}\right)\left[\frac{n_i}{n}\left(1 - \frac{n_i}{n}\right) X'_{n_i} - \frac{n_{i-1}}{n}\left(1 - \frac{n_{i-1}}{n}\right) X'_{n_{i-1}}\right].$$

(K_1, K_2 and K_3 and Y have been defined in (26) and (28).)

Hassanein [29] and Simpson [46] also discuss the optimum choice of quantiles.

9. Tables

Values of $p_Y(y)$ and $F_Y(y)$ as given by (7) and (8) are included in the collection of tables by Owen [34]. Each function is given to four decimal places for

$$y = 0(0.01)1.00(0.05)3.00.$$

(Positive values only of y are needed on account of the symmetry of the distribution.) Inverse tables give (to 4 decimal places) values of y and $p_Y(y)$ for which $F_Y(y)$ takes the values

$$0.50(0.1)0.90(0.005)0.990(0.001)0.999(0.0001)0.9999$$

and some higher values up to 0.999999999.

Tables for use in logit analysis are given by Finney [14] [15]. These tables include the logit transformation ($y = \log [F/(1 - F)]$). There are similar tables (Tables XI and XI_1) in Fisher and Yates [16]. Berkson [4] gives short tables of the 'logit', $\log [p/(1 - p)]$.

For many calculations associated with the use of the logistic distribution, standard tables of hyperbolic functions (sinh, cosh, tanh, etc.) can be used with facility.

10. Applications

Some important uses of the logistic curve or distribution have already been mentioned. These include use in describing growth and as a substitute for the normal distribution. Possibly (see Berkson [2]–[6]), included in the latter is its use in the analysis of quantal response data. This type of analysis has already been described (Chapter 13) in connection with the use of the normal distribution in probit analysis. If a logistic distribution is used, in place of a normal distribution, to represent the population tolerance distribution then the analysis is carried out in terms of logits instead of probits.

The logit, Y, and the corresponding observed proportion P, are connected by the equation

$$P = (1 + e^{-Y}), \qquad \text{that is} \qquad Y = \log [P/(1 - P)].$$

Given observed proportions $P_i = D_i/n_i$ of 'deaths' at 'dosages' x_i ($i = 1,\ldots,k$), the logits $Y_i = \log [P_i/(1 - P_i)]$ are calculated. The estimation of values of the constants α, β in the equation

$$P = (1 + e^{-(\alpha+\beta x)})^{-1}$$

is to be based on the k independent binomial proportions P_i, or equivalently on the k independent logits Y_i. The maximum likelihood estimators $\hat{\alpha}$, $\hat{\beta}$ of α, β respectively, satisfy the equations

$$\sum_{i=1}^{k} n_i(P_i - \hat{P}_i) = 0 = \sum_{i=1}^{k} n_i x_i(P_i - \hat{P}_i) \tag{30}$$

where

$$\hat{P}_i = [1 + \exp\{-(\hat{\alpha} + \hat{\beta}x_i)\}]^{-1}.$$

An iterative method for solving these equations, linked with the idea of fitting a weighted regression of Y_i on x_i, can be constructed in an exactly similar way to that described for probit analysis. In one respect the calculations are simpler, as the weight per unit observation corresponding to P_i is $P_i(1 - P_i)$, which is simpler than the corresponding formula for probit analysis.

As is to be expected from the similarity in shape of the normal and logistic distributions, the results of probit and logit analysis of the same data are usually very similar. Agreement is particularly good in respect of estimates of the median of the tolerance distribution (Finney [14]).

Systems of analysis, using an assumed logistic form for residual variation, have been worked out for 2^n factorial experiments by Dyke and Patterson [12], and for the general linear hypothesis by Grizzle [19]. Multiple comparisons, using the logistic distribution, have been discussed by Reiersøl [41].

11. Related Distributions

The logistic distribution can be regarded as a member of a large class of distributions introduced by Perks [36]. Perks, a British actuary, was primarily interested in obtaining a general function for graduating life-table data, but his formulas are of general applicability.

Perks proposed, as a general form for the probability density function of a random variable Y, the ratio

$$p_Y(y) = \sum_{j=0}^{m} a_j e^{-j\theta y} \Big/ \sum_{j=0}^{m'} b_j e^{-j\theta y}. \tag{31}$$

The quantities a_j, b_j, θ are real parameters. There must be relationships among the values of these parameters to ensure that the conditions

$$p_Y(y) \geq 0 \qquad \text{and} \qquad \int_{-\infty}^{\infty} p_Y(y)\, dy = 1$$

are satisfied. It is always possible to take $\theta = 1$, by a suitable choice of scale, and evidently all a_j's and b_j's can be multiplied by the same (non-zero) constant, without affecting $p_Y(y)$.

A particularly interesting subclass of symmetrical distributions is obtained by taking $m = 1$, $m' = 2$; $a_0 = 0$, $b_0 = b_2$. Then (31) becomes

$$(32) \qquad p_Y(y) = \frac{a_1 e^{-y}}{b_0 + b_1 e^{-y} + b_0 e^{-2y}}$$

$$= \frac{c_1}{e^y + c_2 + e^{-y}} \qquad (\text{with } c_1 = a_1/b_0;\ c_2 = b_1/b).$$

Taking $c_1 > 0$, the condition $p_Y(y) \geq 0$, for all y, requires $c_2 \geq -2$. The condition $\int_{-\infty}^{\infty} p_Y(y)\, dy = 1$ excludes $c_2 = -2$, but, for all $c_2 > -2$, (32) can represent a probability density function.

The logistic distribution is obtained by putting $c_2 = 2$, giving (7).

Another distribution, studied by Talacko [47] is the *hyperbolic secant distribution*, obtained by putting $c_2 = 0$, giving

$$(33) \qquad p_Y(y) = \frac{c_1}{e^y + e^{-y}} = \frac{1}{2} c_1 \operatorname{sech} y.$$

The condition $\int_{-\infty}^{\infty} p_Y(y)\, dy = 1$ requires that $c_1 = 2/\pi$ so that

$$(34) \qquad p_Y(y) = \pi^{-1} \operatorname{sech} y$$

and the cumulative distribution function is

$$(35) \qquad F_Y(y) = \tfrac{1}{2} + \pi^{-1} \tanh^{-1} (\sinh y).$$

Note that if Y has this distribution then e^Y has a half-Cauchy distribution (Chapter 16, Section 6).

The distribution of the sum of n independent random variables, each having the same hyperbolic secant distribution, has been derived by Baten [1].

Returning for a moment to the more general form of distribution (32) the characteristic function, $E(e^{itY})$, is (Talacko [47]).

$$(36) \qquad \frac{\pi}{\cos^{-1} (\frac{1}{2}c_2)} \frac{\sinh [t \cos^{-1} (\frac{1}{2}c_2)]}{\sinh t\pi} \qquad (\text{for } -2 < c_2 < 0;\ 0 < c_2 \leq 2)$$

$$(37) \qquad \frac{\pi}{\cosh^{-1} (\frac{1}{2}c_2)} \frac{\sin [t \cosh^{-1} (\frac{1}{2}c_2)]}{\sinh t\pi} \qquad (\text{for } c_2 > 2).$$

The logistic distribution appears as a limiting case, letting $c_2 \to 2$. The values of $\cos^{-1} (\frac{1}{2}c_2)$ are taken in the range 0 to π. For the hyperbolic secant distribution, the characteristic function is sech $(\frac{1}{2}\pi t)$, and the rth absolute moment about zero is

$$\nu_r' = E(|Y|^r) = \frac{4}{\pi}\int_0^\infty y^r e^{-y}(1+e^{-2y})^{-1}\,dy \tag{38}$$

$$= \frac{4}{\pi}\int_0^\infty \sum_{j=0}^\infty (-1)^j y^r e^{-(2j+1)y}\,dy$$

$$= \frac{4}{\pi}\sum_{j=0}^\infty (-1)^j \int_0^\infty y^r e^{-(2j+1)y}\,dy$$

$$= \frac{4}{\pi}\Gamma(r+1)\sum_{j=0}^\infty (-1)^j (2j+1)^{-(r+1)}.$$

The expected value of Y equals zero; $\text{var}(Y) = \frac{4}{\pi}\cdot 2\cdot\frac{\pi^3}{32} = \frac{1}{4}\pi^2$, $\mu_4(Y) = \frac{5\pi^4}{15}$; $\beta_2(Y) = \alpha_4(Y) = 5$. The mean deviation is $\frac{4}{\pi}c$ (where $c = 0.916$ is Catalan's constant). For this distribution

$$\frac{\text{mean deviation}}{\text{standard deviation}} = \frac{8c}{\pi^2} = 0.742.$$

Harkness and Harkness [27] have investigated the properties of a class of distributions having characteristic functions

$$(\text{sech}\,\theta t)^\rho \qquad (\rho > 0; \theta > 0)$$

which they term *generalized hyperbolic secant* distributions. For integer values of ρ, the distributions are those of sums of independent identically distributed hyperbolic secant random variables. In [27] it is shown that for ρ even $(= 2n)$ the density function is

$$p_X(x) = \left[\frac{4^{n-1}x}{(2n-1)!2\theta^2}\cdot\text{cosech}\,\frac{\pi x}{2\theta}\right]\prod_{j=1}^{n-1}\left(\frac{x^2}{4\theta^2}+j^2\right) \tag{39.1}$$

while for ρ odd $(= 2n+1)$

$$p_X(x) = \left[\frac{2^{2n-1}}{(2n)!\theta}\cdot\text{sech}\,\frac{\pi x}{2\theta}\right]\prod_{j=1}^{n}\left(\frac{x^2}{4\theta^2}+(j-1/2)^2\right). \tag{39.2}$$

Gumbel [20] showed that for a large class of symmetrical continuous distributions with probability density function $\phi(x)$, $[\phi(x) = \phi(-x)]$, the limiting distribution (as $n \to \infty$) of 'reduced mth midrange' defined by ${}_mX = \alpha_m \times$ (mean of mth largest and mth smallest values in a random sample of size n) where

$$\alpha_m = (n/m)\phi(x_{m/(n+1)}); \quad m/(n+1) = \int_{-\infty}^{x_{m/(n+1)}} \phi(x)\,dx$$

has the probability density function

$$p_{_mX}(x) = \frac{(2m-1)!}{[(m-1)!]^2}[e^{-x}(1+e^{-x})^{-2}]^m. \tag{40}$$

These distributions, called *generalized logistic distributions* by Gumbel, belong to the class of Perks' distributions. (For $m = 1$, we have a logistic distribution, cf. Section 2.)

The similarity in general appearance of the logistic and normal distributions (excluding the tails) has been commented upon in Section 5.

Fisk [18] has shown that the Pareto distribution (Chapter 18) can be regarded as a form related to the logistic for certain extreme values of the variable. Making the substitution $e^Y = (T/t_0)^n$ in (8), (with $n > 0, t_0 > 0$), we have

$$\Pr[T < t] = \Pr[(T/t_0)^n < (t/t_0)^n] = (t/t_0)^n[1 + (t/t_0)^n]^{-1}.$$

If t be small compared with t_0, then approximately

$$\Pr[T < t] \propto t^n$$

If t be large compared with t_0, then

$$\Pr[T > t] = (t/t_0)^{-n}[1 + (t/t_0)^{-n}]^{-1}$$

and approximately

$$\Pr[T > t] \propto t^{-n}.$$

Shah and Dave [44] have defined *log-logistic* distributions in a manner analogous to the definition of log-normal distributions (Chapter 14, Section 1). A variable Y is said to be *log-logistic* if $\alpha + \beta \log (Y - \gamma)$ has a standard logistic distribution.

Mixtures of two logistic distributions have been discussed by Shah [42].

A *generalized logistic distribution* has been obtained by Dubey [11] as a mixture of extreme value distributions

$$\Pr[X \le x \mid \eta] = 1 - \exp[-\eta\beta \exp\{-(x-\alpha)/\beta\}] \quad (x > 0; \eta, \beta > 0) \tag{41}$$

with η having a gamma distribution with density function

$$p_\eta(y) = \beta'^b y^{b-1} e^{-\beta' y}/\Gamma/b) \qquad (y, b, \beta' > 0).$$

The resulting density function of X is

$$p_X(x) = b\beta'^b e^{(x-\alpha)/\beta}[\beta' + \beta e^{(x-\alpha)/\beta}]^{-(b+1)}. \tag{42}$$

For $b = 1$ and $\beta' = \beta$, (42) gives the logistic distribution (1), as pointed out in Section 2. The moment generating function of $(X - \alpha)/\beta$ corresponding to (41) is

$$(\eta\beta)^{-t}\Gamma(1 + t)$$

and that corresponding to (42) is

$$\beta^{-t}\Gamma(1+t)E[\eta^{-t}] = (\beta'/\beta)^t\Gamma(1+t)\Gamma(b-t)/\Gamma(b). \tag{43}$$

From (43) we find that, for distribution (42),

$$E[X] = \alpha + \beta[\log(\beta'/\beta) + \psi(1) - \psi(b)] \tag{44.1}$$

$$\kappa_r(X) = \beta^r[\psi^{(r-1)}(1) + (-1)^r\psi^{(r-1)}(b)] \qquad (r \geq 2). \tag{44.2}$$

REFERENCES

[1] Baten, W. D. (1934). The probability law for the sum of n independent variables, each subject to the law $(1/(2h))$ sech $(\pi x/(2h))$, *Bulletin of the American Mathematical Society*, **40,** 284–290.

[2] Berkson, J. (1944). Application of the logistic function to bio-assay, *Journal of the American Statistical Association*, **39,** 357–365.

[3] Berkson, J. (1951). Why I prefer logits to probits, *Biometrics*, **7,** 327–339.

[4] Berkson, J. (1953). A statistically precise and relatively simple method of estimating the bio-assay and quantal response, based on the logistic function, *Journal of the American Statistical Association*, **48,** 565–599.

[5] Berkson, J. (1955). Maximum likelihood and minimum χ^2 estimates of the logistic function, *Journal of the American Statistical Association*, **50,** 130–162.

[6] Berkson, J. (1957). Tables for the maximum likelihood estimate of the logistic function, *Biometrics*, **13,** 28–34.

[7] Berkson, J. and Hodges, J. L. (1961). A minimax estimator for the logistic function, *Proceedings of the Fourth Berkeley Symposium on Mathematical Statistics and Probability*, **4,** 77–86.

[8] Birnbaum, A. and Dudman, J. (1963). Logistic order statistics, *Annals of Mathematical Statistics*, **34,** 658–663.

[9] Blom, G. (1956). On linear estimates with nearly minimum variance, *Arkiv för Matematik*, **3,** 365–369.

[10] Blom, G. (1958). *Statistical Estimates and Transformed Beta Variables*, New York: John Wiley & Sons, Inc.

[11] Dubey, S. D. (1969). A new derivation of the logistic distribution, *Naval Research Logistics Quarterly* **16,** 37–40.

[12] Dyke, G. V. and Patterson, H. D. (1952). Analysis of factorial arrangements when the data are proportions, *Biometrics*, **8,** 1–12.

[13] Erkelens, J. (1968). A method of calculation for the logistic curve, *Statistica Neerlandica*, **22,** 213–217. (In Dutch)

[14] Finney, D. J. (1947). The principles of biological assay, *Journal of the Royal Statistical Society, Series B*, **9,** 46–91.

[15] Finney, D. J. (1952). *Statistical Method in Biological Assay*, New York: Hafner.

[16] Fisher, R. A. and Yates, F. (1957). *Statistical Tables for Biological, Agricultural and Medical Research* (5th edition) London & Edinburgh: Oliver and Boyd, New York: Hafner.

[17] Fisk, P. R. (1961). Estimation of location and scale parameters in a truncated grouped sech square distribution, *Journal of the American Statistical Association*, **56,** 692–702.

[18] Fisk, P. R. (1961). The graduation of income distributions, *Econometrica*, **29,** 171–185.

[19] Grizzle, J. E. (1961). A new method for testing hypotheses and estimating parameters for the logistic model, *Biometrics*, **17,** 372–385.

[20] Gumbel, E. J. (1944). Ranges and midranges, *Annals of Mathematical Statistics*, **15,** 414–422.

[21] Gumbel, E. J. (1961). Bivariate logistic distributions, *Journal of the American Statistical Association*, **56,** 335–349.

[22] Gumbel, E. J. and Keeney, R. D. (1950). The extremal quotient, *Annals of Mathematical Statistics*, **21,** 523–538.

[23] Gumbel, E. J. and Pickands, J. (1967). Probability tables for the extreme quotient, *Annals of Mathematical Statistics*, **38,** 1441–1551.

[24] Gupta, S. S., Qureishi, A. S. and Shah, B. K. (1967). Best linear unbiased estimators of the parameters of the logistic distribution using order statistics, *Technometrics*, **9,** 43–56.

[25] Gupta, S. S. and Shah, B. K. (1965). Exact moments and percentage points of the order statistics and the distribution of the range from the logistic distribution, *Annals of Mathematical Statistics*, **36,** 907–920.

[26] Gupta, S. S. and Gnanadesikan, M. (1966). Estimation of the parameters of the logistic distribution, *Biometrika*, **53,** 565–570.

[27] Harkness, W. L. and Harkness, M. L. (1968). Generalized hyperbolic secant distributions, *Journal of the American Statistical Association*, **63,** 329–337.

[28] Harter, H. L. and Moore, A. H. (1967). Maximum-likelihood estimation, from censored samples, of the parameters of a logistic distribution, *Journal of the American Statistical Association*, **62,** 675–683.

[29] Hassanein, K. M. (1967). Optimum spacing of sample quantiles from the logistic distribution, *Egyptian Journal of Statistics*, **1,** 1–5.

[30] Jung, J. (1956). On linear estimates defined by a continuous weight function, *Arkiv för Matematik*, **3,** 199–209.

[31] Kjelsberg, M. O. (1962). *Estimation of the parameters of the logistics distribution under truncation and censoring*, Ph.D. thesis, University of Minnesota.

[32] Ogawa, J. (1951). Contributions to the theory of systematic statistics, I. *Osaka Journal of Mathematics*, **3,** 175–213.

[33] Oliver, F. R. (1964). Methods of estimating the logistic growth function, *Applied Statistics*, **13,** 57–66.

[34] Owen, D. B. (1962). *Handbook of Statistical Tables*, Reading, Mass.: Addison-Wesley.

[35] Pearl, R. (1940). *Medical Biometry and Statistics*, Philadelphia: W. M. Sanders, Co.

[36] Perks, W. F. (1932). On some experiments in the graduation of mortality statistics, *Journal of the Institute of Actuaries*, **58,** 12–57.

[37] Plackett, R. L. (1959). The analysis of life-test data, *Technometrics*, **1,** 9–19.

[38] Plackett, R. L. (1958). Linear estimation from censored data, *Annals of Mathematical Statistics*, **29,** 131–142.

[39] Rasor, E. A. (1949). The fitting of logistic curves by means of a nomograph, *Journal of the American Statistical Association*, **44,** 548–553.

[40] Reed, L. J. and Berkson, J. (1929). The application of the logistic function to experimental data, *Journal of Physical Chemistry*, **33,** 760–779.

[41] Reiersøl, O. (1961). Linear and non-linear multiple comparisons in logit analysis, *Biometrika*, **48,** 359–365.

[42] Shah, B. K. (1963). A note on method of moments applied to a mixture of two logistic populations, *Journal of the M.S. University of Baroda* (*Science Number*), **12,** 21–22.

[43] Shah, B. K. (1965). Distribution of midrange and semirange from logistic population, *Journal of the Indian Statistical Association*, **3,** 185–188.

[44] Shah, B. K. and Dave, P. H. (1963). A note on log-logistic distribution, *Journal of the M.S. University of Baroda* (*Science Number*), **12,** 15–20.

[45] Silverstone, H. (1957). Estimating the logistic curve, *Journal of the American Statistical Association*, **52,** 567–577.

[46] Simpson, J. S. (1967). *Simultaneous linear estimation of the mean and standard deviation of the normal and logistic distributions by the use of selected order statistics from doubly censored samples*, M.S. thesis, Air Force Institute of Technology, Wright-Patterson Air Force Base, Ohio.

[47] Talacko, J. (1956). Perks' distributions and their role in the theory of Wiener's stochastic variables, *Trabajos de Estadística*, **7,** 159–174.

[48] Tarter, M. E. (1965). *Order statistic moment and product moment relationships for the truncated logistic and other distributions*, Manuscript, University of Michigan.

[49] Tarter, M. E. and Clark, V. A. (1965). Properties of the median and other order statistics of logistic variates, *Annals of Mathematical Statistics*, **36,** 1779–1786.

[50] Tiku, M. L. (1968). Estimating the parameters of normal and logistic distributions from censored samples, *Australian Journal of Statistics*, **10,** 64–74.

[51] Verhulst, P. F. (1845). Recherches mathématiques sur la loi d'accroissement de la population, *Académie de Bruxelles*, **18,** 1–38.

[52] Will, H. S. (1936). On a general solution for the parameters of any function with application to the theory of organic growth, *Annals of Mathematical Statistics*, **7,** 165–190.

23

Laplace Distribution

1. Definition, Genesis and Historical Remarks

The distribution was discovered by Laplace [20] in 1774, as the form of distribution for which the likelihood function is maximized by setting the location parameter equal to the median of the observed values of an odd number of independent identically distributed random variables.

The probability density function of the distribution is

$$p_X(x) = \tfrac{1}{2}\phi^{-1} \exp\left[-\,|x-\theta|/\phi\right] \qquad (\phi > 0). \tag{1}$$

This is known as the *first law of Laplace*. Proceeding further, Laplace replaced the median by the arithmetic mean as the value maximizing the likelihood function, and found that the corresponding distribution is the normal distribution (*second law of Laplace*). Another mode of genesis is as the distribution of the difference of two independent random variables with identical exponential distributions.

The distribution is known under different names. One of the most common is the *double exponential*, even though this is also applied to the extreme value distribution (Chapter 21).* In the Index [12] of the *Annals of Mathematical Statistics* it appears as the *two-tailed exponential*, Feller [8] calls it the *bilateral exponential*, and Weida [32] calls it *Poisson's first law of error*.

2. Moments, Generating Functions and Properties

A standard form of the probability density function (1) is obtained by putting $\theta = 0$, $\phi = 1$, giving

*A distinction in terminology between doubly exponential (extreme value) and double exponential (Laplace) distributions may be helpful.

$$p_X(x) = \tfrac{1}{2}e^{-|x|}. \tag{2}$$

(This is the form which is sometimes called *Poisson's first law of error.*) The characteristic function corresponding to this probability density function is

$$E(e^{itX}) = \tfrac{1}{2}(1 + it)^{-1} + \tfrac{1}{2}(1 - it)^{-1} = (1 + t^2)^{-1} \tag{3}$$

and the moment generating function is $(1 - t^2)^{-1}$. The cumulant generating function is

$$-\log (1 - t^2) \tag{4}$$

and the rth cumulant is

$$\begin{aligned} \kappa_r(X) &= 0, && \text{if } r \text{ is odd,} \\ \kappa_r(X) &= 2[(r-1)!], && \text{if } r \text{ is even.} \end{aligned} \tag{5}$$

The rth central moment is

$$\begin{aligned} \mu_r(X) &= 0, && \text{if } r \text{ is odd,} \\ \mu_r(X) &= r!, && \text{if } r \text{ is even.} \end{aligned} \tag{6}$$

The distribution is symmetrical about $x = 0$; the values of the first two moment ratios are

$$\sqrt{\beta_1} = 0; \qquad \beta_2 = 6. \tag{7}$$

The mean deviation is

$$\nu_1 = E(|X|) = 1 \tag{8}$$

and so, for the Laplace distribution,

$$\frac{\text{mean deviation}}{\text{standard deviation}} = \frac{1}{\sqrt{2}} = 0.707. \tag{9}$$

For the more general distribution (1), the ratios (7) and (9) have the same values as for the standard form (2). The expected value and standard deviation of (1) are θ and $\sqrt{2}\,\phi$ respectively.

The information generating function is

$$\int_{-\infty}^{\infty} (2\phi)^{-u} \exp\{-u|x - \theta|/\phi\}\, dx = (2\phi)^{1-u}u^{-1}. \tag{10}$$

The entropy is $1 + \log(2\phi)$.

The probability density function has a maximum at $x = \theta$, where there is a cusp. The form of the function is sketched in Figure 1. The cumulative distribution function is

$$F_X(x) = \begin{cases} \tfrac{1}{2}\exp[-(\theta - x)/\phi] & (x \leq \theta) \\ 1 - \tfrac{1}{2}\exp[-(x - \theta)/\phi] & (x \geq \theta). \end{cases} \tag{11}$$

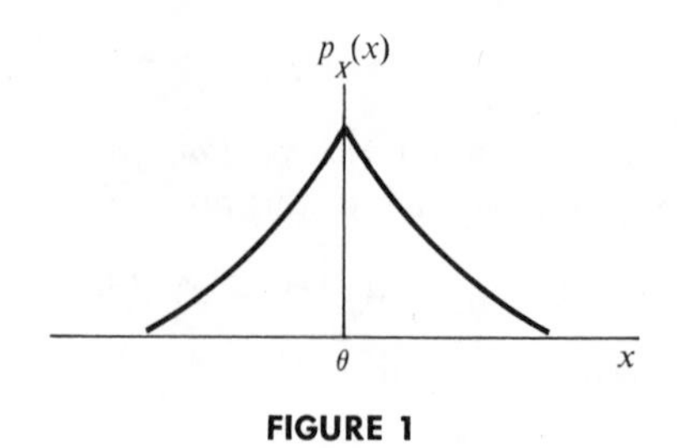

FIGURE 1

Laplace Density Functions

The lower and upper quartiles are $\theta \pm \phi \log_e 2 \doteqdot \theta \pm 0.693\phi$.

The probability density function, expressed in terms of the expected value, ξ, and the standard deviation, σ, is

$$p_X(x) = (\sigma\sqrt{2})^{-1} \exp[-\sqrt{2}\,|x - \xi|/\sigma]. \tag{12}$$

The upper and lower quartiles are $\xi \pm \sigma \cdot 2^{-\frac{1}{2}} \log_e 2 \doteqdot \xi \pm 0.534\sigma$. For the normal distribution, the corresponding values are $\xi \pm 0.674\sigma$. This difference reflects the sharp peak in the Laplace distribution. For quantiles further in the tails the comparison is reversed because the Laplace probability density function decreases as $\exp[-\sqrt{2}\,|x - \xi|/\sigma]$ while the normal decreases as

$$\exp[-\tfrac{1}{2}(x - \xi)^2/\sigma^2].$$

For example the upper and lower 1% points of the Laplace distribution are $\xi \pm 2.722\sigma$, compared with $\xi \pm 2.326\sigma$ for the normal distribution.

Special tables are not needed for numerical calculations connected with the Laplace distribution, as standard tables of the exponential function can be used.

We note that if $\theta = 0$ the probability density function of the arithmetic mean ($\overline{X}$) is

$$\begin{aligned} p_{\overline{X}}(x) &= \frac{n}{\phi^n(n-1)!}\,\frac{d^{n-1}}{dv^{n-1}}\left\{\frac{e^{-vn|x|}}{(1+\phi v)^n}\right\}\Bigg|_{v=\phi^{-1}} \\ &= \frac{(n/\phi)e^{-n|x|/\phi}}{2^{2n-1}(n-1)!}\sum_{j=0}^{n-1}\frac{2^j(2n-j-2)!}{j!(n-j-1)!}\left|\frac{nx}{\phi}\right|^j \end{aligned} \tag{13}$$

(Hausdorff [13], Craig [3], Sassa [27], Weida [32]).

3. Order Statistics

The simple explicit form of $F_X(x)$, as given in (11) leads to simple explicit forms for the distributions of order statistics connected with the Laplace distribution. If $X_1' \leq X_2' \leq \cdots \leq X_n'$ denote the order statistics corresponding to n independent random variables $X_1, X_2, \ldots, X_n$ each having probability density function (1) (so that X'_r is the rth smallest of $X_1, X_2, \ldots, X_n$) then the probability density function of X'_r is

$$(14) \qquad p_{X'_r}(x) = \begin{cases} \dfrac{n!}{(r-1)!(n-r)!} \cdot \dfrac{1}{2\phi} \{1 - \tfrac{1}{2} e^{-(\theta - x)/\phi}\}^{n-r} \\ \qquad \cdot \exp[-(r+1)(\theta - x)/\phi] & \text{for } x \leq \theta \\ \dfrac{n!}{(r-1)!(n-r)!} \cdot \dfrac{1}{2\phi} \{1 - \tfrac{1}{2} e^{-(x-\theta)/\phi}\}^{r-1} \\ \qquad \cdot \exp[-(n-r+1)(x-\theta)/\phi] & \text{for } x \geq \theta. \end{cases}$$

The sth moment of X'_r about θ is

(15)

$$E[(X'_r - \theta)^s] = \phi^s \frac{n!\Gamma(s+1)}{(r-1)!(n-r)!} \times \left[(-1)^s \sum_{j=0}^{n-r} (-1)^j \binom{n-r}{j} 2^{-(r+j+1)} (r+j)^{-(s+1)} + \sum_{j=0}^{r-1} (-1)^j \binom{r-1}{j} 2^{-(n-r+2+j)} (n-r+1+j)^{-(s+1)} \right].$$

In particular, if n is odd, the distribution of the median is obtained by putting $r = \frac{1}{2}(n+1)$. This distribution is symmetrical about θ; the expected value of the median is θ, and the variance is:

(16)

$$\frac{4\phi^2 n!}{\left(\frac{n-1}{2}\right)!} \sum_{j=0}^{\frac{1}{2}(n-1)} (-1)^j \left[j! \left(\frac{n-1}{2} - j\right)! 2^{\frac{1}{2}(n+1)+j} \{\tfrac{1}{2}(n+1) + j\}^3 \right]^{-1}.$$

For any value of n, the expected value of the largest of $X_1, X_2, \ldots, X_n$ is

(17)

$$E(X'_n) = \theta + \phi n \left[\sum_{j=0}^{n-1} (-1)^j \binom{n-1}{j} 2^{-(j+2)} (j+1)^{-2} - 2^{-(1+n)} n^{-2} \right]$$

The expected value of the smallest of $X_1, X_2, \ldots X_n$ is by symmetry $(2\theta - E(X'_n))$, and the expected value of the range $W (= X'_n - X'_1)$ is

$$(18) \qquad E(W) = 2\phi n \left[\sum_{j=0}^{n-1} (-1)^j \binom{n-1}{j} 2^{-(j+2)} (j+1)^{-2} - 2^{-(1+n)} n^{-2} \right] = a_n \phi.$$

Edwards [6] gives the values $a_4 = 2.7708$; $a_5 = 3.1771$. Edwards also gives the following formulas for the cumulative distribution function of range for $n = 4,5$:

(19.1) *For* $n = 4$

$$F_W(w) = 1 + \tfrac{15}{8}e^{-w} - 3e^{-2w} + \tfrac{1}{8}e^{-3w} - \tfrac{3}{4}we^{-w}(4 + e^{-w}) \qquad (w \geq 0)$$

(19.2) *For* $n = 5$

$$F_W(w) = 1 + \tfrac{77}{12}e^{-w} - \tfrac{57}{8}e^{-2w} - \tfrac{1}{4}e^{-3w} - \tfrac{1}{24}e^{-4w} - \tfrac{5}{4}we^{-w}(4 + 3e^{-w}) \qquad (w \geq 0).$$

4. Estimation

4.1 *Point Estimators*

Given observed values of n mutually independent random variables $X_1, \ldots, X_n$, each with probability density function (1), the likelihood function is

$$-n \log (2\phi) - \phi^{-1} \sum_{j=1}^{n} |X_j - \theta|. \tag{20}$$

Whether the value of ϕ is known or not, any value $\hat{\theta}$ minimizing $\sum_{j=1}^{n} |X_j - \theta|$ with respect to θ is a maximum likelihood estimator of θ. If n is odd, then $\hat{\theta}$ is uniquely defined as the median of $X_1, X_2, \ldots, X_n$. This result was obtained by Keynes [18], who also conjectured that this is a characterization of the Laplace distribution (as, indeed, it effectively is). If n is even, then $\hat{\theta}$ can be any value between the $\frac{1}{2}n$th and $(\frac{1}{2}n + 1)$th greatest values inclusive among $X_1, X_2, \ldots, X_n$. The arithmetic mean of these two values is convenient to use, and is an unbiased estimator of θ (as is the median when n is odd).

If ϕ (as well as θ) is unknown, a maximum likelihood estimator of ϕ is

$$n^{-1} \sum_{j=1}^{n} |X_j - \hat{\theta}| \tag{21}$$

(where $\hat{\theta}$ is a maximum likelihood estimator of θ). If θ is known, but ϕ is unknown (most commonly, $\theta = 0$), then the maximum likelihood estimator of ϕ is

$$n^{-1} \sum_{j=1}^{n} |X_j - \theta|. \tag{22}$$

The distribution of the median has been discussed in Section 3. Although the median is a maximum likelihood estimator of θ, and unbiased, it is not a minimum variance unbiased estimator of θ. Indeed, for small values of n (the sample size) it is possible to construct unbiased estimators with smaller variance than the median. Table 1 (from Govindarajulu [11]) gives the coefficients of the best linear unbiased estimators of θ and ϕ in samples of sizes $n = 1(1)10$ (in the rows corresponding to $r = 0$)*. The final column gives the value of ϕ^{-2} times the variances of these estimators. (The original table in Govindarajulu

*Note that for θ, the coefficient of X'_j is the same as that of X'_{n-j+1}; for ϕ, it has the same magnitude but is of opposite sign.

TABLE 1

Coefficients and Variances of Best Linear Estimators of Location (θ) and Scale (ϕ) Parameters.

n	r		Coefficients of X'_n	X'_{n-1}	X'_{n-2}	X'_{n-3}	X'_{n-4}	Variances
2	0	θ	.5000					1.0000
		ϕ	.6667					.7778
3	0	θ	.1481	.7037				.5895
		ϕ	.4444	.0000				.4321
4	0	θ	.0473	.4527				.4155
		ϕ	.3077	.2145				.2986
4	1	θ		.5000				.4201
		ϕ		1.4545				.8512
5	0	θ	.0166	.2213	.5241			.3169
		ϕ	.2331	.2264	.0000			.2290
5	1	θ		.2378	.5244			.3174
		ϕ		.8727	.0000			.4387
6	0	θ	.0063	.1006	.3931			.2548
		ϕ	.1876	.1943	.1132			.1858
6	1	θ		.1069	.3931			.2548
		ϕ		.6135	.1824			.2996
6	2	θ			.5000			.2609
		ϕ			2.2857			.8866
7	0	θ	.0025	.0455	.2386	.4267		.2122
		ϕ	.1572	.1631	.1439	.0000		.1565
7	1	θ		.0480	.2386	.4267		.2122
		ϕ		.4677	.2104	.0000		.2288
7	2	θ			.2862	.4276		.2134
		ϕ			1.3061	.0000		.4468
8	0	θ	.0010	.0208	.1316	.3465		.1814
		ϕ	.1355	.1391	.1391	.0718		.1351
8	1	θ		.0219	.1316	.3465		.1814
		ϕ		.3767	.1910	.0987		.1856
8	2	θ			.1533	.3467		.1816
		ϕ			.1977	.1605		.3020
8	3	θ				.5000		.1873
		ϕ				3.1411		.9078
9	0	θ	.0004	.0097	.0698	.2374	.3654	.1581
		ϕ	.1191	.1211	.1251	.1013	.0000	.1190
9	1	θ		.0101	.0698	.2374	.3654	.1581
		ϕ		.3153	.1643	.1331	.0000	.1562
9	2	θ			.0799	.2374	.3655	.1581
		ϕ			.7023	.1955	.0000	.2295
9	3	θ				.3166	.3668	.1596
		ϕ				1.7451	.0000	.4534
10	0	θ	.0002	.0046	.0364	.1478	.3110	.1399
		ϕ	.1063	.1074	.1110	.1061	.0504	.1062
10	1	θ		.0047	.0364	.1478	.3310	.1399
		ϕ		.2714	.1410	.1347	.0640	.1350
10	2	θ			.0412	.1478	.3110	.1399
		ϕ			.5665	.1854	.0881	.1857
10	3	θ				.1887	.3113	.1403
		ϕ				1.2218	.1448	.3044
10	4	θ					.5000	.1452
		ϕ					4.0125	.9220

TABLE 2

Efficiency of Various Estimators of θ, Relative to Best Linear Unbiased Estimator

Sample Size n		2	3	4	5
Estimator (%)	Arithmetical mean	100.00	88.43	82.80	79.21
	Midrange	100.00	67.90	49.65	38.29
	Median	100.00	92.27	98.90	90.23

Note: Chu and Hotelling [2] showed that var (median) is less than var (arithmetic mean) for $n \geq 7$.

[11] covers values of n up to 20 inclusive.)

Sarhan [25] has compared the variances of the best linear estimator of θ, the median (defined as the arithmetic mean of $X'_{\frac{1}{2}n}$ and $X'_{\frac{1}{2}n+1}$ when n is even), the arithmetic mean $\left(n^{-1}\sum_{j=1}^{n} X_j\right)$, and the midrange $(\frac{1}{2}(X'_1 + X'_n))$. These are all unbiased estimators of θ. Table 2 presents the 'efficiencies' (inverse ratio of variances, expressed as a percentage) of the last three estimators relative to the first. Figures 2a–c represent these values diagrammatically. The irregular appearance of Figure 2c is associated with the different definition of 'median' in samples of odd and even sizes.

We note that the estimator $n^{-1}\sum|X_i - \theta|$ of ϕ (when θ is known), is distributed as $(2n)^{-1}\,\phi \times (\chi^2$ with $2n$ degrees of freedom). The distribution of $n^{-1}\sum|X_i - \hat{\theta}|$, where $\hat{\theta}$ is a median value, has been studied by Karst and Polowy [17].

Table 1 shows the coefficients of the best linear unbiased estimators of θ and ϕ which use symmetrically censored samples. Variances (as multiples of ϕ^2)

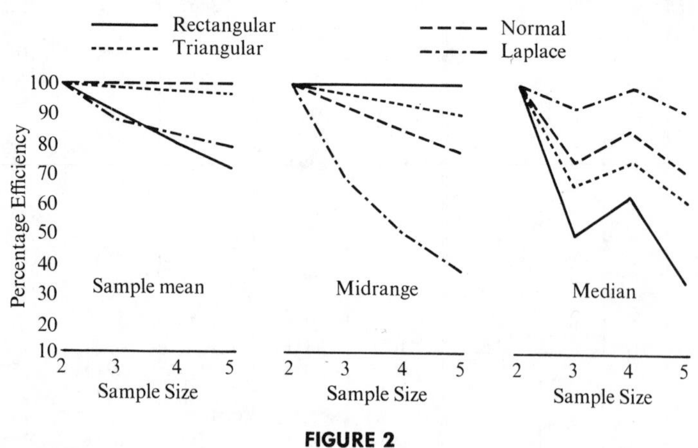

FIGURE 2

Percentage Efficiencies of the Sample Mean, Midrange and Median in Different Populations

of these estimators are also given in this table. Table 3 (Sarhan [26]) gives similar quantities for $n = 3, 4, 5$ for the case of censoring by omission of the r greatest values (note that the scale parameter estimated is the standard deviation, $\phi\sqrt{2}$). This table can also be used if the r least values are omitted by reversing the signs of the observations and estimating $-\theta$ and ϕ.

TABLE 3

Coefficients of Best Linear Unbiased Estimators of Expected Value (θ) and Standard Deviation ($\sigma = \phi\sqrt{2}$) for Samples Censored by Omission of r Largest Observed Values.

n	r	Estimator of	Coefficients of X'_1	X'_2	X'_3	X'_4	Variance $\times\ \sigma^{-2}$	Relative Efficiency
3	1	θ	−0.3300	1.3000				
		σ	−1.3578	1.3578				
4	1	θ	0.0662	0.3333	0.6004		0.1860	(98.23)
		σ	−0.7332	−0.2129	0.9461		0.3339	(89.42)
	2	θ	0.0000	1.0000			0.3335	(62.29)
		σ	−1.2563	1.2563			0.9457	(31.58)
5	1	θ	0.0114	0.2163	0.5243	0.2479	0.1586	(99.88)
		σ	−0.4331	−0.4191	0.0037	0.8484	0.3097	(73.88)
	2	θ	−0.6649	0.1666	0.8998		0.1724	(91.85)
		σ	−0.6655	−0.6233	1.2889		0.4634	(49.37)
	3	θ	−0.5641	1.5641			1.2743	(2.24)
		σ	−1.3925	1.3925			2.8481	(8.03)

Note: In the last column, relative efficiency (inverse ratio of variance to that of best linear unbiased estimator using the complete sample) is shown, as a percentage, in parentheses.

4.2 *Confidence Intervals*

If ϕ is known, confidence limits for θ may be based on the distribution of the median, $\hat{\theta}$. If θ is known, then confidence limits for ϕ may be obtained by using the fact that the distribution of $n^{-1}\sum_{j=1}^{n}|X_j - \theta|$ is that of $(2n)^{-1}\phi \times$ (χ^2 with $2n$ degrees of freedom). The limits of a $100(1-\alpha)\%$ confidence interval for ϕ are then

$$2\sum_{j=1}^{n}|X_j - \theta|/\chi^2_{2n,1-\frac{1}{2}\alpha} \quad \text{and} \quad 2\sum_{j=1}^{n}|X_j - \theta|/\chi^2_{2n,\frac{1}{2}\alpha}. \tag{23}$$

If neither θ nor ϕ is known it would be possible to construct confidence intervals for θ and ϕ, respectively, using the distributions of

$$(\hat{\theta} - \theta)\Big/\sum_{j=1}^{n}|X_j - \hat{\theta}| \quad \text{and} \quad \phi^{-1}\sum_{j=1}^{n}|X_j - \hat{\theta}|,$$

neither of which depend on the actual values of the parameters θ and ϕ ($\hat{\theta}$ is here supposed to be the median, as defined above). However, appropriate tables of these distributions are not at present available.

5. Related Distributions

If X has probability density function (1), then $|X - \theta|$ is distributed exponentially, i.e. as $\frac{1}{2}\phi \times$ (χ^2 with two degrees of freedom). In particular, if $\theta = 0$, then $|X|$ is so distributed. For this reason, if $X_1, X_2, \ldots, X_n$ are independent random variables, each having probability density function (1) with $\theta = 0$, then the distribution of any statistic depending only on the absolute values $|X_1|, |X_2|, \ldots, |X_n|$ can be derived from an initial joint distribution of independent multiples of χ^2 variables. For example, $|X_1|/|X_2|$ is distributed as F with 2,2 degrees of freedom (see Chapter 26).

An interesting connection between the normal and Laplace distributions has been established by Nyquist *et al.* [24]. They showed that if U_1, U_2, U_3 and U_4 are independent unit normal variables then the probability density function of

$$D = \begin{vmatrix} U_1 & U_2 \\ U_3 & U_4 \end{vmatrix} = U_1U_4 - U_2U_3$$

is of form (1) with $\theta = 0$, $\phi = 2$. (The case when the expected values of the U's are not equal to zero leads to a more complicated distribution, and was considered by Nicholson [23].)

The two reciprocal Fourier integrals

$$\tfrac{1}{2}\int_{-\infty}^{\infty} \exp(itx - |x|)\, dx = (1 + t^2)^{-1}$$

and

$$\pi^{-1}\int_{-\infty}^{\infty} (1 + t^2)^{-1} \exp(itx)\, dx = e^{-|x|}$$

represent a formal connection between the Cauchy and Laplace distributions (see also Chapter 16).

Transformed forms of the Laplace distribution have been discussed by Johnson [15]. He considered (by analogy with the lognormal, S_U and S_B systems — see Chapters 14 and 12) distributions of a random variable Y when (with $\delta > 0$)

$$\begin{aligned} X &= \gamma + \delta \log Y, && (S_L' \text{ system}) \\ X &= \gamma + \delta \sinh^{-1} Y, && (S_U' \text{ system}) \\ X &= \gamma + \delta \log[Y/(1 - Y)]; && (S_B' \text{ system}) \end{aligned}$$

and X has the standard Laplace distribution (2).

The (β_1,β_2) points of the S_L' system lie on the line with parametric equations

$$\beta_1(Y) = \frac{4(\delta^2 - 4)(15\delta^2 + 7\delta^2 + 2)^2}{\delta^2(\delta^2 - 9)^2(2\delta^2 + 1)^3} \qquad (\delta > 3); \tag{24.1}$$

$$\beta_2(Y) = \frac{3(\delta^2 - 4)(8\delta^8 + 212\delta^6 + 95\delta^4 + 33\delta^2 + 12)}{\delta^2(\delta^2 - 9)(\delta^2 - 16)(2\delta^2 + 1)^2} \qquad (\delta > 4). \tag{24.2}$$

The (β_1,β_2) points of S'_U lie 'below' this line (i.e. larger values of β_2 for given β_1); those of S'_B lie above it. All possible values of (β_1,β_2) are covered by these three systems combined. For both S'_L and S'_U the rth moment is infinite if $r \geq \delta$.

Asymmetrical Laplace distributions, with probability density functions of form

$$p_X(x) = \begin{cases} (2\phi_1)^{-1} \exp[-|x - \theta|/\phi_1] & (x \geq \theta) \\ (2\phi_2)^{-1} \exp[-|x - \theta|/\phi_2] & (x < \theta) \end{cases} \tag{25}$$

where $\phi_1 \neq \phi_2$ and $\phi_1, \phi_2 > 0$, are sometimes used (see McGill [22]).

Another form of asymmetrical Laplace distribution has probability density function

$$p_X(x) = \begin{cases} p\phi^{-1} \exp[-|x - \theta|/\phi] & (x \geq \theta) \\ (1 - p)\phi^{-1} \exp[-|x - \theta|/\phi] & (x < \theta) \end{cases} \tag{26}$$

with $0 < p < 1$. Holla and Bhattacharya [14] have used this distribution as the compounding distribution of the expected value of a normal distribution. The characteristic function of the resulting compound normal distribution is

$$(1 + t^2\phi^2)^{-1}\{1 + (2p - 1)it\phi\} \exp[it\theta - \tfrac{1}{2}t^2\sigma^2] \tag{27}$$

where σ^2 is the variance of the compounded normal distribution. The probability density function (with argument y) is

$$(\phi\sqrt{2\pi})^{-1}e^{\frac{1}{2}\sigma^2/\phi^2}[pe^{-(y-\theta)/\phi}\{\sqrt{\pi/2} - S_1M(\tfrac{1}{2};\tfrac{3}{2}; -\tfrac{1}{2}S_1^2)\} + (1 - p)e^{(y-\theta)/\phi}\{\sqrt{\pi/2} - S_2M(\tfrac{1}{2};\tfrac{3}{2}; -\tfrac{1}{2}S_2^2)\}] \tag{28}$$

where

$$S_j = (\sigma/\phi) + (-1)^j\{(x - \theta)/\sigma\} \qquad (j = 1,2)$$

and $M(\cdot)$ is the confluent hypergeometric distribution (Chapter 1, Equation (43)).

For the particular case $p = \frac{1}{2}$ we have the distribution

$$\text{Normal}\,(\xi,\sigma) \underset{\xi}{\wedge} \text{Laplace}\,(\theta,\phi).$$

This distribution is symmetrical, with mean θ. The variance is $(\sigma^2 + 2\phi^2)$ and the moment ratio α_4 $(\equiv \beta_2)$ is

$$3 + 12\phi^4(\sigma^2 + 2\phi^2)^{-2}.$$

Holla and Bhattacharya obtain an expression for the distribution of the sum of n independent random variables each having this distribution. They also obtain the following formula for the cumulative distribution function (argument y):

$$(29)\qquad \Phi\left(\frac{y-\theta}{\sigma}\right) - \frac{1}{2}e^{\frac{1}{2}\sigma^2/\phi^2}\sinh\left(\frac{y-\theta}{\sigma}\right) + \frac{1}{2}\frac{1}{\sqrt{2\pi}}e^{\frac{1}{2}\sigma^2/\phi^2}$$
$$\times\{e^{(y-\theta)/\phi}S_2M(\tfrac{1}{2};\tfrac{3}{2};-\tfrac{1}{2}S_2^2) - e^{-(y-\theta)/\phi}S_1M(\tfrac{1}{2};\tfrac{3}{2};-\tfrac{1}{2}S_1^2)\}.$$

Among *compound Laplace distributions*, we note*:

(i) $$\text{Laplace }(\theta,\phi) \underset{\theta}{\wedge} \text{Normal }(\xi,\sigma).$$

The probability density function is

$$(30)\qquad p_X(x) = \frac{1}{2\phi}\{\exp(\tfrac{1}{2}(\sigma/\phi)^2)\}\left[\Phi\left(\frac{x-\xi}{\sigma}-\frac{\sigma}{\phi}\right)\exp\left(-\frac{x-\xi}{\phi}\right)\right.$$
$$\left.+\Phi\left(-\frac{x-\xi}{\sigma}-\frac{\sigma}{\phi}\right)\exp\left(\frac{x-\xi}{\phi}\right)\right]$$

where $\Phi(x) = (2\pi)^{-\frac{1}{2}}\int_{-\infty}^{x}\exp(-\frac{1}{2}t^2)\,dt$.

(ii) $$\text{Laplace }(\theta,\phi) \underset{\phi^{-1}}{\wedge} \text{Gamma }(\alpha,\beta)$$

(The gamma distribution is as given in (23), Chapter 17.)

The probability density function is

$$(31)\qquad p_X(x) = \tfrac{1}{2}\alpha\beta[1+|x-\theta|/\beta]^{-(\alpha+1)}.$$

The relation between distributions of form (31) and Laplace distributions is rather similar to that between Pearson Type VII and normal distributions (Chapter 27). We note, for example, that as β tends to zero and α to infinity, with $\alpha\beta = 1$, $p_X(x) \to \frac{1}{2}\exp(-|x-\theta|)$.

Distribution (31) is symmetrical about θ. Moments of order α or greater do not exist. For r even and less than α

$$(32)\qquad \mu_r = \alpha\beta^r\sum_{j=0}^{r}(-1)^j\binom{r}{j}(\alpha+j-r)^{-1}.$$

In particular, the variance is

$$(33.1)\qquad \sigma^2 = \frac{2\beta^2}{(\alpha-1)(\alpha-2)}\qquad(\alpha>2)$$

and also

$$(33.2)\qquad \beta_2 = \frac{6(\alpha-1)(\alpha-2)}{(\alpha-3)(\alpha-4)}\qquad(\alpha>4).$$

The mean deviation is

$$(33.3)\qquad \nu_1' = \frac{\beta}{\alpha-1}$$

*In accordance with the notation introduced in Chapter 8, θ is here a random variable which is normally distributed with expected value ξ and standard deviation σ.

and so

$$\frac{\text{mean deviation}}{\text{standard deviation}} = \sqrt{\frac{2(\alpha-2)}{\alpha-1}}.$$

The cumulative distribution function is of very simple form:

$$F_X(x) = \begin{cases} \tfrac{1}{2}[1 + \{1 + |x-\theta|/\beta\}^{-\alpha}] & (x \geq \theta) \\ \tfrac{1}{2}[1 - \{1 + |x-\theta|/\beta\}^{-\alpha}] & (x \leq \theta). \end{cases}$$

Subbotin [31], on the basis of certain broad requirements for 'error distributions', obtained the class

$$(34) \quad p_X(x) = [2^{\frac{1}{2}\delta+1}\Gamma(\tfrac{1}{2}\delta+1)]^{-1}\phi^{-1}\exp\left[-\frac{1}{2}\left|\frac{x-\theta}{\phi}\right|^{2/\delta}\right] \qquad (\delta,\phi > 0).$$

(See also Fréchet [9], where Subbotin's arguments are criticized.) This class of distributions includes Laplace ($\delta = 2$) normal ($\delta = 1$), and, as a limiting ($\delta \to 0$) case, rectangular distributions. It is symmetrical about θ, and has finite moments of all positive orders. The rth central moment is

$$(35) \qquad \mu_r = \begin{cases} 0 & (r \text{ odd}) \\ \phi^r 2^{\frac{1}{2}r\delta}\Gamma\left(\dfrac{(r+1)\delta}{2}\right)\Big/\Gamma\left(\dfrac{\delta}{2}\right) & (r \text{ even}). \end{cases}$$

The variance is

$$(36.1) \qquad \sigma^2 = \frac{2^\delta\Gamma\left(\dfrac{3\delta}{2}\right)}{\Gamma\left(\dfrac{\delta}{2}\right)}\phi^2$$

and the mean deviation is

$$(36.2) \qquad \nu_1' = \frac{2^{\frac{1}{2}\delta}\Gamma(\delta)}{\Gamma\left(\dfrac{\delta}{2}\right)}\phi$$

so

$$(36.3) \qquad \frac{\text{mean deviation}}{\text{standard deviation}} = \frac{\Gamma(\delta)}{\left[\Gamma\left(\dfrac{\delta}{2}\right)\Gamma\left(\dfrac{3\delta}{2}\right)\right]^{\frac{1}{2}}}.$$

Also

$$(36.4) \qquad \beta_2 = \frac{\Gamma\left(\dfrac{5\delta}{2}\right)\Gamma\left(\dfrac{\delta}{2}\right)}{\left[\Gamma\left(\dfrac{3\delta}{2}\right)\right]^2}.$$

Some values of β_2 and the ratio (36.3) are given in Table 3.

The cumulative distribution function corresponding to (34) can be expressed in terms of incomplete gamma functions. Maximum likelihood estimation of

TABLE 3

Ratio of Mean Deviation (m.d.) to Standard Deviation (s.d.), and β_2 for Subbotin Distributions

δ	m.d./s.d.	β_2
0 (uniform)	0.866	1.800
0.25	0.858	1.923
0.5	0.841	2.188
0.75	0.815	2.548
1 (normal)	0.798	3.000
1.5	0.757	4.222
2 (Laplace)	0.707	6.000
3	0.623	12.257
4	0.548	25.200
5	0.481	51.951

the parameters has been discussed by Diananda [4].

The classes of distributions (31) and (34) have been used by Box and Tiao [1] as prior distributions for certain Bayesian statistical analyses. Distribution (34) provides a convenient set of alternatives to normality, if symmetry can be assumed.

Krysicki [19] has given formulas for estimating the parameters in a mixture of two Laplace distributions, each having $\theta = 0$.

Śródka [30] has discussed the distributions obtained if ϕ^{-1} is supposed to have a generalized gamma distribution (as defined in (65) of Chapter 17).

REFERENCES

[1] Box, G. E. P. and Tiao, G. C. (1962). A further look at robustness via Bayes's theorem, *Biometrika*, **49,** 419–432.

[2] Chu, J. T. and Hotelling, H. (1955). The moments of the sample median, *Annals of Mathematical Statistics*, **26,** 593–606.

[3] Craig, A. T. (1932). On the distribution of certain statistics, *American Journal of Mathematics*, **54,** 353–366.

[4] Diananda, P. H. (1949). Note on some properties of maximum likelihood estimates, *Proceedings of the Cambridge Philosophical Society*, **45,** 536–544.

[5] Dwinas, S. (1948). A deduction of the Laplace-Gauss law of errors, *Revista Matemática Hispano-Americana*, (4), **8,** 12–18. (In Spanish)

[6] Edwards, L. (1948). The use of normal significance limits when the parent population is of Laplace form, *Journal of the Institute of Actuaries Students' Society*, **8,** 87–99.

[7] Farison, J. B. (1965). On calculating moments for some common probability laws, *IEEE Transactions on Information Theory*, **11,** 568–589.

[8] Feller, W. (1966). *An Introduction to Probability Theory and Its Applications*, Vol. **II,** New York: John Wiley & Sons, Inc.

[9] Fréchet, M. (1924). Sur la loi des erreurs d'observation, *Matematicheskii Sbornik*, **32,** 1–8.

[10] Fréchet, M. (1928). Sur l'hypothèse de l'additivité des erreurs partielles, *Bulletin des Sciences et Mathématiques, Paris*, **63,** 203–206.

[11] Govindarajulu, Z. (1966). Best linear estimates under symmetric censoring of the parameters of a double exponential population, *Journal of the American Statistical Association*, **61,** 248–258.

[12] Greenwood, J. A., Olkin, I. and Savage, I. R. (1962). Index to *Annals of Mathematical Statistics*, Volumes 1–31, 1930–1960. University of Minnesota, North Central Publishing Company.

[13] Hausdorff, F. (1901). Beiträge zur Wahrscheinlichkeitsrechnung, *Verhandlungen der Konigliche Sächsischen Gesellschaft der Wissenschaften, Leipzig, Mathematisch-Physische Classe*, **53,** 152–178.

[14] Holla, M. S. and Bhattacharya, S. K. (1968). On a compound Gaussian distribution, *Annals of the Institute of Statistical Mathematics, Tokyo*, **20,** 331–336.

[15] Johnson, N. L. (1954). Systems of frequency curves derived from the first law of Laplace, *Trabajos de Estadística*, **5,** 283–291.

[16] Kacki, K. and Krysicki, W. (1967). Die Parameterschätzung einer Mischung von zwei Laplaceschen Verteilungen (im allgemeinen Fall), *Roczniki Polskiego Towarzystwa Matematycznego, Seria I: Prace Matematyczne*, **11,** 23–31.

[17] Karst, O. J. and Polowy, H. (1963). Sampling properties of the median of a Laplace distribution, *American Mathematical Monthly*, **70,** 628–636.

[18] Keynes, J. M. (1911). The principal averages and the laws of error which lead to them, *Journal of the Royal Statistical Society, Series A*, **74,** 322–328.

[19] Krysicki, W. (1966). Zastosowanie metody momentów do estymacji parametrów mieszaniny dwóch rozkladów Laplace'a, *Zeszyty Naukowe Politechniki Lódzkiej*, **59,** 5–13.

[20] Laplace, P. S. (1774). Mémoire sur la probabilité des causes par les évènemens, *Mémoires de Mathématique et de Physique*, **6,** 621–656.

[21] Mantel, N. and Pasternack, B. S. (1966). Light bulb statistics, *Journal of the American Statistical Association*, **61,** 633–639.

[22] McGill, W. J. (1962). Random fluctuations of response rate, *Psychometrika*, **27,** 3–17.

[23] Nicholson, W. L. (1958). On the distribution of 2×2 random normal determinants, *Annals of Mathematical Statistics*, **29,** 575–580.

[24] Nyquist, H., Rice, S. O. and Riordan, J. (1954). The distribution of random determinants, *Quarterly of Applied Mathematics*, **42,** 97–104.

[25] Sarhan, A. E. (1954). Estimation of the mean and standard deviation by order statistics, Part I, *Annals of Mathematical Statistics*, **25,** 317–328.

[26] Sarhan, A. E. (1955). *Ibid.*, Part III, *Annals of Mathematical Statistics*, **26,** 576–592.

[27] Sassa, H. (1968). The probability density of a certain statistic in one sample from the double exponential population, *Bulletin*, *Tokyo Gakugei* University, **19,** 85–89. (In Japanese)

[28] Smith, J. H. (1947). Estimation of linear functions of cell proportions, *Annals of Mathematical Statistics*, **18,** 231–254.

[29] Śródka, T. (1964). Estymatory i przedzialy ufności odchylenia standardowego w rozkladzie Laplace'a, *Zeszyty Naukowe Politechniki Lódzkiej*, **57,** 5–9.

[30] Śródka, T. (1966). Zloženie rozkladu Laplace'a z pewnym uogólnionym rozkladem gamma, Maxwella i Weibulla, *Zeszyty Naukowe Politechniki Lódzkiej*, **59,** 21–28.

[31] Subbotin, M. T. (1923). On the law of frequency of errors, *Matematicheskii Sbornik*, **31,** 296–301.

[32] Weida, F. M. (1935). On certain distribution functions when the law of the universe is Poisson's first law of error, *Annals of Mathematical Statistics*, **6,** 102–110.

[33] Wilson, E. B. (1923). First and second law of errors, *Journal of the American Statistical Association*, **18,** 841–851.

24

Beta Distribution

1. Definition

The family of beta distributions is composed of all distributions with probability density functions of form:

$$p_Y(y) = \frac{1}{B(p,q)} \frac{(y-a)^{p-1}(b-y)^{q-1}}{(b-a)^{p+q-1}} \qquad (a \le y \le b) \tag{1}$$

with $p > 0$, $q > 0$.

This will be recognized as a Pearson Type I (or II) distribution (see Chapter 12, Section 4.1). If $q = 1$, the distribution is sometimes called a *power-function* distribution.

If we make the transformation

$$X = (Y - a)/(b - a)$$

we obtain the probability density function

$$p_X(x) = \frac{1}{B(p,q)} x^{p-1}(1-x)^{q-1} \qquad (0 \le x \le 1). \tag{2}$$

This is the *standard form* of the beta distribution with *parameters* p, q. It is the form which will be used in most of this chapter. (The *standard power-function* density is

$$p_X(x) = px^{p-1} \qquad (0 \le x \le 1).) \tag{2$'$}$$

The probability integral of the distribution (2) up to x is called the *incomplete beta function ratio* and is denoted by $I_x(p,q)$, so that

(3) $$I_x(p,q) = \frac{1}{B(p,q)} \int_0^x t^{p-1}(1-t)^{q-1}\,dt.$$

The word 'ratio' which distinguishes (3) from the *incomplete beta function*

(4) $$B_x(p,q) = \int_0^x t^{p-1}(1-t)^{q-1}\,dt$$

is often omitted.

A description of the properties of $I_x(p,q)$ is contained in Chapter 1 (and also Chapter 3, Section 8).

2. Genesis

In 'normal theory', the beta distribution arises naturally as the distribution of $V^2 = X_1^2/(X_1^2 + X_2^2)$ where X_1^2, X_2^2 are independent random variables, and X_j^2 is distributed as χ^2 with ν_j degrees of freedom ($j = 1,2$). The distribution of V^2 is then a standard beta distribution, as in (2), with $p = \frac{1}{2}\nu_1$, $q = \frac{1}{2}\nu_2$.

Notice that, in this situation, V^2 and $(X_1^2 + X_2^2)$ are mutually independent. An extension of this result is that if $X_1^2, X_2^2, \ldots, X_k^2$ are mutually independent with X_j^2 distributed as χ^2 with ν_j degrees of freedom ($j = 1,2,\ldots,k$) then,

$$\begin{aligned} V_1^2 &= X_1^2/(X_1^2 + X_2^2) \\ V_2^2 &= (X_1^2 + X_2^2)/(X_1^2 + X_2^2 + X_3^2) \\ &\vdots \\ V_{k-1}^2 &= (X_1^2 + \cdots + X_{k-1}^2)/(X_1^2 + \cdots + X_k^2) \end{aligned}$$

are mutually independent random variables, each with a beta distribution, the values of p, q for V_j^2 being $\frac{1}{2}\sum_{i=1}^{j} \nu_i$, $\frac{1}{2}\nu_{j+1}$ respectively. Under these conditions, the product of any successive set of V_j^2's also has a beta distribution (see, for example, Jambunathan [25]). This property also holds when the ν's are any positive numbers (not necessarily integers). Kotlarski [31] has investigated general conditions under which products of independent variables have a beta distribution.

Another way in which the beta distribution arises is as the distribution of an ordered variable from a rectangular distribution (Chapter 25). If $Y_1, Y_2, \ldots, Y_n$ are independent random variables each having the standard rectangular distribution, so that

$$p_{Y_j}(y) = 1 \qquad (0 \le y \le 1)$$

and the corresponding order statistics are $Y_1' \le Y_2' \le \cdots \le Y_n'$, the sth order statistic Y_s' has the beta distribution

(5) $$p_{Y_s'}(y) = [B(r, n-r+1)]^{-1} y^{r-1}(1-y)^{n-r} \qquad (0 \le y \le 1).$$

This result may be used to generate beta distributed random variables from standard rectangularly distributed variables. Using this method, only integer values can be obtained for n and $(n - s)$. A method applicable for fractional values of n and $(n - s)$, has been constructed by Jöhnk [26]. He has shown that if X and Y are independent standard rectangular variables then the *conditional* distribution of $X^{1/n}$ *given that* $X^{1/n} + Y^{1/r} \leq 1$, is a standard beta distribution with parameters n, $r + 1$; and the conditional distribution of $Y^{1/r}$ is beta with parameters $n + 1$ and r.

This process involves the calculation of $X^{1/n}$ and $Y^{1/r}$ which may be awkward. Bánkövi [7] has suggested a method whereby these calculations may be avoided if n and r are both rational. This consists of selecting integers $a_1, a_2, \ldots, a_M$, $b_1, b_2, \ldots, b_N$ such that

$$n = \sum_{j=1}^{M} a_j^{-1}; \qquad r = \sum_{j=1}^{N} b_j^{-1}$$

and then using the fact that, if $X_1, X_2, \ldots, X_M, Y_1, \ldots, Y_N$ are independent standard rectangular variables, $\max(X_1^{a_1}, X_2^{a_2}, \ldots, X_M^{a_M})$ and $\max(Y_1^{b_1}, \ldots, Y_N^{b_N})$ are distributed as $X^{1/n}$, $Y^{1/r}$ respectively.

If n (or r) is not a rational fraction, it may be approximated as closely as desired by such a fraction. Bánkövi has investigated the effects of such approximation on the desired beta variates.

Yet another way in which the beta distribution can arise has been described by Murthy [37]. He has shown that if $X_1, \ldots, X_{2M+1}$ are independent unit normal variables and $r_L = \left(\sum_{j=1}^{2M+1} X_j X_{j+L}\right) \Big/ \left(\sum_{j=1}^{2M+1} X_j^2\right)$ — (the Lth lag correlation) — with X_{2M+1+i} defined equal to X_i, then $(2M + 1)^{-1} \times \sum_{j=1}^{2M+1} r_j$ has a standard beta distribution.

The special standard beta distribution with $p = q = \frac{1}{2}$ (known as the *arc-sine distribution* because $\Pr[X \leq x] = (2/\pi)\sin^{-1}\sqrt{x}$ for $0 \leq x \leq 1$) arises in an interesting way in the theory of 'random walks'. Suppose a particle moves along the real line by steps of unit length, starting from zero, it being equally likely that a step will be to the left (decreasing) or right (increasing). Let the random variable T_{2n} denote the number of times in the first $2n$ steps for which the point is in the interval 0 to $2n$ inclusive at the conclusion of a step. Then

$$\Pr[T_{2n} = 2k] = \binom{2k}{k}\binom{2n - 2k}{n - k} 2^{-2n} \qquad (k = 0, 1, \ldots, n).$$

The ratio $T_{2n}/(2n)$ can be regarded as 'the fraction of time spent on the positive part of the real line'. As n tends to infinity, the limiting distribution of $T_{2n}/(2n)$ is the arc-sine distribution, that is,

$$(6) \qquad \lim_{n\to\infty} \left\{\sum_{k \leq nx} \Pr[T_{2n} = 2k]\right\} = \frac{1}{\pi}\int_0^x t^{-\frac{1}{2}}(1 - t)^{-\frac{1}{2}}\,dt = (2/\pi)\sin^{-1}\sqrt{x}.$$

Standard beta distributions with $p + q = 1$, but $p \neq \frac{1}{2}$ are sometimes called *generalized arc-sine* distributions.

A beta distribution can also be obtained as the limiting distribution of eigenvalues in a sequence of random matrices. Suppose A_n to be a symmetric $n \times n$ matrix whose elements $a_{ij}(i \leq j)$ are independent random variables, all a_{ij}'s with $i \neq j$ having a common distribution, and all a_{ii}'s another common distribution, both distributions being symmetrical about zero with variance σ^2 and with all absolute moments finite. Under these conditions Wigner [52] has shown that the proportion of eigenvalues of the "normalized" matrix $(2\sigma\sqrt{n})^{-1}A_n$ which are less than x tends to the limit

$$2\pi^{-1}\int_{-1}^{x} \sqrt{1-t^2}\, dt$$

as $n \to \infty$. This is of form (1) with $a = -1$, $b = 1$, $p = q = \frac{3}{2}$.

Arnold [2] has shown that this result holds under much weaker conditions on the distributions of the a's.

3. Properties

If X has the standard beta distribution (2), its rth moment about zero is

$$\mu'_r = \frac{B(p+r,q)}{B(p,q)} = \frac{\Gamma(p+r)\Gamma(p+q)}{\Gamma(p)\Gamma(p+q+r)} \tag{7}$$

$$= \frac{p^{[r]}}{(p+q)^{[r]}} \qquad \text{(if } r \text{ is an integer)}$$

where $y^{[r]} = y(y+1)\ldots(y+r-1)$ is the ascending factorial. In particular

$$E(X) = p/(p+q); \tag{8.1}$$

$$\operatorname{var}(X) = pq(p+q)^{-2}(p+q+1)^{-1}. \tag{8.2}$$

Also

(8.3)
$$\alpha_3(X) = \sqrt{\beta_1(X)} = 2(q-p)\sqrt{p^{-1}+q^{-1}+(pq)^{-1}}\cdot(p+q+2)^{-1}$$

$$\alpha_4(X) = \beta_2(X) = 3(p+q+1)\{2(p+q)^2 + pq(p+q-6)\} \times [pq(p+q+2)(p+q+3)]^{-1}. \tag{8.4}$$

The moment generating function can be expressed as a confluent hypergeometric function (Equation (45), Chapter 1):

$$E[e^{tX}] = M(p;p+q;t) \tag{9}$$

and of course the characteristic function is $M(p;p+q;it)$.

The mean deviation is

$$\frac{2p^p q^q}{(p+q)^{p+q}} \cdot \frac{\Gamma(p+q)}{\Gamma(p)\Gamma(q)}. \tag{10}$$

For p and q large, using Stirling's approximation to the gamma function, the mean deviation is approximately

$$(11)\qquad \sqrt{\frac{2pq}{\pi(p+q)}}\cdot\frac{1}{p+q}\,(1+\tfrac{1}{12}(p+q)^{-1}-\tfrac{1}{12}p^{-1}-\tfrac{1}{12}q^{-1})$$

and

$$\frac{\text{mean deviation}}{\text{standard deviation}} \doteqdot \sqrt{\frac{2}{\pi}}\,(1+\tfrac{7}{12}(p+q)^{-1}-\tfrac{1}{12}p^{-1}-\tfrac{1}{12}q^{-1}).$$

If $p > 1$ and $q > 1$ then $p_X(x) \to 0$ as $x \to 0$ or $x \to 1$; if $0 < p < 1$, $p_X(x) \to \infty$ as $x \to 0$ and if $0 < q < 1$, $p_X(x) \to \infty$ as $x \to 1$. If $p = 1(0)$, $p_X(x)$ tends to a finite non-zero value as $x \to 0(1)$.

If $p > 1$ and $q > 1$, the density function has a single mode at $x = (p-1)/(p+q-2)$. If $p < 1$ and $q < 1$ there is an antimode (minimum value) of $p_X(x)$ at this value of x. Such distributions are called U-shaped beta (or Type I or II) distributions. If $(p-1)(q-1)$ is not positive the probability density function does not have a mode or an antimode for $0 < x < 1$. Such distributions are called *J-shaped* beta (or Type I) distributions.

If $p = q$, the distribution is symmetrical about $x = \frac{1}{2}$.

For all positive values of p and q, there are points of inflexion at

$$(12)\qquad \frac{p-1}{p+q-2} \pm \frac{1}{p+q-2}\sqrt{\frac{(p-1)(q-1)}{p+q-3}}$$

provided these values are real and lie between 0 and 1. Note that, as for all Pearson curves, the points of inflexion are equidistant from the modes.

The expected value $p/(p+q)$ depends on the ratio p/q. If this ratio is kept constant, but p and q both increased, the variance decreases, and the (standardized) distribution tends to the unit normal distribution.

Some of the properties of beta distributions, described in this section, are exhibited in Figure 1. Note that if the values of p and q are interchanged, the distribution is 'reflected' about $x = \frac{1}{2}$.

4. Estimation

Estimation of all four parameters in distribution (1) can be effected by equating sample and population values of the first four moments. Calculation of a, b, p and q from the mean μ_1' and central moments μ_2, μ_3, μ_4 is effected using the following formulas (Elderton and Johnson [16]). Putting

$$r = 6(\beta_2 - \beta_1 - 1)/(6 + 3\beta_1 - 2\beta_2)$$

then

$$(13)\qquad p, q = \tfrac{1}{2}r\{1 \pm (r+2)\sqrt{\beta_1\{(r+2)^2\beta_1 + 16(r+1)\}^{-1}}\}$$

with $p \lesseqgtr q$ according $\alpha_3 = \sqrt{\beta_1} \gtreqless 0$. Also

$$(14)\qquad (p-1)/(q-1) = [\text{mode }(Y) - a]/[b - \text{mode }(Y)]$$

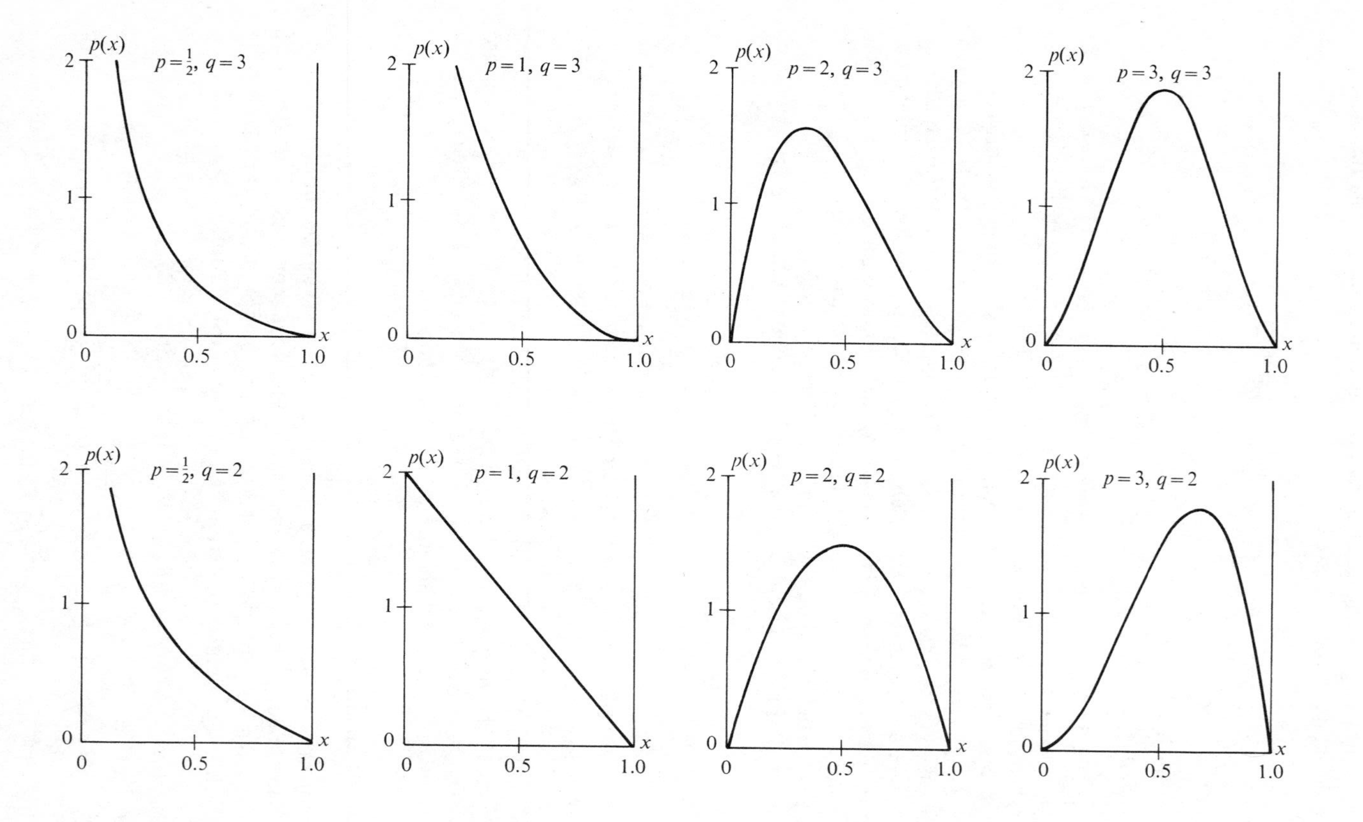
p(x)
p = 1/2, q = 3
p = 1, q = 3
p = 2, q = 3
p = 3, q = 3
p = 1/2, q = 2
p = 1, q = 2
p = 2, q = 2
p = 3, q = 2
x
0
0.5
1.0
1
2

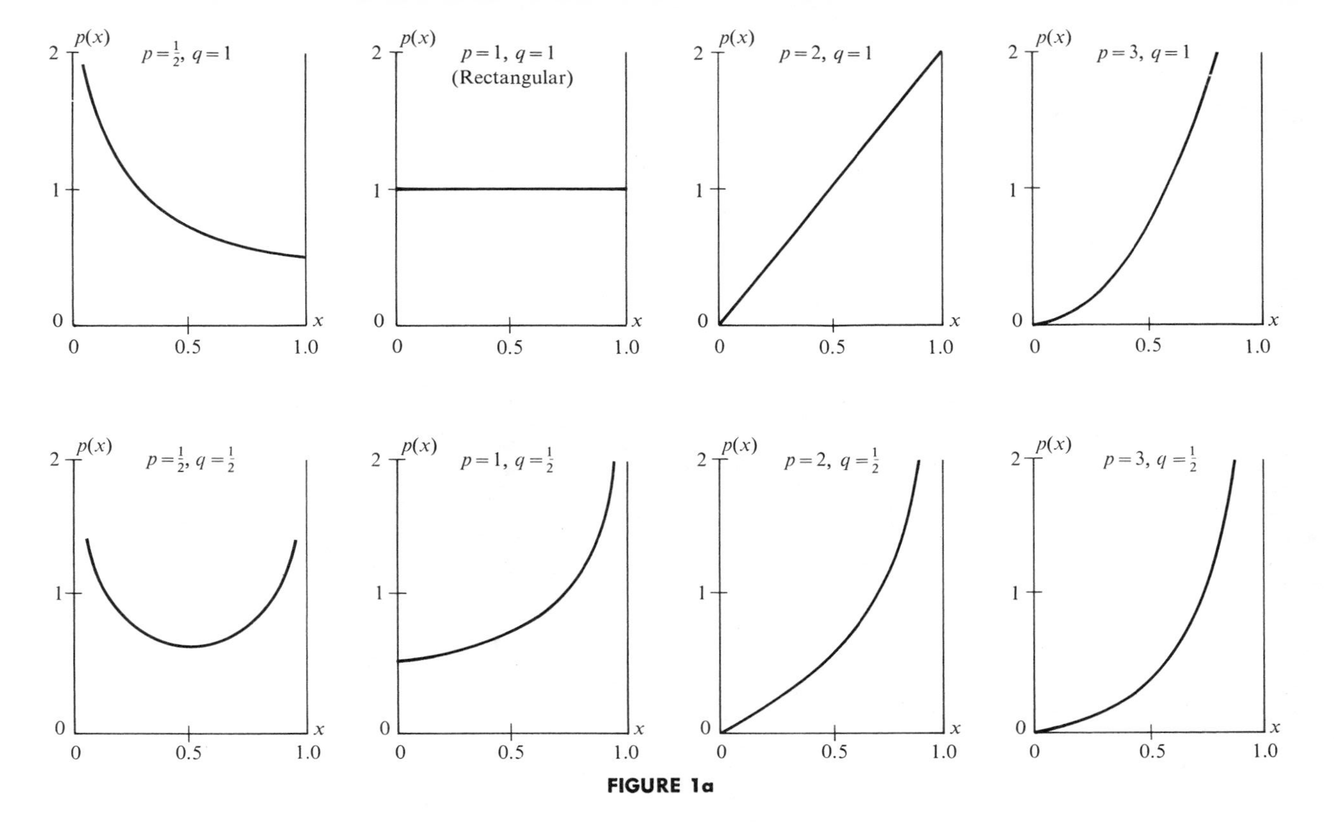

FIGURE 1a

Beta Density Functions

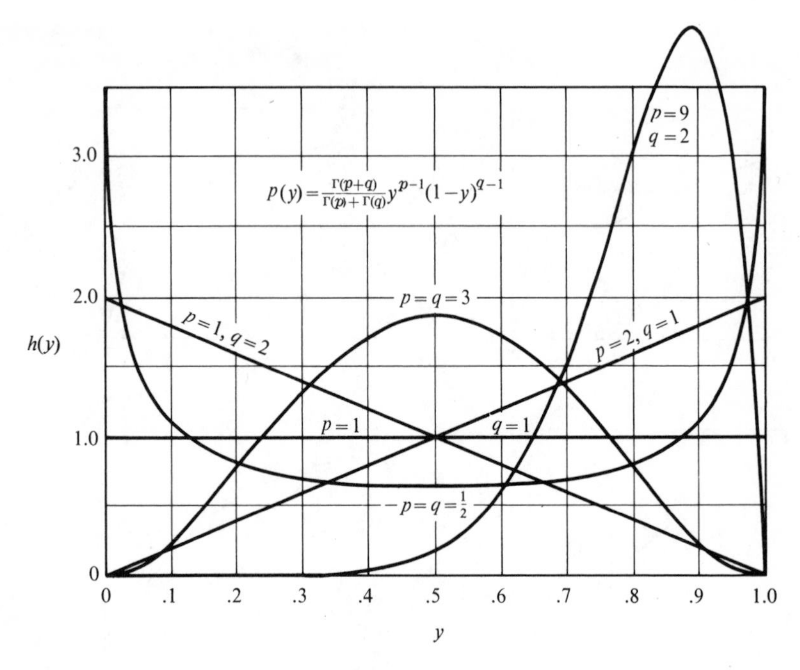

FIGURE 1b

Beta Density Functions

(where mode $(Y) = a + (b - a)(p - 1)/(p + q - 2)$) and

$$b - a = \tfrac{1}{2}\sqrt{\mu_2}\sqrt{(r + 2)^2\beta_2 + 16(r + 1)}. \tag{15}$$

If the values of a and b are known then only the first and second moments need be used, giving

$$\begin{aligned} \mu_1' &= a + (b - a)p/(p + q) \\ \mu_2 &= (b - a)^2 pq(p + q)^{-2}(p + q + 1)^{-1} \end{aligned} \tag{16}$$

whence

$$\frac{\mu_1' - a}{b - a} = \frac{p}{p + q} \tag{17}$$

$$\frac{\mu_2}{(b - a)^2} = \frac{p}{p + q}\left(1 - \frac{p}{p + q}\right)\frac{1}{p + q + 1}\,. \tag{18}$$

Thus

$$p + q = \frac{\mu_1' - a}{b - a}\left(1 - \frac{\mu_1' - a}{b - a}\right)\Big/\left(\frac{\mu_2}{(b - a)^2}\right) - 1, \tag{19}$$

$$p = \left(\frac{\mu_1' - a}{b - a}\right)^2\left(1 - \frac{\mu_1' - a}{b - a}\right)\left(\frac{\mu_2}{(b - a)^2}\right)^{-1} - \frac{\mu_1' - a}{b - a}\,. \tag{20}$$

If a and b are known and $Y_1, Y_2, \ldots, Y_n$ are independent random variables each having distribution (1), the maximum likelihood equations for estimators $\hat{p}$, $\hat{q}$ of p, q respectively are

$$\psi(\hat{p}) - \psi(\hat{p} + \hat{q}) = n^{-1} \sum_{j=1}^{n} \log \left(\frac{Y_j - a}{b - a} \right) \tag{21.1}$$

$$\psi(\hat{q}) - \psi(\hat{p} + \hat{q}) = n^{-1} \sum_{j=1}^{n} \log \left(\frac{b - Y_j}{b - a} \right) \tag{21.2}$$

where $\psi(\cdot)$ is the digamma function (Equation (32), Chapter 1). Equations (21.1) and (21.2) must be solved by trial and error. If $\hat{p}$ and $\hat{q}$ are not too small, the approximation

$$\psi(t) \doteqdot \log (t - \tfrac{1}{2})$$

may be used. Then approximate values of $(\hat{p} - \frac{1}{2})/(\hat{p} + \hat{q} - \frac{1}{2})$ and $(\hat{q} - \frac{1}{2})/(\hat{p} + \hat{q} - \frac{1}{2})$ can be obtained from (21.1) and (21.2), whence follow, as first approximations to p and q:

(22.1)

$$\hat{p} \doteqdot \frac{1}{2}\left\{1 - \prod_{j=1}^{n} \left(\frac{b - Y_j}{b - a}\right)^{\frac{1}{n}}\right\} \Big/ \left\{1 - \prod_{j=1}^{n} \left(\frac{Y_j - a}{b - a}\right)^{\frac{1}{n}} - \prod_{j=1}^{n} \left(\frac{b - Y_j}{b - a}\right)^{\frac{1}{n}}\right\};$$

(22.2)

$$\hat{q} \doteqdot \frac{1}{2}\left\{1 - \prod_{j=1}^{n} \left(\frac{Y_j - a}{b - a}\right)^{\frac{1}{n}}\right\} \Big/ \left\{1 - \prod_{j=1}^{n} \left(\frac{Y_j - a}{b - a}\right)^{\frac{1}{n}} - \prod_{j=1}^{n} \left(\frac{b - Y_j}{b - a}\right)^{\frac{1}{n}}\right\}.$$

Starting from these values, solutions of (21.1) and (21.2) can be obtained by an iterative process. Gnanadesikan *et al.* [17] give exact numerical solutions for a few cases.

The asymptotic covariance matrix of $\sqrt{n}\,\hat{p}$ and $\sqrt{n}\,\hat{q}$ (as $n \to \infty$) is

$$[\psi'(p)\psi'(q) - \psi'(p + q)\{\psi'(p) + \psi'(q)\}]^{-1} \times \begin{pmatrix} \psi'(q) - \psi'(p + q) & \psi'(p + q) \\ \psi'(p + q) & \psi'(p) - \psi'(p + q) \end{pmatrix}. \tag{23}$$

Introducing approximations for $\psi'(\cdot)$ we have, for large values of p and q

$$\operatorname{var}(\hat{p}) \doteqdot p(2p - 1)n^{-1}; \quad \operatorname{var}(\hat{q}) \doteqdot q(2q - 1)n^{-1} \tag{24}$$
$$\operatorname{corr}(\hat{p},\hat{q}) \doteqdot \sqrt{(1 - 2p^{-1})(1 - 2q^{-1})}.$$

If a and b are unknown, and maximum likelihood estimators of a, b, p and q are required, the above procedure can be repeated using a succession of trial values of a and b, until the pair (a,b) for which the maximized likelihood (given a and b) is as great as possible, is attained. This procedure is practicable with mechanical calculating aids (even desk calculators) but is likely to be long and expensive, so it is not generally employed. Fortunately, in many cases a and b, or at least one of these parameters, can be assigned known values.

If only the r smallest values $X'_1, X'_2, \ldots, X'_r$ are available, the maximum likelihood equations are:

$$\frac{r}{n}\log\left[\left(\prod_{j=1}^{r} X'_j\right)^{1/r}\right] = \psi(\hat{p}) - \psi(\hat{p}+\hat{q}) - \left(1-\frac{r}{n}\right)\frac{\partial}{\partial\hat{p}}\log\left[\int_{X'_r}^{1} t^{\hat{p}-1}(1-t)^{\hat{q}-1}\,dt\right]$$

and

$$\frac{r}{n}\log\left[\left(\prod_{j=1}^{r}(1-X'_j)\right)^{1/r}\right] = \psi(\hat{q}) - \psi(\hat{p}+\hat{q}) - \left(1-\frac{r}{n}\right)\frac{\partial}{\partial\hat{q}}\log\left[\int_{X'_r}^{1} t^{\hat{p}-1}(1-t)^{\hat{q}-1}\,dt\right]$$

(Gnanadesikan *et al.* [17]).

If one of the values p and q is known, the equations are much simpler to solve. In particular, for the standard power-function distribution ($q = 1$), the maximum likelihood estimator of p is

$$\hat{p} = \left[n^{-1}\sum_{j=1}^{n}\log X_j\right]^{-1}$$

and we have

$$n \operatorname{var} \hat{p} \doteqdot p^2.$$

A moment estimator of p in this case is

$$\tilde{p} = \bar{X}(1-\bar{X})^{-1}$$

for which

$$n \operatorname{var} \tilde{p} \doteqdot p(p+1)^2(p+2)^{-1}.$$

Note that $(\operatorname{var}\hat{p})/(\operatorname{var}\tilde{p}) \doteqdot p(p+2)(p+1)^{-2}$. The asymptotic relative efficiency of $\tilde{p}$ increases with p; it is as high as 75% for $p = 1$ and tends to 100% as $p \to \infty$, but tends to zero as $p \to 0$. Further discussion of power-function distributions will be found in Section 7, Chapter 19.

It is of interest to note that Guenther [19] has shown that for the special case when q is known to be 1, and $a = 0$, $b = 1$, so that

$$p_{X_j}(x_j) = p x_j^{p-1} \quad (0 < x_j < 1; j = 1, \ldots, n),$$

the minimum variance unbiased estimator of p is $-(n-1)\left[\sum_{j=1}^{n}\log X_j\right]^{-1}$. Its variance is $p^2(n-2)^{-1}$, while the Cramér-Rao lower bound (Chapter 1, Section 6) is p^2n^{-1}.

5. Applications

The beta distribution is one of the most frequently employed to fit theoretical distributions. Usually the range of variation (a,b) of such distributions is known, and fitting is effected by equating the first and second moments of the theoretical and fitted curve. No random sample values enter into this calculation, so that maximum likelihood methods are inapplicable, and *à fortiori* arguments based on asymptotic efficiency are irrelevant.

An example of some importance is the use of beta distributions to fit the distribution of certain likelihood ratios used in statistical tests, or some function thereof. Usually the range of variation of the likelihood ratio is known to be from zero to one, and that of any monotonic function of the likelihood ratio can be derived from this knowledge. If the likelihood ratio is based on n independent identically distributed random variables, it is often found that a usefully good fit can be obtained by supposing

$$(\text{likelihood ratio})^{2/n}$$

to have a beta distribution with $a = 0$, $b = 1$. Use of the power $2n^{-1}$ is suggested by Wilks' theorem that, under certain fairly broad conditions, $-2n^{-1}$ log (likelihood ratio) has an asymptotic χ^2 distribution (as $n \to \infty$). (See also Chapter 28, Section 9 where some additional cases are discussed.) Of course a general power c might be used, and c, as well as p and q, fitted, if a substantially improved fit could be expected by using this method. This would be equivalent to fitting a 'generalized' beta distribution, to be described in Section 7.

A standard Type I distribution has been found to give good approximation when fitted (by equation of first two moments) to relative frequencies of a binomial distribution. (Benedetti [10].) If the binomial parameters are N, ω then the approximate value for the probability that the binomial variable is less than r is

$$I_{(r-\frac{1}{2})}((N-1)\omega, (N-1)(1-\omega))$$

as compared with the exact value

$$I_{1-\omega}(N-r+1, r).$$

(See also Equation (34), Chapter 3.) Numerical comparisons given in Benedetti [10], and also in Johnson [27], show that except in the 'tails' (probabilities not between 0.05 and 0.95) a good practical approximation is obtained for $N \geq 50$ and $0.1 \leq \omega \leq 0.9$.

Of recent years, a fashionable use for the beta distribution has been as a 'prior' distribution for a binomial proportion (see Section 2(x), Chapter 8). While this leads to some conveniently simple mathematics, and the beta distribution is often referred to as a 'natural' prior distribution for the binomial parameter p (in that the posterior distribution obtained by its use is also of the same form) there seems to be little definite evidence in its favor (see, for example, Barnard [8] in the discussion of Horsnell [24]).

A direct application of beta distribution to the analysis of Markov processes with "uncertain" transition probabilities was developed in a monograph by Silver [47].

6. Approximations and Tables

6.1 *Approximations*

A number of approximations to the incomplete beta function ratio $I_x(p,q)$ have been described in Section 8 of Chapter 3. Relevant references, given at the end of this chapter for convenience, include Aroian [3], [4], Cadwell [12], Hartley and Fitch [21], Nair [38], Pearson and Pearson [42], Thompson [50], Wise [53], [54]. Here we add only three further approximations which came to our attention subsequent to the completion of Volume I of this book. The first approximation is one of a number proposed by Peizer and Pratt [43] [44]. For the incomplete beta function ratio, their approximations are:

$$I_x(p,q) \doteqdot \Phi(Z)$$

where Z is of form

$$Z = \frac{d}{|q - \frac{1}{2} - n(1-x)|}\left\{\frac{2}{1+(6n)^{-1}}\left[(q-\tfrac{1}{2})\log\left\{\frac{q-\frac{1}{2}}{n(1-x)}\right\} + (p-\tfrac{1}{2})\log\left\{\frac{p-\frac{1}{2}}{nx}\right\}\right]\right\}^{\frac{1}{2}} \tag{25}$$

with $n = p + q - 1$; and the value of d is either

(a) $d = q - \frac{1}{3} - (n + \frac{1}{3})(1 - x)$

or

(b) $d = q - \frac{1}{3} - (n + \frac{1}{3})(1 - x) + \frac{1}{5}(x/q) - (1 - x)/p + (x - \frac{1}{2})/(p + q)$.

Value (b) generally gives the more accurate results. Using this value of d, the error in $I_x(p,q)$ is less than 0.001 for $p, q \geq 2$ and less than 0.01 for $p, q \geq 1$. Limits on the *proportional* error are

$$\frac{|\Phi(Z) - I_x(p,q)|}{I_x(p,q)} < \begin{cases} 0.01 \text{ if } p, q \geq 3 \text{ and } 0.2 \leq R \leq 5.0 \\ 0.02 \text{ if } p, q \geq 1.75 \text{ and } 0.125 \leq R \leq 8 \\ 0.03 \text{ if } p, q \geq 1.5 \text{ and } 0.1 \leq R \leq 10 \end{cases} \tag{26}$$

where

$$R = \frac{(q - \frac{1}{2})x}{(p - \frac{1}{2})(1 - x)}.$$

The second approximation is more of a general theoretical nature: Let $p(x;kn,ln)$ be the density of beta distribution with parameters kn, ln, where k and l are fixed positive constants and let $n \to \infty$. It is known (see e.g. Chapter 13, Section 2) that the normalized density $p_1(x) = \sigma p(a + \sigma x,kn,ln)$ converges to the standardized normal density $N(0,1)$. Krolikówska [32] investigated the behavior of the main term of the deviation $|p_1(x) - N(0,1)|$ when $n \to \infty$. She found that if $k \neq l$ this deviation is of order $n^{-\frac{1}{2}}$, if $k = l$ the deviation is of order n^{-1}.

Molina [36] has obtained the following approximation to the incomplete beta function:

$$I_x(p,q) \doteqdot \sum_{j=0}^{6} \frac{A_j}{j!} \left(\frac{z}{N}\right)^{q+j} D(q + j,z) \tag{27}$$

where

$$N = p + \tfrac{1}{2}q - \tfrac{1}{2}; \quad z = -N \log x;$$
$$A_0 = 1; \quad A_1 = A_3 = A_5 = 0;$$
$$A_2 = \tfrac{1}{12}(q - 1); \quad A_4 = \tfrac{1}{240}(q - 1)(5q - 7);$$
$$A_6 = \tfrac{1}{4032}(q - 1)(35q^2 - 112q + 93);$$

and

$$D(a,b) = \int_0^1 t^{a-1}e^{-bt}\, dt = b^{-a}\Gamma_b(a),$$

where $\Gamma_b(a)$ is an incomplete gamma function, as defined in Chapter 17. (See also appendix in [17], where more computational details are given.)

Woods and Posten [56] have constructed computer programs based on the Fourier expansion

$$I_x(p,q) = 1 - \theta\pi^{-1} - \sum_{j=1}^{\infty} b_j \sin j\theta$$

where

$$\theta = \cos^{-1}(2x - 1)$$
$$b_1 = 2\pi^{-1}(p - q)(p + q)^{-1}$$
$$b_2 = \pi^{-1}(p + q)^{-1}(p + q + 1)^{-1}\{2(p - q)^2 - (p + q)(p + q - 1)\}$$
$$(j + 2)b_{j+2} = (j + p + q + 1)^{-1}\{2(p - q)(j + 1)b_{j+1} + (j + 1 - p - q)jb_j\}.$$

They found that (for m sufficiently large) if the infinite series be terminated at the term containing b_m the error is less than

$$\tfrac{1}{2}m|b_m| \{\min(p,q)\}^{-1} \qquad \text{if } p \neq q,$$
$$\tfrac{1}{4}m|b_m| \{\min(p,q)\}^{-1} \qquad \text{if } p = q \text{ and } m \text{ is even.}$$

In some special cases there are simple explicit formulas for the b's.

(i) $p = q$:

$$b_{2j-1} = 0$$
$$b_{2j} = (j\pi)^{-1} \prod_{i=1}^{j} \left(\frac{2j - 1 - 2p}{2j - 1 + 2p}\right);$$

(ii) $q = \frac{1}{2}$:

$$b_j = 2(j\pi)^{-1} \prod_{i=1}^{j} \left(\frac{2p - 2j + 1}{2p + 2j - 1}\right).$$

If both p and q have fractional part equal to one-half, then $b_j = 0$ for $j \geq p + q$.

Kalinin [29] has obtained the expansion

$$[B(p,q)]^{-1}x^{p-1}(1-x)^{q-1} = (pq)^{-\frac{1}{2}}(p+q)^{\frac{3}{2}}Z(y)\exp\left[\sum_{j=1}^{\mu-1} W_j p^{-\frac{1}{4}j} + R_\mu\right]$$

where $y = (p+q)(p^{-1}+q^{-1})^{\frac{1}{2}}\{x - p/(p+q)\}$, $R_\mu = O(p^{-\frac{1}{2}\mu})$, and

$$W_j = \frac{y^j}{j}\left(\frac{p}{p+q}\right)^{j/2}\left[\left(\frac{p}{q}\right)^{j/2} - \left(\frac{q}{p}\right)^{j/2}\right]$$
$$- \frac{y^{j+2}}{j+2}\left(\frac{p}{p+q}\right)^{j/2+1}\left[\left(\frac{p}{q}\right)^{j/2} - \left(\frac{q}{p}\right)^{j/2+1}\right], \quad j \text{ odd};$$

$$W_{2k} = \frac{y^{2k}}{2k}\left(\frac{p}{p+q}\right)^k\left[\left(\frac{p}{q}\right)^k + \left(\frac{q}{p}\right)^k\right]$$
$$- \frac{y^{2k+2}}{2k+2}\left(\frac{p}{p+q}\right)^{k+1}\left[\left(\frac{p}{q}\right)^k + \left(\frac{q}{p}\right)^{k+1}\right]$$
$$+ \frac{B_{k+1}}{k(k+1)}\left[\left(\frac{p}{p+q}\right)^k - \left(\frac{p}{q}\right)^k - 1\right].$$

Kalinin also gives similar (but rather more complicated) expansions for the density functions of gamma, F, and t distributions. Because of their complexity, we have not reproduced these in the appropriate chapters.

6.2 *Tables*

The first edition of Pearson's tables [41] included values of $I_x(p,q)$ to 7 decimal places for

$$p, q = 0.5(0.5)11.0(1)50 \text{ with } p \geq q$$
$$x = 0.00(0.01)1.00.$$

The second edition also includes values of $I_x(p,q)$ to 7 decimal places for

$$p = 11.5(1.0)14.5 \text{ with } q = 0.5$$
$$x = 0.00(0.01)1.00;$$

and to 8 decimal places for $p = 0.5(0.5)11.0(1)16$; $q = 0.5$, $x = 0.988(0.0005)0.9985$, $0.9988(0.0001)0.9999$ and for $q = 1.0(0.5)3.0$, $x = 0.988(0.001)0.999$. Values to 7 decimal places are also given for $x = 0.975, 0.985$.

Further values have been calculated by Osborn and Madey [39]. These cover values of p, q in a region where interpolation using Pearson's tables is difficult. Values of $B_x(p,q)$ and $I_x(p,q)$ are to 5 significant figures for

$$p, q = 0.50(0.05)2.00$$
$$x = 0.10(0.01)1.00$$

The formulas used for calculation were

$$B_x(p,q) = x^p\left[\frac{1}{p} + \frac{1-q}{p+1}x + \frac{(1-q)(2-q)}{2!(p+2)}x^2 + \cdots\right]$$

for $0 < x \leq \frac{1}{2}$, and

$$B_x(p,q) = B_{0.5}(p,q) + \frac{1-w^q}{(q)2^q} + \frac{(1-p)(1-w^{q+1})}{1!(q+1)2^{q+1}}$$
$$+ \frac{(1-p)(2-p)(1-w^{q+2})}{2!(q+2)2^{q+2}} + \cdots$$

with $w = 2(1-x)$ for $\frac{1}{2} < x < 1$.

Percentage points of the beta distribution have been tabulated by Thompson [49], Clark [14], Harter [20] and Vogler [51]. Thompson gave values of $X(P;p;q)$, where

$$I_{X(P;p;q)}(p,q) = P$$

to 5 significant figures for

$$p = 0.5(0.5)15.0,\ 20,\ 30,\ 60$$
$$q = 0.5(0.5)5.0,\ 6,\ 7.5,\ 10,\ 12,\ 15,\ 20,\ 30,\ 60$$
$$P = 0.50,\ 0.25,\ 0.10,\ 0.05,\ 0.025,\ 0.01,\ 0.005$$

These tables are included in [40], the third edition of which contains also values for $P = 0.0025$ and 0.001, calculated by Amos [1]. Harter [20] gives $X(P;p;q)$ to seven significant figures for $p, q = 1(1)40$; $P = 0.0001, 0.0005, 0.001, 0.005, 0.01, 0.025, 0.05, 0.1(0.1)0.5$. Vogler [51] gives $X(P;p;q)$, and also $B(p,q)$, to six significant figures for

$$p = 0.50(0.05)1.00,\ 1.1,\ 1.25(0.25)2.50,\ 3.0(0.5)5.0,\ 6,\ 7.5,\ 10,\ 12,\ 15,\ 20,\ 30,\ 60$$
$$q = 0.5(0.5)5.0,\ 6,\ 7.5,\ 10,\ 12,\ 15,\ 20,\ 30,\ 60$$
$$P = 0.0001,\ 0.001,\ 0.005,\ 0.01,\ 0.025,\ 0.05,\ 0.1,\ 0.25,\ 0.5.$$

7. Related Distributions

If X has distribution (2), then by the transformation

$$T = X/(1-X)$$

we obtain a distribution with probability density function

$$p_T(y) = \frac{1}{B(p,q)}\left(\frac{t}{1+t}\right)^{p-1}\left(\frac{1}{1+t}\right)^{q-1}\frac{1}{(1+t)^2} \tag{28}$$
$$= \frac{1}{B(p,q)}\frac{t^{p-1}}{(1+t)^{p+q}} \qquad (t > 0)$$

This is a standard form of Pearson Type VI distribution, sometimes called a *beta-prime* distribution (Keeping [30]). This relationship between the Type VI and beta distributions is exploited in Chapter 26 to express the probability integral of the central F distribution in terms of an incomplete beta function ratio.

Little work has been done on 'Weibullized' beta distributions, obtained by supposing a random variable Z to be such that (for some c) Z^c has a standard beta distribution. It is, of course, easy to write down the moments of such a distribution, since

$$\mu_r'(Z) = E[Z^r] = E[(Z^c)^{r/c}]$$

and so $\mu_r'(Z)$ is the (r/c)-th moment of the corresponding beta distribution.

If X has a power-function distribution (Section 1) then X^{-1} has a Pareto distribution (Chapter 19).

Compound beta distributions may be formed by ascribing distributions to some or all of the parameters p, q, a and b of distribution (1). However, such distributions have not been used much in applied statistical work. Continuous distributions for p and q usually present analytical difficulties, owing to the presence of the beta function $B(p,q)$ in (1) (or (2)).

As a matter of interest, we may note that if we suppose that p and q are positive integers and that for $p + q = s$ (≥ 2) fixed, p is equally likely to take values 1, 2 . . . $(s - 1)$ then the probability density function of X, given $p + q = s$ is

$$p_X(x \mid s) = \frac{(s-1) \sum_{p-1=0}^{s-2} \binom{s-2}{p-1} x^{p-1}(1-x)^{s-2-(p-1)}}{(s-1)} = 1 \qquad (0 < x < 1),$$

that is, the distribution is rectangular (Chapter 25).

It follows that *whatever the distribution of* $(p + q)$, the compound distribution is rectangular if the conditional distribution of p, given $(p + q)$ is discrete rectangular as described above.

Johnson [27] has considered the distribution of $\log [X/(1 - X)]$ when X has distribution (2). The moment generating function of $\log [X/(1 - X)]$ is

$$E[X^t(1-X)^{-t}] = B(p+t, q-t)/B(p,q) = \frac{\Gamma(p+t)\Gamma(q-t)}{\Gamma(p)\Gamma(q)} \tag{29}$$

whence the rth cumulant is

$$\kappa_r = \psi^{(r-1)}(p) + (-1)^r\psi^{(r-1)}(q). \tag{30}$$

Approximating to the polygamma functions we obtain (for p,q large)

$$\begin{aligned} \alpha_3^2 &= \beta_1 \doteqdot p^{-1} + q^{-1} - 4(p+q)^{-1} \\ \alpha_4 &= \beta_2 \doteqdot 3 + 2p^{-1} + 2q^{-1} - 6(p+q)^{-1}. \end{aligned} \tag{31}$$

These may be compared with the approximations (derived from (8.3) and

(8.4)) for the moment ratios of X,

$$\beta_1(X) \doteqdot 4(p^{-1} + q^{-1}) - 16(p + q)^{-1} \tag{32}$$
$$\beta_2(X) \doteqdot 3 + 6(p^{-1} + q^{-1}) - 30(p + q)^{-1}.$$

We have already remarked (Section 2) that beta distributions can be generated as the distributions of ratios $X_1/(X_1 + X_2)$ where X_1, X_2 are independent random variables having chi-squared distributions.

If one or both of X_1, X_2 have *non-central* χ^2 distributions the distribution of the ratio is called a *non-central beta distribution* [23], [46]. These distributions are evidently related to singly or doubly non-central F distributions — and will be discussed in Chapter 30 — in the same way as beta distributions are related to central F distributions (see earlier part of this section).

REFERENCES

[1] Amos, D. E. (1963). Additional percentage points for the incomplete beta distribution, *Biometrika*, **50,** 449–457.

[2] Arnold, L. (1967). On the asymptotic distribution of the eigenvalues of random matrices, *Journal of Mathematical Analysis and Applications*, **20,** 262–268.

[3] Aroian, L. A. (1941). Continued fractions for the incomplete beta function, *Annals of Mathematical Statistics*, **12,** 218–223. (Correction: **30,** 1265).

[4] Aroian, L. A. (1950). On the levels of significance of the incomplete beta function and the *F*-distributions, *Biometrika*, **37,** 219–223.

[5] Bancroft, T. A. (1945). Note on an identity in the incomplete beta function, *Annals of Mathematical Statistics*, **16,** 98–99.

[6] Bancroft, T. A. (1949). Some recurrence formulae in the incomplete beta function ratio, *Annals of Mathematical Statistics*, **20,** 451–455.

[7] Bánkővi, G. (1964). A note on the generation of beta distributed and gamma distributed random variables, *Publications of the Mathematical Institute, Hungarian Academy of Sciences, Series A*, **9,** 555–562.

[8] Barnard, G. A. (1957). (Discussion of reference [24]), *Journal of the Royal Statistical Society, Series A*, **120,** 148–191.

[9] Békéssy, A. (1964). Remarks on beta distributed random numbers, *Publications of the Mathematical Institute, Hungarian Academy of Sciences, Series A*, **9,** 565–571.

[10] Benedetti, C. (1956). Sulla rappresentabilità di una distribuzione binomiale mediante une distribuzione B e vice versa, *Metron*, **18,** 121–131.

[11] Bol'shev, L. N. (1964). Some applications of Pearson transformations, *Review of the International Statistical Institute*, **32,** 14–15.

[12] Cadwell, J. H. (1952). An approximation to the symmetrical incomplete beta function, *Biometrika*, **39,** 204–207.

[13] Champernowne, D. G. (1953). The economics of sequential sampling procedures for defectives, *Applied Statistics*, **2,** 118–130.

[14] Clark, R. E. (1953). Percentage points of the incomplete beta function, *Journal of the American Statistical Association*, **48,** 831–843.

[15] Dodge, H. F. and Romig, H. G. (1959). *Sampling Inspection Tables, Single and Double Sampling*, 2nd Edition, New York: John Wiley & Sons, Inc. (Originally published in *Bell System Technical Journal*, **20,** (1941)).

[16] Elderton, W. P. and Johnson, N. L. (1969). *Systems of Frequency Curves*, London: Cambridge University Press.

[17] Gnanadesikan, R., Pinkham, R. S. and Hughes, L. P. (1967). Maximum likelihood estimation of the parameters of the beta distribution from smallest order statistics, *Technometerics*, **9,** 607–620.

[18] Govindarajulu, Z. and Hubacker, N. W. (1962). *Percentile points of order statistics in samples from beta, normal, chi (1 d.f.) populations*, Case Institute of Technology, Statistical Laboratory Publication No. 101. Cleveland, Ohio.

[19] Guenther, W. C. (1967). A best statistic with variance not equal to the Cramér-Rao lower bound, *American Mathematical Monthly*, **74,** 993–994.

[20] Harter, H. L. (1964). *New Tables of the Incomplete Gamma-Function Ratio and of Percentage Points of the Chi-square and Beta Distributions*, Aerospace Research Laboratories, Wright-Patterson Air Force Base, Ohio.

[21] Hartley, H. O. and Fitch, E. R. (1951). A chart for the incomplete beta-function and the cumulative binomial distribution, *Biometrika*, **38,** 423–426.

[22] Heselden, G. P. M. (1955). Some inequalities satisfied by incomplete beta functions, *Skandinavisk Aktuarietidskrift*, **38,** 192–200.

[23] Hodges, J. L., Jr. (1955). On the noncentral beta-distribution, *Annals of Mathematical Statistics*, **26,** 648–653.

[24] Horsnell, G. (1957). Economical acceptance sampling schemes, *Journal of the Royal Statistical Society, Series A*, **120,** 148–191.

[25] Jambunathan, M. V. (1954). Some properties of beta and gamma distributions, *Annals of the Mathematical Statistics*, **25,** 401–405.

[26] Jöhnk, M. D. (1964). Erzeugung von beta-verteilten und gamma-verteilten Zufallszahlen, *Metrika*, **8,** 5–15.

[27] Johnson, N. L. (1949). Systems of frequency curves generated by methods of translation, *Biometrika*, **36,** 149–176.

[28] Johnson, N. L. (1960). An approximation to the multinomial distribution: Some properties and applications, *Biometrika*, **47,** 93–102.

[29] Kalinin, V. M. (1968). Limit properties of probability distributions, *Akademia Nauk SSSR*, Steklov Institute, **104,** 88–134. (English translation to appear.)

[30] Keeping, E. S. (1962). *Introduction to Statistical Inference*, New York: D. Van Nostrand.

[31] Kotlarski, I. (1962). On groups of n independent random variables whose product follows the beta distribution, *Colloquium Mathematicum*, **9,** 325–332.

[32] Królikowska, T. (1966). O pewnej wlasności granicznej rozkladu beta, *Zeszyty Naukowe Politechniki Lódzkiej*, **77,** 15–20.

[33] Laha, R. G. (1964). On a problem connected with beta and gamma distributions. *Transactions of the American Mathematical Society*, **113,** 287–298.

[34] Massonie, J. P. (1965). Utilization d'une statistique linéaire correcte pour estimer la vitesse d'évasion des étoiles dans le plan galactique, *Publications de l'Institut de Statistique de l'Université de Paris*, **14,** 1–62.

[35] Mendenhall, W. and Lehman, E. H. (1960). An approximation to the negative moments of the positive binomial useful in life testing, *Technometrics*, **2,** 227–242.

[36] Molina, E. C. (1932). An expansion for Laplacian integrals in terms of incomplete gamma functions, and some applications, *Bell System Technical Journal*, **11,** 563–575.

[37] Murthy, V. K. (1960). On the distribution of the sum of circular serial correlation coefficients and the effect of non-normality on its distribution, (Abstract), *Annals of Mathematical Statistics*, **31,** 239–240.

[38] Nair, K. R. (1948). The Studentized form of the extreme mean square test in the analysis of variance, *Biometrika*, **35,** 16–31.

[39] Osborn, D. and Madey, R. (1968). The incomplete beta function and its ratio to the complete beta function, *Mathematics of Computation*, **22,** 159–162.

[40] Pearson, E. S. and Hartley, H. O. (1954). *Biometrika Tables for Statisticians,* **1,** London: Cambridge University Press (Second edition, 1958; Third edition, 1966.)

[41] Pearson, K. (Ed.) (1968). *Tables of the Incomplete Beta Function*, 2nd edition, London: Cambridge University Press. (1st edition, 1934)

[42] Pearson, K. and Pearson, Margaret V. (1935). On the numerical evaluation of high order incomplete Eulerian integrals, *Biometrika*, **27,** 409–412.

[43] Peizer, D. B. and Pratt, J. W. (1968). A normal approximation for binomial, *F*, Beta, and other common, related tail probabilities, I, *Journal of the American Statistical Association*, **63,** 1416–1456.

[44] Pratt, J. W. (1968). A normal approximation for binomial, *F*, Beta, and other common, related tail probabilities, II, *Journal of the American Statistical Association*, **63,** 1457–1483.

[45] Scheffé, H. (1944). Note on the use of the tables of percentage points of the incomplete beta function to calculate small sample confidence intervals for a binomial p, *Biometrika*, **33,** 181.

[46] Seber, G. A. F. (1963). The non-central chi-squared and beta distributions, *Biometrika*, **50,** 542–544.

[47] Silver, E. A. (1963). *Markovian decision processes with uncertain transition probabilities or rewards*, Operations Research Center, M.I.T., Technical Report, No. **1.**

[48] Stuart, A. (1962). Gamma-distributed products of independent random variables, *Biometrika*, **49,** 564–565.

[49] Thompson, Catherine M. (1941). Tables of percentage points of the incomplete beta-function, *Biometrika*, **32,** 151–181.

[50] Thomson, D. H. (1947). Approximate formulae for the percentage points of the incomplete beta function and of the χ^2 distribution, *Biometrika*, **34,** 368–372.

[51] Vogler, L. E. (1964). *Percentage Points of the Beta Distribution*, Technical Note No. 215, National Bureau of Standards, Colorado.

[52] Wigner, F. (1958). On the distribution of the roots of certain symmetric matrices, *Annals of Mathematics*, **67,** 325–327.

[53] Wise, M. E. (1950). The incomplete beta function as a contour integral and a quickly converging series for its inverse, *Biometrika*, **37,** 208–218.

[54] Wise, M. E. (1960). On normalizing the incomplete beta-function for fitting to dosage-response curves, *Biometrika*, **47,** 173–175.

[55] Wishart, J. (1927). On the approximate quadrature of certain skew curves, with an account of the researches of Thomas Bayes, *Biometrika*, **19,** 1–38.

[56] Woods, J. D. and Posten, H. O. (1968). *Fourier series and Chebyshev polynomials in statistical distribution theory*, Research Report No. 37, Department of Statistics, University of Connecticut, Storrs, Connecticut.

25

Uniform or Rectangular Distribution

1. Definition

The *uniform* or *rectangular* distribution is the special case of the beta distribution (Equation (1), Chapter 24) obtained by putting the exponents p and q each equal to 1. The name is based on the appearance of a graph of its probability density function (Figure 1a). The function shown here has the analytic expression

$$p_Y(y) = (2h)^{-1} \qquad (a - h \leq y \leq a + h; \quad h > 0). \tag{1}$$

The cumulative distribution function, on the other hand, has the appearance shown in Figure 1b, which represents

$$F_Y(y) = \begin{cases} 0 & y < a - h \\ (2h)^{-1}(y - a - h) & a - h \leq y \leq a + h \\ 1 & y > a + h. \end{cases} \tag{2}$$

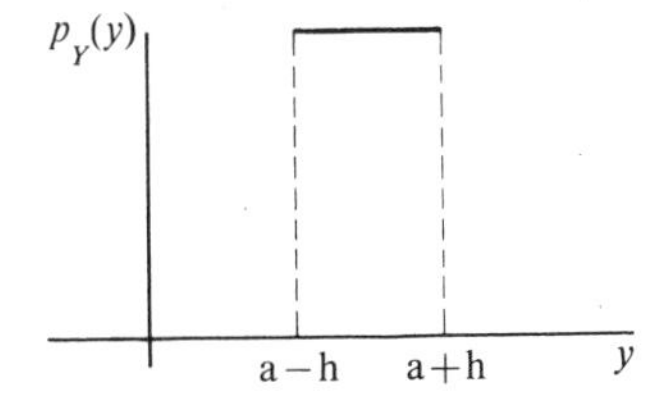

FIGURE 1a

Rectangular Density Function

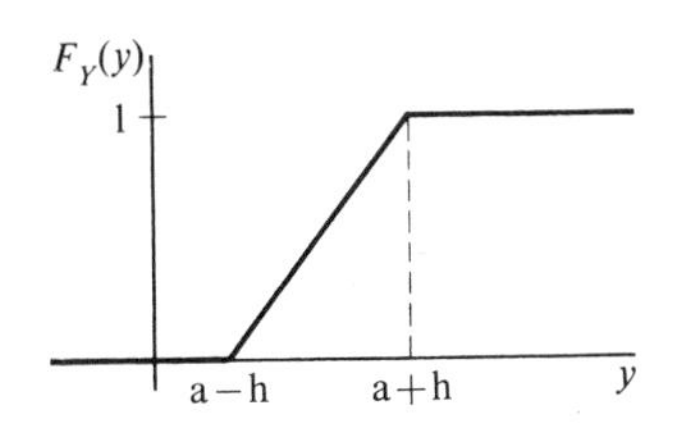

FIGURE 1b

Rectangular Cumulative Distribution Function

A general formula, alternative to (1), is

$$p_Y(y) = (\beta - \alpha)^{-1} \qquad (\alpha \leq y \leq \beta). \tag{1)'$$

Standard forms are obtained by taking $\alpha = 0$; or $\alpha = 0$, $\beta = 1$ (the *unit uniform* distribution); or $\alpha = -\frac{1}{2}$, $\beta = \frac{1}{2}$. We will usually take the form (1).

2. Genesis

A uniform distribution, with $a = 0$, $h = \frac{1}{2} \times 10^{-k}$ is often used to represent the distribution of "rounding-off" errors in values tabulated to the nearest k decimal places. Of course, if the rounding were applied to figures expressed in a binary scale, we would have $h = 2^{-(k+1)}$.

Nagaev and Mukhin [28] have investigated conditions under which a rectangular distribution of rounding-off errors is to be expected. In particular they have shown that if $X_1, X_2, \ldots$ are independent random variables with characteristic functions $E[e^{itX_j}] = \phi_j(t)$, then for any positive integer a, a necessary and sufficient condition for

$$\lim_{n\to\infty} \Pr\left\{\sum_{j=1}^{n} X_j - a\left[\sum_{j=1}^{n} X_j/a\right] \leq x\right\} = x/a \qquad (0 \leq x < a)$$

is that

$$\prod_{j=1}^{\infty} \phi_j(2\pi k/a) = 0 \qquad \text{for } k = \pm 1, \pm 2, \ldots.$$

$\left[\sum_{j=1}^{n} X_j/a\right]$ in the above expression means "integral part of $\sum_{j=1}^{n} X_j/a$", so that $\sum_{j=1}^{n} X_j - a\left[\sum_{j=1}^{n} X_j/a\right]$ is the "rounding-off" error of $\sum_{j=1}^{n} X_j$ in units of a. The condition $\prod_{j=1}^{n} \phi_j(2\pi k/a) = 0$ is certainly satisfied if all X's have the same distribution and

$$|\phi_j(2\pi k/a)| < \eta_k < 1 \quad \text{for infinitely many } j \text{ and some } \eta_k < 1.$$

Holewijn [16] has shown that if

$$\lim_{n\to\infty} n^{-1} \sum_{j=1}^{n} \phi_j(2\pi k) = 0 \qquad (k = 1, 2, \ldots)$$

then the fractional part sequence is uniformly distributed "almost certainly" — that is, nearly all the variables $\{X_j - [X_j]\}$ have unit uniform distributions.

A rectangular distribution also arises as a result of the *probability integral transformation.* If X is a continuous variable and $\Pr[X \leq x] = F(x)$, then $F(X)$ is distributed according to (1) with $a = h = \frac{1}{2}$. This result (first employed by

Fisher [12]) has been applied in a number of ways (Durbin [10], Pearson [30], Stephens [45]) in various techniques for combining results of statistical tests (see Section 8).

3. Historical Remarks

The uniform distribution is so natural a one that it has probably been in use far more than can be inferred from printed records. Among such records we may mention in particular descriptions of the use of the distribution by Bayes [3] and Laplace [23].

Some particular historical interest attaches to the distribution of the sum of independent random variables each having the *same* rectangular distribution. Seal [42] gives an extensive bibliography on this subject.

4. Generating Functions and Moments

The expected value of a random variable Y with probability density function (1) is a. The distribution is symmetrical, and all odd central moments are zero. If r is even, the rth central moment of Y is

$$\mu_r(Y) = (r+1)^{-1}h^r. \tag{3}$$

It follows that $\operatorname{var}(Y) = \frac{1}{3}h^2$ and that $\sqrt{\beta_1} = \alpha_3 = 0$ and $\beta_2 = \alpha_4 = 1.8$.

Formula (3) gives the values of the rth *absolute* central moment for all positive r. In particular, the mean deviation is $\frac{1}{2}h$. Hence, for this distribution

$$\frac{\text{mean deviation}}{\text{standard deviation}} = \frac{\sqrt{3}}{2} = 0.866.$$

The characteristic function is $E(e^{itY}) = e^{ita}\left(\frac{\sin(th)}{th}\right)$. The moment generating function is $E(e^{\tau Y}) = e^{\tau a}\left(\frac{\sinh(\tau h)}{\tau h}\right)$. The cumulants are

$$\begin{cases} \kappa_1 = a; \quad \kappa_r = 0 & (r > 1 \text{ and odd}) \\ \kappa_r = 2^{r-1}h^r B_r/r & (r \text{ even}) \end{cases} \tag{4}$$

(B_r is the rth Bernoulli number — Chapter 1, Section 3).

The information generating function (($u-1$)-th frequency-moment) is

$$T(u) = (2h)^{1-u}. \tag{5}$$

The entropy is $-T'(1) = \log_e(2h)$.

5. Estimation of Parameters

If observations in a random sample be represented by independent random variables $Y_1, Y_2, \ldots, Y_n$ each with distribution (1), the likelihood function is equal to $(2h)^{-n}$ for $a - h \leq \min(Y_1,\ldots,Y_n) \leq \max(Y_1,\ldots,Y_n) \leq a + h$. This likelihood is maximized by making h as small as possible, i.e. the maximum likelihood estimator of h is

$$\hat{h} = \tfrac{1}{2}[\text{range}\,(Y_1, Y_2,\ldots,Y_n)].$$

The maximum likelihood estimator of a is therefore

$$\begin{aligned}\hat{a} &= \tfrac{1}{2}[\min(Y_1,\ldots,Y_n) + \max(Y_1,\ldots,Y_n)]\\ &= \text{mid-range}\,(Y_1,\ldots,Y_n).\end{aligned}$$

In fact, the best linear unbiased estimators of h and a are

$$(n+1)(n-1)^{-1}\hat{h} \text{ and } \hat{a}$$

respectively. The variances of these estimators are

$$2h^2(n-1)^{-1}(n+2)^{-1} \qquad \text{and} \qquad 2h^2(n+1)^{-1}(n+2)^{-1}$$

respectively.

The estimators $\hat{a}$ and $\hat{h}$ are uncorrelated, but not independent. In fact their joint probability density function is

$$p_{\hat{a},\hat{h}}(a',h') = \left(\frac{2}{h}\right)^{n-1} n(n-1)h'^{n-2} \qquad (6)$$

$$\left\{\begin{matrix} 0 \leq h' \leq h; \\ 0 \leq a' - a - h' \leq a' - a + h' \leq 2h \end{matrix}\right\}.$$

The distribution of $\hat{h}$ alone is

$$p_{\hat{h}}(h') = 2\left(\frac{2}{h}\right)^{n-1} n(n-1)h'^{n-2}(h-h') \qquad (0 \leq h' < h). \qquad (7)$$

The distribution of $\hat{a}$ alone is

$$p_{\hat{a}}(a') = \left(\frac{2}{h}\right)^{n-1} n(\tfrac{1}{2} - |a' - a - \tfrac{1}{2}|)^{n-2} \qquad (8)$$

$$(a - h \leq a' \leq a + h).$$

Cumulative probabilities $\Pr[\hat{h} < H]$, $\Pr[\hat{a} < A]$ are easily evaluated from these probability density functions. The arithmetic mean $\bar{Y}$ and the median $\tilde{Y}$ are also unbiased estimators of the parameter a.

It was noted by Carlton [7] that

$$\text{var}(\hat{a})/\text{var}(\bar{Y}) = 6n/(n+1)(n+2)$$

and

$$\text{var}(\tilde{Y})/\text{var}(\bar{Y}) = 3n/(n+2).$$

If n is allowed to increase, $\text{var}(\hat{a})/\text{var}(\bar{Y})$ tends to zero and $\text{var}(\tilde{Y})/\text{var}(\bar{Y})$ tends o 3, so that the efficiency of the mean is zero, and the median is only one-third s efficient as the mean. (However, as $\hat{a}$ does not have a normal limiting distribution, we should not strictly apply the concept of efficiency here.)

It will be noted that these estimators are functions of order statistics (in fact of the smallest and greatest values). The theory of order statistics for random amples from rectangular distributions is remarkably simple. This has led to he proliferation of methods of estimation based on various combinations of order statistics. We will discuss some of these in the next section.

Sometimes the initial value $(a - h)$ is known, and it is desired to estimate $2h$. Such a case, for example, is arrived at as a continuous approximation to a problem proposed by Schrödinger (referred to in [13]), calling for estimation of a number N (positive integer) given n independent integers each equally likely to be $1, 2, 3, \ldots N$. Geary [13], by regarding N as a parameter which could take any positive value, and assuming the observed values Y_i are uniformly distributed continuous variables, used the model

$$p_{Y_i}(y) = N^{-1} \qquad (0 \leq y \leq N) \tag{9}$$

which is (1) with $a = \frac{1}{2}N$, $h = \frac{1}{2}N$.

Johnson [19] discussed four estimators of N, each depending only on $\max(Y_1, Y_2, \ldots, Y_n) = Y'_n$. They were

(i) the maximum likelihood estimator Y'_n,
(ii) the minimum mean square error estimator $\hat{N}' = (n + 2)Y'_n/(n + 1)$,
(iii) the unbiased estimator $\hat{N}'' = (n + 1)Y'_n/n$
(iv) the closest estimator $\hat{N}''' = 2^{1/n}Y'_n$.

The following Table 1 (from [19] shows how the method of comparison affects the assessment of "relative merit" of estimators. Column (a) shows the ratio $\dfrac{\text{mean square error of } \hat{N}'''}{\text{mean square error of } \hat{N}'}$; column ($b$) shows the "closeness criterion"

$$\Pr[|\hat{N}''' - N| < |\hat{N}' - N|].$$

TABLE 1

n	(a)	(b)
1	1.333	0.571
2	1.029	0.530
3	1.001	0.505
4	1.002	0.509
5	1.008	0.519
10	1.036	0.542
20	1.061	0.556
∞	1.094	0.571

6. Estimation Using Order Statistics — Censored Samples

The variances and covariances of the ordered variables $Y_1' \leq Y_2' \leq \cdots \leq Y_n'$ corresponding to a random sample of size n from (1) are

$$\begin{aligned} \text{var}(Y_r') &= 4h^2 r(n - r + 1)(n + 1)^{-2}(n + 2)^{-1} \\ \text{var}(Y_r', Y_s') &= 4h^2 r(n - s + 1)(n + 1)^{-2}(n + 2)^{-1} \qquad (r \leq s). \end{aligned} \tag{10}$$

Together with the formula

$$E(Y_r') = r/(n + 1) \tag{11}$$

these formulas make it possible to obtain best linear unbiased estimators of a and h. Such estimators were discussed by Lloyd [24], Sarhan [38] and Sarhan and Greenberg [39]. The more important results are summarized below.

If the smallest r_1 and largest r_2 values (out of a random sample of total size n) are omitted, the best linear unbiased estimator of a is

$$\begin{aligned} \hat{a}^* = \tfrac{1}{2}[&(n - 2r_2 - 1)(\text{least observed value}) \\ &+ (n - 2r_1 - 1)(\text{greatest observed value})](n - r_1 - r_2 - 1)^{-1}. \end{aligned} \tag{12}$$

The best linear unbiased estimator of h is

$$\begin{aligned} \hat{h}^* = \tfrac{1}{2}&(n + 1)[(\text{greatest observed value}) - (\text{least observed value})] \\ &\times (n - r_1 - r_2 - 1)^{-1}. \end{aligned} \tag{13}$$

The variances of these estimators are

$$\begin{aligned} \text{var}(\hat{a}^*) = h^2&[(r_1 + 1)(n - 2r_2 - 1) + (r_2 + 1)(n - 2r_1 - 1)] \\ &\times (n + 1)^{-1}(n + 2)^{-1}(n - r_1 - r_2 - 1)^{-1} \end{aligned} \tag{14}$$

$$\text{var}(\hat{h}^*) = h^2(r_1 + r_2 + 2)(n + 2)^{-1}(n - r_1 - r_2 - 1)^{-1}. \tag{15}$$

The correlation between $\hat{a}^*$ and $\hat{h}^*$ is

$$(r_2 - r_1)\left[\frac{n + 1}{(r_1 + r_2 + 2)[(r_1 + 1)(n - 2r_2 - 1) + (r_2 + 1)(n - 2r_1 - 1)]}\right]^{\frac{1}{2}}.$$

It will be noted that we again use only the least and greatest of all the observed values (just as when all the sample values are available).

By putting $r_1 = 0$, the special case of censoring only the largest r_2 values is obtained. In this case

$$\begin{aligned} \hat{a}^* = \tfrac{1}{2}[&(n - 2r_2 - 1)(\text{least observed value}) \\ &+ (n - 1)(\text{greatest observed value})](n - r_2 - 1)^{-1} \end{aligned} \tag{16}$$

and

$$\begin{aligned} \hat{h}^* = \tfrac{1}{2}&(n + 1)[(\text{greatest observed value}) \\ &- (\text{least observed value})](n - r_2 - 1)^{-1}. \end{aligned} \tag{17}$$

(The corresponding estimator of the lower limit of the range of variation is

$$\hat{a}^* - \hat{h}^* = \tfrac{1}{2}[2(n - r_2)(\text{least observed value}) - 2(\text{greatest observed value})](n - r_2 - 1)^{-1}.)$$

Putting $r_2 = 0$, the special case of censoring only the r_1 smallest values is obtained.

7. Tables of Random Numbers

A set of numbers 0 to 9 chosen independently of each other with each number equally likely to be any one of the ten digits 0–9 is known as a *table of random numbers*. Although the distribution corresponding to the individual recorded number is a discrete rectangular distribution (Chapter 10, Section 2), good approximations to samples from (continuous) rectangular distributions are obtainable by combining several integers together and applying an appropriate linear transformation. For example, taking groups of four numbers, adjoining a 5 at the right-hand end, and dividing by 100,000 gives a good approximation to random samples from a standard uniform distribution (over the interval 0 to 1). The best known sets of random numbers are (in chronological order):

1927	Tippett [48]
1938, 1940	Kendall and Babington Smith [21], [22]
1955	RAND [32]
1966	Clark [8]

We may also note that Kadyrov [20] published tables of random numbers in 1936, which were found unsatisfactory by Bol'shev [5] in 1964.

By combining uniformly distributed random variables in various ways, a number of other distributions can be built up. For example (see beginning of Section 8), if Y has a standard rectangular distribution (over the interval 0 to 1) then $-2 \log Y$ is distributed as χ^2 with two degrees of freedom. More sophisticated examples will be found in Marsaglia [25].

8. Related Distributions

The rectangular distribution is a special form of beta distribution, as mentioned in Chapter 24.

If X is rectangularly distributed according to (1) with $a = h = \tfrac{1}{2}$, then $Z = -2 \log X$ has the exponential distribution (Chapter 18)

$$p_Z(z) = \tfrac{1}{2}e^{-z/2} \qquad (z > 0). \tag{18}$$

That is, Z is distributed as χ^2 with two degrees of freedom. This relationship is used in the construction of certain methods for the combination of tests. Values of independent random variables $Z_1, Z_2, \ldots, Z_k$ obtained from k independent tests may be combined by checking the value of $\sum_{i=1}^{k} Z_i$ against a χ^2 distribution with $2k$ degrees of freedom.

The distribution of the arithmetic mean of n independent random variables, each having the same rectangular distribution, can be expressed in terms of a number (n, in fact) of polynomial arcs. If S_n denotes the sum of n independent random variables each having probability density function (1) with $a = h = \frac{1}{2}$, then

$$\Pr[S_n \le x] = \sum_j (-1)^j \binom{n}{j} (x - j)^n/n! \qquad (0 \le x \le n) \tag{19}$$

where the summation $\sum_j$ is over all $j < x$. The remarkable history of investigation of this distribution has been mentioned in Section 3. When $n = 2$, the distribution of $\overline{X}_2 = \frac{1}{2}S_2$ is a symmetrical *triangular distribution*:

$$p_{\overline{X}_2}(x) = \begin{cases} (x - a + h)/h^2 & (a - h \le x \le a) \\ (a + h - x)/h^2 & (a \le x \le a + h). \end{cases} \tag{20}$$

The standard triangular distribution (see Ayyangar [2]) is represented (possibly after linear transformation) by a probability density function of form

$$p_X(x) = \begin{cases} 2x/H & 0 \le x \le H \\ 2(1 - x)/(1 - H) & H \le x \le 1 \end{cases}$$

The graph of $p_X(x)$, sketched in Figure 2, indicates why the name *triangular* is given to these distributions. If $H = \frac{1}{2}$, the distribution is symmetrical. Symmetrical triangular distributions have been called *tine distributions* (Schmidt [41]).

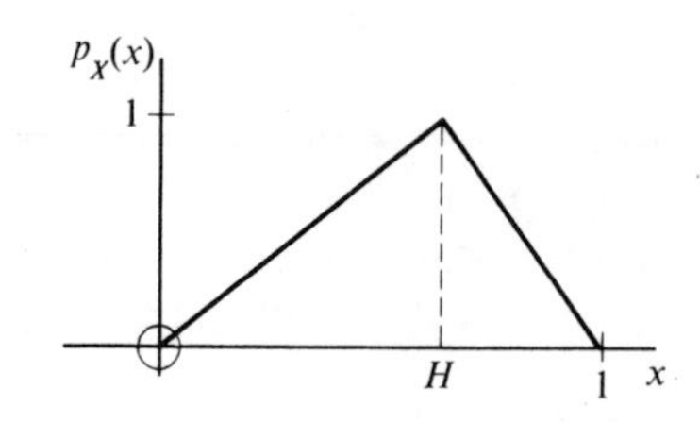

FIGURE 2

Triangular Density Function

The rth moment about H is

$$\begin{aligned} {}_H\mu_r(X) &= E[(X - H)^r] \\ &= 2(-1)^r \int_0^H (H - x)^r (x/H)\,dx + 2\int_H^1 (x - H)^r \{(1 - x)/(1 - H)\}\,dx \\ &= \frac{2[(-1)^r H^{r+1} + (1 - H)^{r+1}]}{(r + 1)(r + 2)}. \end{aligned}$$

The expected value is

$$\begin{aligned} H + {}_H\mu_1(X) &= H + \tfrac{1}{3}\{-H^2 + (1 - H)^2\} \\ &= \tfrac{1}{3}(1 + H) \end{aligned}$$

and the variance is

$$
\begin{aligned}
{}_H\mu_2(X) - [{}_H\mu_1(X)]^2 &= \tfrac{1}{6}\{H^3 + (1 - H)^3\} - \tfrac{1}{9}(1 - 2H)^2 \\
&= \tfrac{1}{18}(1 - H + H^2).
\end{aligned}
$$

The median is at $\sqrt{\frac{1}{2}}\max(H,1 - H)$. The mean deviation is

$$\tfrac{2}{81}H^{-1}(1 + H)^3 \qquad \text{if } H > \tfrac{1}{2}$$

and

$$\tfrac{2}{81}(1 - H)^{-1}(2 - H)^3 \qquad \text{if } H < \tfrac{1}{2}.$$

The ratio (mean deviation)/(standard deviation) has the following values:

H	0.5	0.6	0.7	0.8	0.9
Ratio	0.816	0.820	0.827	0.833	0.837

When each X_j can have different values of a_j and h_j, the distribution of S_n is much more complicated. Tach [47] gives 5 decimal place tables of the cumulative distribution function of S_n for $n = 2, 3, 4$ with $a_j = 0$ for all j, and various h_j (subject to $\sum_{j=1}^{n} h_j = 1$).

The joint distribution of the differences of successive order statistics is a *Dirichlet distribution* (Chapter 40). Thus if $Y_1' \leq Y_2' \leq \cdots \leq Y_n'$ are defined as in Section 6 with $a = h$, and $V_i = Y_i' - Y_{i-1}'$ $(i = 1,\ldots,(n + 1))$; $Y_0' = a - h$; $Y_{n+1}' = a + h$ then

$$p_{V_1,\ldots,V_n}(v_1,\ldots,v_n) = n!(2h)^{-n} \qquad \left(0 \leq v_i;\ \sum_{i=1}^{n} v_i \leq 2h\right). \tag{21}$$

Putting $h = \frac{1}{2}$ (corresponding to a range of 1) we have

$$p_{V_1,\ldots,V_n}(v_1,\ldots,v_n) = n! \qquad \left(0 \leq v_i;\ \sum_{i=1}^{n} v_i \leq 1\right). \tag{22}$$

More insight into the nature of this joint distribution is gained by noting that the same distribution (22) would be obtained if

$$V_i = W_i \Big/ \sum_{j=1}^{n+1} W_j \qquad (i = 1,2,\ldots,(n + 1))$$

where $W_1, W_2, \ldots, W_{n+1}$ are mutually independent random variables each distributed as χ^2 with two degrees of freedom. From this point of view it is clear that the ratio $(Y_{n-s+1}' - Y_s')/(Y_{n-r+1}' - Y_r')$, with $s > r$, has a beta distribution with parameters $(n - 2s + 1)$, $2(s - r)$ (Chapter 24, Section 2). Criteria of this kind were suggested by David and Johnson [9] as tests for kurtosis, when the probability integral transformation can be used. As another example, noting that the range $(Y_n' - Y_1')$ is equal to $\sum_{i=2}^{n} V_i$, it is clear that range has a beta distribution with parameters $(n - 1)$, 2 as can be confirmed from (7).

The distribution of the ratio (U) of ranges calculated from independent random samples of sizes n', n'' from distributions (1) with the same value of h respectively has been studied by Rider [34] [35]. The probability density function of this ratio (sample size n' in numerator) is

$$p_U(u) = \begin{cases} C[(n' + n'')u^{n'-2} - (n' + n'' - 2)u^{n'-1}] & 0 \leq u \leq 1 \\ C[(n' + n'')u^{-n''} - (n' + n'' - 2)u^{-n''-1}] & 1 \leq u \end{cases} \tag{23}$$

with $C = n'(n' - 1)n''(n'' - 1)/[(n' + n'')(n' + n'' - 1)(n' + n'' - 2)]$. Tables of upper 10%, 5% and 1% points are given in [34]; more extensive tables are given in [35].

The distribution of the ratio (V) of maximum values when both random samples come from distributions with $a = h$ has been studied by Murty [27]. He found that

$$p_V(v) = \begin{cases} \dfrac{n'n''}{n' + n''} v^{n'-1} & \text{for } 0 \leq v \leq 1; \\ \dfrac{n'n''}{n' + n''} v^{-n''-1} & \text{for } v \geq 1. \end{cases} \tag{24}$$

Murty gives tables of the upper 5% point of the distribution of max (V, V^{-1}) for $n', n'' = 2(1)20$, (n' = sample size with *greater* maximum value).

These last two criteria might be used in testing for identity of two rectangular distributions with respect to changes in range of variation (with, in the second case, known initial point). A further criterion introduced by Hyrenius [18] uses "cross-ranges". If we denote by L', L'' the smallest observations in the two samples, and U', U'' the largest, then the cross ranges are $U'' - L'$, $U' - L''$. For $V = (U'' - L')/(U' - L'')$, Hyrenius obtained the probability density function

$$p_V(v) = \begin{cases} \dfrac{(n' - 1)n''}{n' + n'' - 1} v^{n''-1} & \text{for } 0 \leq v \leq 1; \\ \dfrac{(n' - 1)n''}{n' + n'' - 1} v^{-n'} & \text{for } v \geq 1. \end{cases} \tag{25}$$

He also considers $T = \dfrac{L'' - L'}{U' - L'} = V +$ (ratio of ranges of the two samples).

Among other distributions derived (at least in part) from the rectangular distribution there may be mentioned:

(a) The Student's t (Chapter 27) when the variables $X_1, \ldots, X_n$ from which the t is calculated are independent and have a common rectangular distribution. Rider [33] showed that for samples of size $n = 2$, the probability density function is of form $\frac{1}{2}(1 + |t|)^{-2}$. Perlo [31] derived the distribution for samples of size $n = 3$ and Siddiqui [44] obtained a general formal result for any size of sample, and gives some inequalities for the cumulative distribution function. The cumulative probability for samples of size 3 is

$$\Pr[t \le t_0] = \begin{cases} 1 + \sqrt{3}\,\dfrac{t_0\sqrt{1-t_0^2}}{4-t_0^2} - 3\sqrt{3}\,\dfrac{t_0}{(4-t_0^2)^2}\sqrt{1-t_0^2} & (0 \le t_0 \le \tfrac{1}{2}); \\ \dfrac{3\sqrt{3}\,t_0}{(4-t_0^2)^{3/2}}\tan^{-1}\dfrac{\sqrt{t_0^2-4}}{\sqrt{3}\,(t_0+2)} - \dfrac{9}{2(t_0+1)(t_0^2-4)} & (t_0 \ge \tfrac{1}{2}). \end{cases} \tag{26}$$

(b) Various distributions arising from the construction of tests of adherence to a specified distribution, using the probability integral transformation. Pearson [30] pointed out that if Y has distribution (1) with $a = h$, so do $2h - Y$, $2|Y - h|$, etc. Durbin [10] showed that if the differences $V_1, V_2, \ldots, V_{n+1}$ (as defined in the beginning of this section) are arranged in ascending order of magnitude $V_1' \le V_2' \le \cdots \le V_{n+1}'$, the joint probability density function of $V_1', \ldots, V_n'$ is (for $h = \frac{1}{2}$):

$$p_{V_1',\ldots,V_n'}(y_1,y_2,\ldots,y_n) = (n+1)!n! \qquad \left(0 \le y_1 \le \cdots \le y_n;\ \sum_{i=1}^n y_i \le 1\right). \tag{27}$$

Also if

$$G_j = (n+2-j)(V_j' - V_{j-1}')$$

(Sukhatme's [46] transformation) then

$$p_{G_1,\ldots,G_n}(g_1,g_2,\ldots,g_n) = n! \qquad \left(g_j \ge 0;\ 1 \ge \sum_{j=1}^n g_j\right) \tag{28}$$

and hence the quantities

$$W_r' = \sum_{j=1}^r G_j \qquad (r = 1,\ldots,n)$$

have the *same* joint distribution as the original ordered variables Y_r', and so any function of the W''s has the same distribution as the corresponding function of the Y_r''s.

Durbin gives a considerable number of references to other work on "random division of an interval", i.e. on uniform distributions.

(c) The distribution of the ratio of a rectangular variable (Y) to an independent normal variable (Z) is described by Broadbent [6]. Calculations are facilitated by noting that, for example, with $a = 0$ in (1),

$$\begin{aligned} \Pr[Y/Z \le K(>0) \mid Z > 0] &= \Pr[Y \le KZ \mid Z > 0] \\ &= \int_0^\infty p_Z(z)F_Y(Kz)\,dz \Big/ \int_0^\infty p_Z(z)\,dz. \end{aligned} \tag{29}$$

(d) There is a connection between a uniform angular distribution around a semicircle and a Cauchy distribution on a line. If, in Figure 3, θ has the uniform probability density

$$p_\theta(t) = \pi^{-1} \qquad \left(-\frac{\pi}{2} \leq t \leq \frac{\pi}{2}\right) \tag{30}$$

then the probability density function of $X = PQ \left(\angle OPQ = \frac{\pi}{2}\right)$ is

$$p_X(x) = \pi^{-1}\frac{d}{dx}\tan^{-1}\frac{x}{|OP|} = \frac{1}{\pi|OP|}\frac{1}{[1 + (x/|OP|)^2]}. \tag{31}$$

Hence X has a Cauchy distribution (Chapter 16).

FIGURE 3

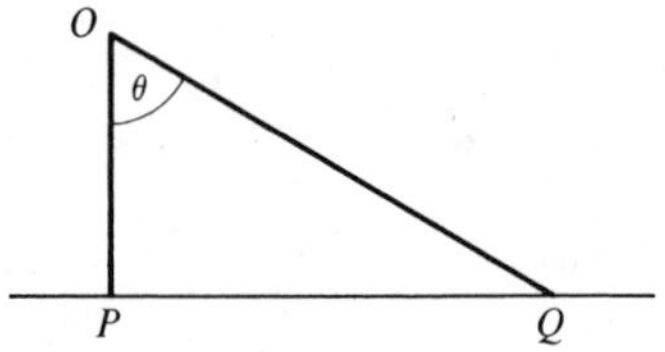

9. Applications

The distribution finds innumerable uses in the construction of mathematical models of physical, biological and social phenomena. In this section we give a few examples.

9.1 *Corrections for Grouping*

The uses of the rectangular distribution to represent the distribution of rounding-off errors, and in connection with the probability integral transformation, have already been mentioned. This distribution also plays a central role in the derivation of *Sheppard's corrections* [43], which adjust the values of sample moments to allow for (on the average) effects of grouping.

Suppose that the probability density function of X is $p_X(x)$. If the observed value is not in fact X, but the nearest value ($\tilde{X}$) to X in the set of values $\{\alpha + jh\}$ where j can take any (positive, negative or zero) integer value, then

$$\Pr[\tilde{X} = \alpha + jh] = \int_{\alpha+(j-\frac{1}{2})h}^{\alpha+(j+\frac{1}{2})h} p_X(x)\, dx. \tag{32}$$

We now seek to find an "average" relation between the cumulants of X and those of $\tilde{X}$. If we assume that $(\tilde{X} - X)$ has a uniform distribution then

$$\tilde{X} = X + Y$$

where Y has a uniform distribution with a probability density function

$$p_Y(y) = h^{-1} \qquad (-\tfrac{1}{2}h \leq y \leq \tfrac{1}{2}h).$$

Also X and Y are mutually independent, so

$$\kappa_r(\tilde{X}) = \kappa_r(X) + \kappa_r(Y). \tag{33}$$

Using (4) we find

$$\begin{aligned} \kappa_1(X) &= \kappa_1(\tilde{X}) \\ \kappa_2(X) &= \kappa_2(\tilde{X}) - \tfrac{1}{12}h^2 \\ \kappa_3(X) &= \kappa_3(\tilde{X}) \\ \kappa_4(X) &= \kappa_4(\tilde{X}) + \tfrac{1}{120}h^4 \end{aligned} \tag{34}$$

The last equation implies $\mu_4(X) = \mu_4(\tilde{X}) - \tfrac{1}{2}h^2\mu_2(\tilde{X}) + \tfrac{7}{240}h^4$.

9.2 *Life Testing*

We give the distribution of a statistic based on the r smallest of n independent observation from a unit uniform distribution. In life testing terminology, this statistic includes as special cases (i) the sum of the r earliest failure times, (ii) the total observed life up to the rth failure, and (iii) the sum of all n failure times.

The statistic is given by

$$T_{r,m}^{(n)} = t_1 + t_2 + \cdots + t_r + (m - r)t_r \tag{35}$$

where $t_i = t_i^{(n)}$ is the ith smallest of n independent observations and m is greater than $r - 1$ but is not necessarily an integer. For the case where $m = n$ this statistic can be interpreted as the total observed life in a life-testing experiment without replacement. The results given are due to Gupta and Sobel [15].

The density function and the distribution function of $T_{r,m}^{(n)}$ are given by:

$$p_{T_{r,m}^{(n)}}(t) = A_{r-1,m}^{(n,n)}(m - t)$$

and

$$F_{T_{r,m}^{(n)}}(t) = 1 - \frac{1}{n+1} A_{r-1,m}^{(n,n+1)}(m - t)$$

respectively, where $0 \leq t \leq m$ and

$$A_{r-1,m}^{(n,n)}(m - t) = \frac{n}{(r-1)!}\left\{\binom{r-1}{0}\frac{(m-t)^{n-1}}{m^{n-r+1}} - \binom{r-1}{1} \times \frac{(m-1-t)^{n-1}}{(m-1)^{n-r+1}} + \cdots\right\}.$$

From the above we get as special cases the densities and cumulative distribution functions of (i) $T_{r,r}^{(n)}$, (ii) $T_{r,n}^{(n)}$, and (iii) $T_{n,n}^{(n)}$.

The first two moments and the variance of $T_{r,m}^{(n)}$ are given by the following formulas:

$$E[T_{r,m}^{(n)}] = \frac{r(2m - r + 1)}{(2n + 1)}$$

$$E[(T_{r,m}^{(n)})^2] = \frac{r(r + 1)}{12(n + 1)(n + 2)}[12m^2 - 12m(r - 1) + (r - 1)(3r - 2)]$$

$$\text{Var}[T_{r,m}^{(n)}] = \frac{r(n - r + 1)(2m - r + 1)^2}{4(n + 1)^2(n + 2)} + \frac{r(r + 1)(r - 1)}{12(n + 1)(n + 2)}.$$

The statistic $T_{r,m}^{(n)}$ is asymptotically normal for $r = \lambda n$, $m = \gamma n$ (γ and λ fixed with $0 < \lambda \leq 1$ and $\lambda \leq \gamma < \infty$) and $n \to \infty$.

9.3 *Traffic Flow Applications*

Allan [1] has applied the rectangular distribution to form a model of the distribution of traffic along a straight road. The road is divided into intervals each of length a, and it is supposed that for each interval there are probabilities p of 1 vehicle being in the interval, q of no vehicles being there. (For purposes of this model, the possibility of two vehicles being in the same interval is neglected. So are the lengths of the vehicles.) It is further supposed that given that a vehicle is in an interval, its position is uniformly distributed over the interval. From Figure 4, it can be seen that the distance from a vehicle A to the next vehicle B, ahead of it, is distributed as

$$L = aY + X_1 + X_2$$

where Y represents the number of empty intervals between A and B, and has the geometric distribution (Chapter 5)

$$P[Y = y] = q^y p \qquad (y = 0,1,\ldots)$$

and X_1, X_2 are independent random variables each having the distribution

$$p_X(x) = a^{-1} \qquad (0 \leq x \leq a).$$

The distribution of $S = X_1 + X_2$ has density function

$$p_S(s) = a^{-2}(a - |s - a|) \qquad (0 \leq s \leq 2a)$$

(cf. Equation (20)). The distribution of $T = aY + S$ has density function

$$(36) \quad p_T(t) = \begin{cases} pa^{-2}t & (0 \leq t \leq a) \\ pa^{-2}\{q^{k-1}[(k + 1)a - t] + q^k(t - ka)\} & \\ = pq^{k-1}a^{-2}\{(1 + kp)a - pt\} & (ka \leq t \leq (k + 1)a;\ k \leq 1). \end{cases}$$

FIGURE 4

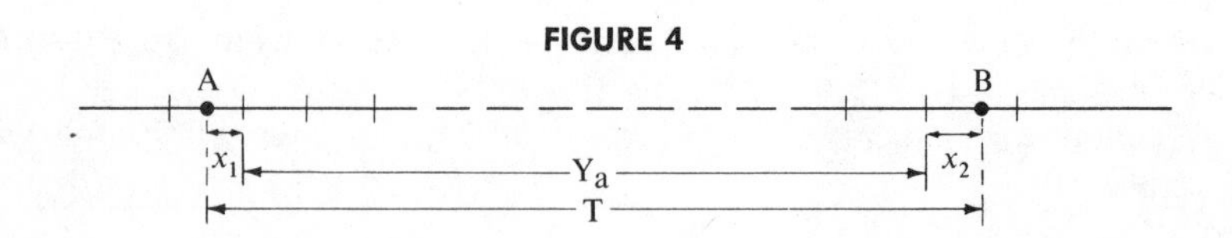

Allan calls this distribution the *binomial-uniform* distribution. It should not be confused with the binomial-beta distribution of Chapter 8 (Section 2(x)). The graph of the density function (36) consists of a succession of straight lines. Figure 5, taken from [1] shows some examples.

Since Y, X_1 and X_2 are independent it is easy to find the moments of T. The rth cumulant is

$$\kappa_r(T) = a^r\kappa_r(Y) + 2\kappa_r(X_1)$$

whence we find

$$E(T) = ap^{-1} \tag{37.1}$$

$$\text{var}(T) = (ap^{-1})^2(q + \tfrac{1}{6}p^2) \tag{37.2}$$

$$\alpha_3(T) = q(1 + q)(q + \tfrac{1}{6}p^2)^{-3/2} \tag{37.3}$$

$$\alpha_4(T) = 3 + (6q^2 + qp^2 - \tfrac{1}{60}p^4)(q + \tfrac{1}{6}p^2)^{-2}. \tag{37.4}$$

Allan [1] also obtains the distribution of the sum of an independent identically distributed binomial-uniform variables and provides tables of the cumulative distribution function to 4 decimal places for $p = 0.4(0.1)0.9$; $t/a = 0.5(0.5)25$; $n = 1(1)20$.

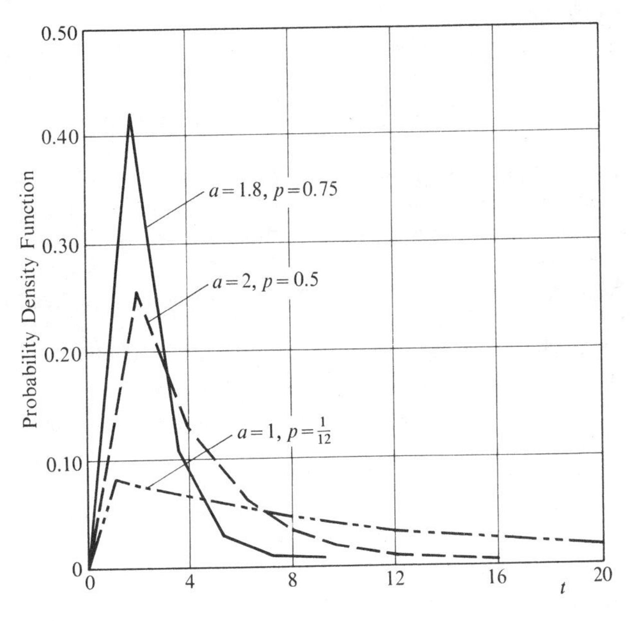

FIGURE 5

The Binomial-Uniform Headway Distribution

REFERENCES

[1] Allan, R. R. (1966). Extension of the binomial model of traffic flow to the continuous case, *Proceedings of the Third Conference of the Australian Road Research Board*, **3,** 276–316.

[2] Ayyangar, A. A. K. (1941). The triangular distribution, *Mathematics Student*, **9,** 85–87.

[3] Bayes, T. (1763). An essay towards solving a problem in the Doctrine of Chances, *Philosophical Transactions of the Royal Society of London*, **53,** 370–418.

[4] Bhate, D. H. (1951). A note on the significance level of the distribution of the means of a rectangular population. *Bulletin of the Calcutta Statistical Association*, **3,** 172–173.

[5] Bol'shev, L. N. (1964). On M. Kadyrov's random numbers in J. Janko's statistical tables, *Teoriya Veroyatnostei i ee Primeneniya*, **9,** 152–154. (In Russian)

[6] Broadbent, S. R. (1954). The quotient of a rectangular or triangular and a general variate, *Biometrika*, **41,** 330–337.

[7] Carlton, A. G. (1946). Estimating the parameters of a rectangular distribution, *Annals of Mathematical Statistics*, **17,** 355–358.

[8] Clark, C. E. (1966). *Random Numbers in Uniform and Normal Distribution*, San Francisco: Chandler.

[9] David, F. N. and Johnson, N. L. (1956). Some tests of significance with ordered variables, *Journal of the Royal Statistical Society, Series B*, **18,** 1–20; Discussion, 20–31.

[10] Durbin, J. (1961). Some methods of constructing exact tests, *Biometrika*, **48,** 41–55.

[11] Eisenhart, C., Deming, L. S. and Martin, C. S. (1963). *Tables describing small-sample properties of the mean, median, standard deviation, and other statistics in sampling from various distributions.* U. S. National Bureau of Standards, Technical Note 191.

[12] Fisher, R. A. (1932). *Statistical Methods for Research Workers*, London: Oliver & Boyd.

[13] Geary, R. C. (1944). Comparison of the concepts of efficiency and closeness for consistent estimates of a parameter, *Biometrika*, **33,** 123–128.

[14] Graybill, F. A. and Connell, T. L. (1964). Sample size required to estimate the parameter in the uniform density within *d* units of the true value, *Journal of the American Statistical Association*, **59,** 550–556.

[15] Gupta, S. S. and Sobel, M. (1958). On the distribution of a statistic based on ordered uniform chance variables, *Annals of Mathematical Statistics*, **29,** 274–281.

[16] Holewijn, P. J. (1969). Note on Weyl's criterion and the uniform distribution of independent random variables, *Annals of Mathematical Statistics*, **40,** 1124–1125.

[17] Hull, T. E. and Dobell, A. R. (1962). Random number generators, *Journal of the Society of Industry and Applied Mathematics*, **4,** 230–254.

[18] Hyrenius, H. (1953). On the use of ranges, cross-ranges and extremes in comparing small samples, *Journal of the American Statistical Association*, **48,** 534–545, (Correction, **48,** 907).

[19] Johnson, N. L. (1950). On the comparison of estimators, *Biometrika*, **37,** 281–287.

[20] Kadyrov, M. (1936). *Tables of Random Numbers*, Tashkent: SGU, (Mid-Asian State University).

[21] Kendall, M. G. and Babington Smith, B. (1938). Randomness and random sampling numbers, *Journal of the Royal Statistical Society, Series A*, **120.**

[22] Kendall, M. G. and Babington Smith, B. (1940). *Tables of Random Sampling Numbers*, Tracts for Computers, **24,** London: Cambridge University Press.

[23] Laplace, P. S. (1812). *Théorie Analytique des Probabilitiés*, 1st edition, *Paris.*

[24] Lloyd, E. H. (1952). Least-squares estimation of location and scale parameters using order statistics, *Biometrika*, **39,** 88–95.

[25] Marsaglia, G. (1961). Expressing a random variable in terms of uniform random variables, *Annals of Mathematical Statistics*, **32,** 894–898.

[26] Morimoto, H. and Sibuya, M. (1967). Sufficient statistics and unbiased estimation of restricted selection parameters, *Sankhyā, Series A*, **29,** 15–40.

[27] Murty, V. N. (1955). The distribution of the quotient of maximum values in samples from a rectangular distribution, *Journal of the American Statistical Association*, **50,** 1136–1141.

[28] Nagaev, S. V. and Mukhin, A. B. (1966). On a case of convergence to a uniform distribution on an interval; pp. 113–116 in *Limit Theorems and Statistical Inference.* (S. H. Siraždinov, Ed.). Tashkent: FAN. (In Russian)

[29] Packer, L. R. (1950). The distribution of the sum of *n* rectangular variates, I. *Journal of the Institute of Actuaries Students' Society*, **10,** 52–61.

[30] Pearson, E. S. (1938). Tests based on the probability integral transformation, *Biometrika*, **30,** 134–148.

[31] Perlo, V. (1933). On the distribution of Student's ratio for samples of three drawn from a rectangular distribution, *Biometrika*, **25,** 203–204.

[32] RAND Corporation (1955). *A Million Random Digits and 100,000 Normal Deviates*, Glencoe, Illinois: Free Press.

[33] Rider, P. R. (1929). On the distribution of the ratio of mean to standard deviation in small samples from non-normal universes, *Biometrika*, **21,** 124–143.

[34] Rider, P. R. (1951). The distribution of the quotient of ranges in samples from a rectangular population, *Journal of the American Statistical Association*, **46,** 502–507.

[35] Rider, P. R. (1963). *Percentage Points of the Ratio of Ranges of Samples from a Rectangular Distribution*, ARL 63–194, Aerospace Research Laboratories, U. S. Air Force, Wright-Patterson Air Force Base, Ohio.

[36] Roach, S. A. (1963). The frequency distribution of the sample mean where each member of the sample is drawn from a different rectangular distribution, *Biometrika*, **50,** 508–513.

[37] Sakamoto, H. (1943). On the distribution of the product and the quotient of the independent and uniformly distributed random variables, *Tôhoku Mathematical Journal*, **49,** 243–260.

[38] Sarhan, A. E. (1955). Estimation of the mean and standard deviation by order statistics, III. *Annals of Mathematical Statistics*, **26,** 576–592.

[39] Sarhan, A. E. and Greenberg, B. G. (1959). Estimation of location and scale parameters for the rectangular population from censored samples, *Journal of the Royal Statistical Society, Series B*, **21,** 356–363.

[40] Sarhan, A. E. and Greenberg, B. G. (Ed.) (1961). *Order Statistics*, New York: John Wiley & Sons, Inc.

[41] Schmidt, R. (1934). Statistical analysis of one-dimensional distributions, *Annals of Mathematical Statistics*, **5,** 30–43.

[42] Seal, H. L. (1950). Spot the prior reference, *Journal of the Institute of Actuaries Students' Society*, **10,** 255–256.

[43] Sheppard, W. F. (1907). Calculation of moments of a frequency distribution, *Biometrika*, **5,** 450–459.

[44] Siddiqui, M. M. (1964). Distribution of Student's t in samples from a rectangular universe, *Revue de l'Institut International de Statistique*, **32,** 242–250.

[45] Stephens, M. A. (1966). Statistics connected with the uniform distribution: percentage points and application to testing for randomness of directions, *Biometrika*, **53,** 235–238.

[46] Sukhatme, P. V. (1937). Tests of significance for samples from the χ^2 population with two degrees of freedom, *Annals of Eugenics, London*, **8,** 52–56.

[47] Tach, L. T. (1958). *Tables of Cumulative Distribution Function of a Sum of Independent Random Variables*, Convair Aeronautics Report No. ZU-7-119-TN, San Diego, California.

[48] Tippett, L. H. C. (1927). Random sampling numbers, *Tracts for Computers*, **XV,** London: Cambridge University Press.

26

F-Distribution

1. Introduction

If X_1, X_2 are independent random variables distributed as $\chi^2_{\nu_1}$, $\chi^2_{\nu_2}$ respectively, then the distribution of

$$(X_1/\nu_1)(X_2/\nu_2)^{-1}$$

is the *F-distribution with ν_1, ν_2 degrees of freedom.* We will use the symbol F_{ν_1,ν_2} generically to denote a random variable having this distribution. The phrase "F_{ν_1,ν_2}-distribution" can be used as an abbreviation for "*F*-distribution with ν_1, ν_2 degrees of freedom".

Note that the *order* ν_1, ν_2 is important. From the definition it is clear that the random variables F_{ν_1,ν_2} and $F^{-1}_{\nu_2,\nu_1}$ have identical distributions. In particular, using the suffix α to denote "lower $100\alpha\%$ point",

$$F_{\nu_1,\nu_2,\alpha} = F^{-1}_{\nu_2,\nu_1,1-\alpha} \tag{1}$$

(since $\Pr[F_{\nu_1,\nu_2} \leq K] = \Pr[F_{\nu_2,\nu_1} \geq K^{-1}]$).

The importance of the *F*-distribution in statistical theory derives mainly from its applicability to the distribution of ratios of independent 'estimators of variance'. If $\{X_{ti}\}$ $(t = 1,2; i = 1,2,\ldots,n_t; n_t \geq 2)$ denote independent random variables each normally distributed, and the expected value and standard deviation of X_{ti} are ξ_t, σ_t respectively (not depending on i) then

$$S_t^2 = (n_t - 1)^{-1} \sum_{i=1}^{n_t} (X_{ti} - \overline{X}_{t.})^2 \qquad \left(\text{with } \overline{X}_{t.} = n_t^{-1} \sum_{i=1}^{n_t} X_{ti}\right)$$

is distributed as $\chi^2_{n_t-1}\sigma_t^2(n_t - 1)^{-1}$ $(t = 1,2)$. The ratio (S_1^2/S_2^2) is therefore distributed as:

$$\frac{\chi^2_{n_1-1}\sigma_1^2(n_1 - 1)^{-1}}{\chi^2_{n_2-1}\sigma_2^2(n_2 - 1)^{-1}}, \qquad \text{that is, as } \left(\frac{\sigma_1}{\sigma_2}\right)^2 F_{n_1-1,n_2-1}.$$

The statistic $(S_1/S_2)^2$ is used in testing the hypothesis of equality between σ_1 and σ_2. The hypothesis is rejected if either

$$(S_1/S_2)^2 \leq F_{n_1-1,n_2-1,\alpha_1} \qquad \text{or} \qquad (S_1/S_2)^2 \geq F_{n_1-1,n_2-1,1-\alpha_2}$$

with $\alpha_1 + \alpha_2 < 1$. The significance level of the test is $(\alpha_1 + \alpha_2)$, and the power (with respect to a specified value of the ratio σ_1/σ_2) is

$$(2) \qquad 1 - \Pr[F_{n_1-1,n_2-1,\alpha_1} < (S_1/S_2)^2 < F_{n_1-1,n_2-1,1-\alpha_2} \mid \sigma_1/\sigma_2] \\ = 1 - \Pr[(\sigma_2/\sigma_1)^2 F_{n_1-1,n_2-1,\alpha_1} < F_{n_1-1,n_2-1} \\ < (\sigma_2/\sigma_1)^2 F_{n_1-1,n_2-1,1-\alpha_2}].$$

Since

$$\Pr[(\sigma_1/\sigma_2)^2 F_{n_1-1,n_2-1,\alpha_1} < (S_1/S_2)^2 < (\sigma_1/\sigma_2)^2 F_{n_1-1,n_2-1,1-\alpha_2} \mid \sigma_1/\sigma_2] \\ = 1 - \alpha_1 - \alpha_2$$

the values

$$\left(\frac{S_1}{S_2}\right)^2 (F_{n_1-1,n_2-1,1-\alpha_2})^{-1}; \quad \left(\frac{S_1}{S_2}\right)^2 (F_{n_1-1,n_2-1,\alpha_1})^{-1}$$

enclose a confidence interval for $(\sigma_1/\sigma_2)^2$, with confidence coefficient $(1 - \alpha_1 - \alpha_2)$. For a given value, α, say of $(\alpha_1 + \alpha_2)$, the length of this interval is minimized by choosing α_1, α_2 (subject to $\alpha_1 + \alpha_2 = \alpha$) to minimize

$$(F_{n_1-1,n_2-1,\alpha_1})^{-1} - (F_{n_1-1,n_2-1,1-\alpha_2})^{-1} \\ = F_{n_2-1,n_1-1,1-\alpha_1} - F_{n_2-1,n_1-1,\alpha_2}.$$

This is achieved by making the ordinates of the probability density function of F_{n_2-1,n_1-1} equal at the values $F_{n_2-1,n_1-1,1-\alpha_1}$ and F_{n_2-1,n_1-1,α_2} if $n_2 \geq 3$; if $n_2 < 3$, then α_1 must be taken equal to α, and $\alpha_2 = 0$, so that

$$F_{n_2-1,n_1-1,\alpha_2} = 0.$$

It must be noted that minimization of the length of confidence interval for $(\sigma_1/\sigma_2)^2$ does not usually minimize the length of interval for $(\sigma_2/\sigma_1)^2$ (or for σ_1/σ_2 or σ_2/σ_1). It is more natural to minimize the length of interval for $\log(\sigma_1/\sigma_2)$ — this will minimize the length for $\log(\sigma_2/\sigma_1)$, or indeed for $\log[(\sigma_1/\sigma_2)^r]$ for any $r \neq 0$. To do this we need to choose α_1, α_2 (subject to $\alpha_1 + \alpha_2 = \alpha$) so that the probability density function of F_{n_1-1,n_2-1} has the same value at $f = F_{n_1-1,n_2-1,1-\alpha_2}$ and at $f = F_{n_1,n_2-1,\alpha_1}$. Tables giving appropriate values for $\alpha = 0.01, 0.05, 0.10, 0.25$ and $n_1, n_2 = 6(1)31$ (with $n_1 \geq n_2$) have been constructed by Tiao and Lochner [56].

Further applications for the F-distribution are described in Section 5.

Figure 1 (reproduced from Hald [30]) contains graphs of the probability density function for $\nu_1 = 10$ and a number of values of ν_2. It can be seen that the graphs appear to approach a limiting form. This is, in fact the distribution of $\chi^2_{\nu_1}/\nu_1$ with $\nu_1 = 10$. (Sections 2 and 5.)

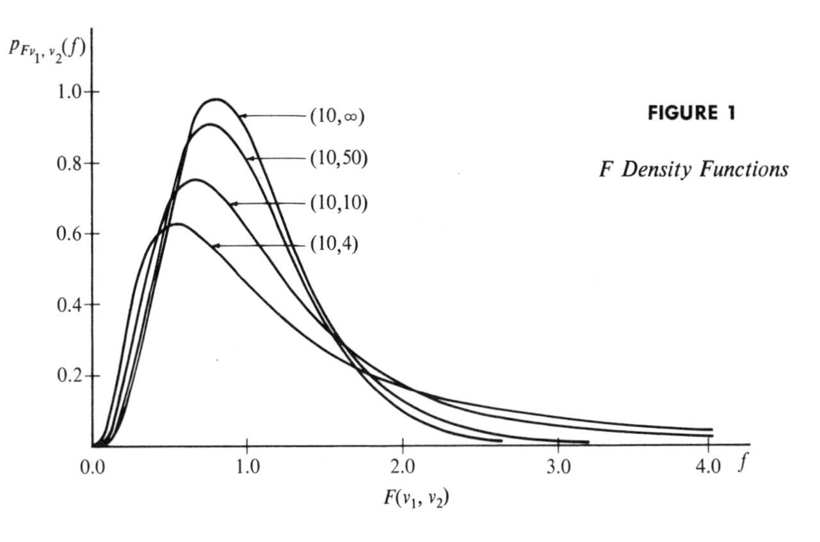

FIGURE 1

F Density Functions

2. Properties

If X_1, X_2 are distributed as described in Section 1, then the probability density function of $G = X_1/X_2$ is

$$p_G(g) = \frac{1}{B(\frac{1}{2}\nu_1, \frac{1}{2}\nu_2)} \frac{g^{\frac{1}{2}\nu_1 - 1}}{(1 + g)^{\frac{1}{2}(\nu_1+\nu_2)}} \qquad (0 < g). \tag{3}$$

This is a Pearson Type VI distribution (Chapter 12, Section 4).

The probability density function of $F = \nu_2 G/\nu_1$ is*

$$p_F(f) = \frac{\nu_1^{\frac{1}{2}\nu_1} \nu_2^{\frac{1}{2}\nu_2}}{B(\frac{1}{2}\nu_1, \frac{1}{2}\nu_2)} \frac{f^{\frac{1}{2}\nu_1 - 1}}{(\nu_2 + \nu_1 f)^{\frac{1}{2}(\nu_1+\nu_2)}} \qquad (0 < f). \tag{4}$$

As f tends to infinity, $p_F(f)$ tends to zero; this is also the case as f tends to zero, provided $\nu_1 > 2$. In this case there is a single mode, at

$$f = [\nu_2(\nu_1 - 2)][\nu_1(\nu_2 + 2)]^{-1}.$$

If $\nu_1 = 2$, there is a mode at $f = 0$; if $\nu_1 = 1$, $p_F(f) \to \infty$ as $f \to 0$.

The rth moment of F about zero is

$$\begin{aligned} \mu_r' &= \left(\frac{\nu_2}{\nu_1}\right)^r E[(\chi^2_{\nu_1})^r] E[(\chi^2_{\nu_2})^{-r}] \\ &= \left(\frac{\nu_2}{\nu_1}\right) \frac{\nu_1(\nu_1 + 2) \cdots (\nu_1 + 2 \cdot \overline{r - 1})}{(\nu_2 - 2)(\nu_2 - 4) \cdots (\nu_2 - 2r)}. \end{aligned} \tag{5}$$

Note that if $r \geq \frac{1}{2}\nu_2$, μ_r' is infinite.

*Note that the subscripts ν_1, ν_2 have been omitted for convenience.

In particular

$$E(F) = \frac{\nu_2}{\nu_2 - 2} \qquad (\nu_2 > 2) \tag{6.1}$$

$$\text{var}(F) = \frac{2\nu_2^2(\nu_1 + \nu_2 - 2)}{\nu_1(\nu_2 - 2)^2(\nu_2 - 4)} \qquad (\nu_2 > 4) \tag{6.2}$$

$$\sqrt{\beta_1} = \sqrt{\frac{8(\nu_2 - 4)}{(\nu_1 + \nu_2 - 2)\nu_1}} \cdot \frac{(2\nu_1 + \nu_2 - 2)}{(\nu_2 - 6)} \qquad (\nu_2 > 6) \tag{6.3}$$

(6.4)

$$\beta_2 = 3 + \frac{12\{(\nu_2 - 2)^2(\nu_2 - 4) + \nu_1(\nu_1 + \nu_2 - 2)(5\nu_2 - 22)\}}{\nu_1(\nu_2 - 6)(\nu_2 - 8)(\nu_1 + \nu_2 - 2)} \qquad (\nu_2 > 8)$$

(Wishart [59]).

The characteristic function of F_{ν_1,ν_2} is the confluent hypergeometric function

$$M(\tfrac{1}{2}\nu_1; -\tfrac{1}{2}\nu_2; -(\nu_2/\nu_1)it). \tag{7}$$

(Chapter 1, Equation (43).)

From (3) it can be seen that $G(1 + G)^{-1} = \nu_1 F(\nu_2 + \nu_1 F)^{-1}$ has a standard beta distribution (as defined in Chapter 24) with parameters $\frac{1}{2}\nu_1$, $\frac{1}{2}\nu_2$. It follows that

$$\Pr[F \leq K] = I_{K_1}(\tfrac{1}{2}\nu_1, \tfrac{1}{2}\nu_2) \qquad \text{where } K_1 = \nu_1 K/(\nu_2 + \nu_1 K). \tag{8}$$

Random variates following F distributions are easily constructed from beta variates, which can be obtained by methods described in Chapter 24, Section 2.

As ν_2 increases, the value of $\chi^2_{\nu_2}/\nu_2$ tends to 1 with probability 1. If ν_1 remains constant, the distribution of F_{ν_1,ν_2} tends to that of $\chi^2_{\nu_1}/\nu_1$ as ν_2 tends to infinity.

For purposes of approximation it is often convenient to study the distribution of $z_{\nu_1,\nu_2} = \frac{1}{2}\log F_{\nu_1,\nu_2}$* rather than that of F_{ν_1,ν_2} itself. (This variable was, in fact, used by Fisher [22] in tables published in 1924.) Approximations themselves will be discussed in the next section, but here we give formulas for the moments of z_{ν_1,ν_2}. Dropping the suffices for convenience, we have the moment generating function

$$E[e^{tz}] = E[F^{\frac{1}{2}t}] = \left(\frac{\nu_2}{\nu_1}\right)^{\frac{1}{2}t} \frac{\Gamma(\frac{1}{2}(\nu_1 + t))\Gamma(\frac{1}{2}(\nu_2 - t))}{\Gamma(\frac{1}{2}\nu_1)\Gamma(\frac{1}{2}\nu_2)}. \tag{9}$$

The cumulants of z are

$$\kappa_1(z) = \tfrac{1}{2}[\log(\nu_2/\nu_1) + \psi(\tfrac{1}{2}\nu_1) - \psi(\tfrac{1}{2}\nu_2)] \tag{10.1}$$

$$\kappa_r(z) = 2^{-r}[\psi^{(r-1)}(\tfrac{1}{2}\nu_1) + (-1)^r\psi^{(r-1)}(\tfrac{1}{2}\nu_2)], \qquad (r \geq 2). \tag{10.2}$$

Note that all moments of z are finite.

*Although this is a random variable, it is usually denoted in the literature by a small z.

For $r \geq 2$, an alternative formula is:

$$(10.2)' \qquad \kappa_r(z) = (r-1)! \sum_{j=0}^{\infty} [(-1)^r(\nu_1 + 2j)^{-r} + (\nu_2 + 2j)^{-r}].$$

There are also a number of special expressions for $\kappa_r(z)$ depending on the parity of ν_1 and ν_2. We now give these in summary form (see Aroian [3], Wishart [60]).

ν_1 and ν_2 both even

$$(11.1) \qquad E(z) = \tfrac{1}{2}[\log (\nu_2/\nu_1) - \sum_j{}^* j^{-1}]$$

where $\sum_j^*$ denotes summation from $j = \frac{1}{2} \min(\nu_1,\nu_2)$ to $(\frac{1}{2} \max(\nu_1,\nu_2) - 1)$ inclusive and $\sum_j^* j^{-1} = 0$ if $\nu_1 = \nu_2$.

$$(11.2) \qquad \text{var}(z) = 0.822467 - \frac{1}{4}\left(\sum_{j=1}^{\frac{1}{2}\nu_1 - 1} j^{-2} + \sum_{j=1}^{\frac{1}{2}\nu_2 - 1} j^{-2}\right)$$

and generally (for $r \geq 2$)

$$(11.3) \quad \kappa_r(z) = 2^{-r}(r-1)!\left[\left(S_r - \sum_{j=1}^{\frac{1}{2}\nu_1 - 1} j^{-r}\right) + (-1)^r\left(S_r - \sum_{j=1}^{\frac{1}{2}\nu_2 - 1} j^{-r}\right)\right],$$

where

$$S_r = \sum_{j=1}^{\infty} j^{-r}.$$

ν_1 and ν_2 both odd

$$(12.1) \quad \kappa_r(z) = (r-1)!\left[\left(T_r - \sum_{j=0}^{\frac{1}{2}(\nu_2 - 3)} (2j+1)^{-r}\right) + (-1)^{-r}\left(T_r - \sum_{j=0}^{\frac{1}{2}(\nu_1 - 3)} (2j+1)^{-r}\right)\right]$$

for $r \geq 2$, where

$$T_r = S_r(1 - 2^{-r}) = \sum_{j=0}^{\infty} (2j+1)^{-r}.$$

In particular

$$(12.2) \quad \text{var}(z) = 2.467401 - \frac{1}{4}\left(\sum_{j=0}^{\frac{1}{2}(\nu_2 - 3)} (j + \tfrac{1}{2})^{-2} + \sum_{j=0}^{\frac{1}{2}(\nu_1 - 3)} (j + \tfrac{1}{2})^{-2}\right).$$

Also

$$(12.3) \qquad E(z) = \tfrac{1}{2} \log (\nu_2/\nu_1) - \sum_j{}^* (2j+1)^{-r}$$

where the limits of summation in $\sum_j^*$ are *now* from $j = \frac{1}{2} \min(\nu_1,\nu_2) - \frac{1}{2}$ to $j = \frac{1}{2} \max(\nu_1,\nu_2) - \frac{3}{2}$.

ν_1 even, ν_2 odd

(13)
$$\kappa_r(z) = (r-1)!\times\left[\left(T_r - \sum_{j=0}^{\frac{1}{2}(\nu_2-3)}(2j+1)^{-r}\right) + (-\tfrac{1}{2})^r\left(S_r - \sum_{j=1}^{\frac{1}{2}\nu_1-1} j^{-r}\right)\right] \quad \text{for } r \geq 2.$$

ν_2 even, ν_1 odd

(14)
$$\kappa_r(z) = (r-1)!\left[2^{-r}\left(S_r - \sum_{j=0}^{\frac{1}{2}\nu_2-1} j^{-r}\right) + (-1)^r\left(T_r - \sum_{j=0}^{\frac{1}{2}(\nu_1-3)}(2j+1)^{-r}\right)\right] \quad \text{for } r \geq 2.$$

It is helpful, in remembering these formulas, to think of the relation

(14)′
$$\kappa_r(z) = 2^{-r}[\kappa_r(\log \chi^2_{\nu_1}) + (-1)^r\kappa_r(\log \chi^2_{\nu_2})] \qquad (\text{for } r \geq 2).$$

3. Tables

In view of relation (8), tables of the incomplete beta function ratio can be used to evaluate the cumulative distribution function of F_{ν_1,ν_2}. In suitable cases, of course, tables of binomial (or negative binomial) probabilities can be used. This has been noted by a number of authors (e.g. Bizley [9], Mantel [41], Johnson [34]). Similarly from tables of percentile points of the beta distribution, corresponding points of F distributions can be obtained with little computational effort. It is, however, convenient to have available tables giving percentile points of F distributions (i.e. values of $F_{\nu_1,\nu_2,\alpha}$) directly for given values of ν_1, ν_2 and α. Such tables are commonly available. Here we give only the more notable sources, excluding reproductions (in part or whole) in text-books. It is to be noted that it is customary to give only *upper* percentiles (i.e. $F_{\nu_1,\nu_2,\alpha}$ with $\alpha \geq 0.5$). Lower percentiles are easily obtained from the formula

$$F_{\nu_1,\nu_2,1-\alpha} = F^{-1}_{\nu_2,\nu_1,\alpha}.$$

Greenwood and Hartley [29] classify tables of $F_{\nu_1,\nu_2,1-\alpha}$ according to two broad categories:

(A) with $\alpha = 0.005, 0.001, 0.025, 0.05, 0.1$ and 0.25
(B) with $\alpha = 0.001, 0.01, 0.05$ and 0.20.

Merrington and Thompson [43] give tables of type (A) to 5 significant figures for $\nu_1 = 1(1)10, 12, 15, 20, 24, 30, 40, 60, 120, \infty$ and $\nu_2 = 1(1)30, 40, 60, 120, \infty$. (The higher values are chosen to facilitate harmonic interpolation with $120\nu^{-1}$ as the variable). Fisher and Yates [24] give tables of type (B) to 2 decimal places for $\nu_1 = 1(1)6, 8, 12, 24, \infty$ and $\nu_2 = 1(1)30, 40, 60, 120, \infty$. More extensive tables (though only to 3 significant figures) have been given by Hald [30]. He gives values of $F_{\nu_1,\nu_2,1-\alpha}$ for $\alpha = 0.0005, 0.0001, 0.1, 0.3$ and 0.5 with $\nu_1 = 1(1)10(5)20, 30, 50, 100, 200, 500, \infty$ and $\nu_2 = 1(1)20(2)30(5)50,$

60(20)100, 200, 500, ∞, and also for $\alpha = 0.005, 0.01, 0.025, 0.05$ with $\nu_1 = 1(1)20(2)30(5)50$, 60(20)100, 200, 500, ∞ and $\nu_2 = 1(1)30(2)50(5)$ 70(10)100(25)150, 200, 300, 500, 1000, ∞.

Most other tables of $F_{\nu_1,\nu_2,\alpha}$ are derived from one or more of these tables.

However, the earliest tables (as has already been mentioned in Section 2) gave values, not of $F_{\nu_1,\nu_2,\alpha}$, but of $z_{\nu_1,\nu_2,\alpha} = \frac{1}{2}\log F_{\nu_1,\nu_2,\alpha}$. For large values of ν_1 or ν_2, interpolation with respect to numbers of degrees of freedom is much easier for z than for F. This does not appear to be the original reason for introducing z (see Fisher [22]), but it is now the main reason for using such tables (Fisher and Yates [24]). Of course, for occasional interpolation it is possible to use tables of the F-distribution, and to calculate the needed values of

$$z_{\nu_1,\nu_2,\alpha} = \tfrac{1}{2}\log F_{\nu_1,\nu_2,\alpha}.$$

Zinger [63] has proposed the following method for interpolating in tables of percentile points of F_{ν_1,ν_2} in order to evaluate $\Pr[F_{\nu_1,\nu_2} > K]$. One seeks to find tabled values $F_{\nu_1,\nu_2,1-\alpha}$ and $F_{\nu_1,\nu_2,1-m\alpha}$ such that $F_{\nu_1,\nu_2,1-m\alpha} \leq K \leq F_{\nu_1,\nu_2,1-\alpha}$. Then

$$\Pr[F_{\nu_1,\nu_2} > K] \doteqdot 1 - \alpha m^k$$

where k satisfies the equation

$$kF_{\nu_1,\nu_2,1-m\alpha} + (1-k)F_{\nu_1,\nu_2,1-\alpha} = K.$$

Laubscher [39] has given examples showing the accuracy of harmonic interpolation (with arguments ν_1^{-1}, ν_2^{-1}) for either univariate or bivariate interpolation with respect to ν_1 and ν_2, where α is fixed. Bol'shev *et al.* [11] devised auxiliary tables for accurate computations of the beta and z-distribution functions.

4. Approximations and Nomograms

Because of the relation (8) between the probability integral of the F-distribution and the incomplete beta function ratio, approximations to the latter can be applied to the former. Such approximations have already been described in Chapters 3 and 24. Here we will describe more particularly some approximations to the F-distribution. They also can be used as approximations to the incomplete beta function ratio. In fact, the section may be regarded as an extension of the discussions of Chapter 3, Section 8, and Chapter 24, Section 6.

We have already noted that $z = \frac{1}{2}\log F$ has a more nearly normal distribution than does F itself. A number of approximations are based on either a normal approximation to the distribution of z or some modification thereof—for example, that provided by using a Cornish-Fisher expansion.

For large values of both ν_1 and ν_2 the distribution of z may be approximated by a normal distribution with expected value $\frac{1}{2}(\nu_2^{-1} - \nu_1^{-1})$ $(= \delta)$ and variance $\frac{1}{2}(\nu_1^{-1} + \nu_2^{-1})$ $(= \sigma^2)$. This leads to the simple approximate formula

$$z_{\nu_1,\nu_2,\alpha} \doteqdot \tfrac{1}{2}(\nu_2^{-1} - \nu_1^{-1}) + U_\alpha\sqrt{\tfrac{1}{2}(\nu_1^{-1} + \nu_2^{-1})} \tag{15}$$

suggested by Fisher [22]. Fisher also suggested that replacement of ν_1^{-1}, ν_2^{-1} by $(\nu_1 - 1)^{-1}$, $(\nu_2 - 1)^{-1}$ might improve accuracy.

More elaborate approximations can be obtained in expansions of Cornish-Fisher type [18]. One such approximation (Aroian [3], Fisher and Cornish [23], Wishart [61]) using *approximate* formulas for the cumulants is

$$
\begin{aligned}
(16) \qquad z_{\nu_1,\nu_2,\alpha} \doteqdot\ & U_\alpha\sigma + \frac{1}{3}\,\delta(U_\alpha^2 + 2) \\
& + \sigma\left\{\frac{\sigma^2}{12}(U_\alpha^3 + 3U_\alpha) + \frac{1}{36}\left(\frac{\delta}{\sigma}\right)^2(U_\alpha^3 + 11U_\alpha)\right\} \\
& + \frac{1}{30}\,\delta\sigma^2(U_\alpha^4 + 9U_\alpha^2 + 8) \\
& - \frac{1}{810}\frac{\delta^3}{\sigma^2}(3U_\alpha^4 + 7U_\alpha^2 - 16) + \cdots.
\end{aligned}
$$

The first two terms of approximation (16) can be written

$$\delta + U_\alpha\sigma + \tfrac{1}{3}\delta(U_\alpha^2 - 1) = \delta[1 + \tfrac{1}{3}(U_\alpha^2 - 1)] + U_\alpha\sigma.$$

Fisher [22] suggested the approximation

$$(17) \qquad z_{\nu_1,\nu_2,\alpha} \doteqdot \delta[1 + \tfrac{1}{3}(U_\alpha^2 - 1)] + U_\alpha\sigma(1 - \sigma^2)^{-\frac{1}{2}}$$

which makes some allowance for later terms in (16). This was improved by Cochran [16] to the formula

$$(18) \qquad z_{\nu_1,\nu_2,\alpha} \doteqdot \delta[1 + \tfrac{1}{3}(U_\alpha^2 - 1)] + U_\alpha\sigma[1 - \tfrac{1}{6}(U_\alpha^2 + 3)\sigma^2]^{-\frac{1}{2}}$$

which differs from the sum of the first three terms of (16) by approximately $\frac{1}{36}(U_\alpha^3 + 11U_\alpha)(\delta^2/\sigma)$.

Carter [15] obtained another formula using more accurate expressions for the cumulants of z derived by Wishart, together with certain modifications in the expansions. His formula for ν_1, ν_2 large is

$$
\begin{aligned}
(19) \qquad z_{\nu_1,\nu_2,\alpha} \doteqdot\ & \frac{U_\alpha\{\frac{1}{6}U_\alpha^2 - \frac{1}{2} + 2\nu_1'\nu_2(\nu_1' + \nu_2')^{-1}\}^{\frac{1}{2}}}{2\nu_1'\nu_2'(\nu_1' + \nu_2')^{-1}} \\
& - \frac{1}{6}\left(\frac{1}{\nu_1'} - \frac{1}{\nu_2'}\right)\left(U_\alpha^2 + 2 - \frac{2(\nu_1' + \nu_2')}{\nu_1'\nu_2'}\right)
\end{aligned}
$$

where

$$\nu_j' = \nu_j - 1 \qquad (j = 1,2).$$

These formulas were further modified by Aroian [5], who gave a number of interesting numerical comparisons of the accuracy of various approximations. His conclusions were that a better approximation, than that provided by expansions of the type we have so far discussed (except (19)), is provided (at least for $\nu_1, \nu_2 > 20$, say) by an approximation suggested by Paulson [46].

This is based on the Wilson-Hilferty approximation to the distribution of χ^2, the remarkable accuracy of which has been described in Chapter 17 (Section 5). If the distributions of X_1 and X_2 (at the beginning of this chapter) are each approximated in this way, we see that the distribution of $F^{\frac{1}{3}}_{\nu_1,\nu_2}$ is approximated by the distribution of the ratio of two independent normal variables. In fact, $F^{\frac{1}{3}}_{\nu_1,\nu_2}$ is approximately distributed as

$$\frac{1 - \dfrac{2}{9\nu_1} + U_1\sqrt{\dfrac{2}{9\nu_1}}}{1 - \dfrac{2}{9\nu_2} + U_2\sqrt{\dfrac{2}{9\nu_2}}}$$

where U_1, U_2 are independent unit normal variables. Using a further approximate formula for the distribution of this ratio (Chapter 13, Section 6.3) we are led to the approximation of taking

$$z = \left[\left(1 - \frac{2}{9\nu_2}\right)F^{\frac{1}{3}}_{\nu_1,\nu_2} - \left(1 - \frac{2}{9\nu_1}\right)\right]\left(\frac{2}{9\nu_2}F^{\frac{2}{3}}_{\nu_1,\nu_2} + \frac{2}{9\nu_1}\right)^{-\frac{1}{2}} \tag{20}$$

to have a unit normal distribution.

This approximation is also remarkably accurate for $\nu_2 \geq 10$. Smillie and Anstey [52] have utilized it in a computer routine.

For $\nu_2 \leq 10$, values of upper percentage points $F_{\nu_1,\nu_2,\alpha}$ calculated from (20) can be improved by using the formula

$$\text{(improved value)} = m \times \text{(calculated value)} + c$$

where m and c depend on ν_2 and the percentile, but not on ν_1. Ashby [1] gives values of m and c for $\alpha = 0.95, 0.99, 0.99$; $\nu_2 = 1(1)10$ with which accuracy to three significant figures is attained.

For small values of $\nu_2 \leq 3$, Kelley [36] recommends replacing z by

$$z' = z(1 + 0.08z^4\nu_2^{-3}).$$

A computer program for calculating $\Pr[F > F_0]$ using this corrected formula has been published by Jaspen [33]. He gives comparisons of exact and calculated values for $\nu_1 = 2, \nu_2 = 2$. Further comparisons, for $\nu_1, \nu_2 = 1, 2, 4, 10, 20, 60, 1000$ but only for $F_0 = F_{\nu_1,\nu_2,0.95}$, have been made by Golden *et al.* [28]. It appears from [33] that accuracy is better in the lower, than the upper, tail of the F-distribution.

A similar approximation can be obtained using Fisher's approximation to the χ^2 distribution [$\sqrt{2\chi^2_\nu} - \sqrt{2\nu - 1}$ approximately a unit normal variable] in place of the Wilson-Hilferty approximation.

The resulting approximation is to regard

$$\left[\sqrt{1 - \frac{1}{2\nu_1}}F^{\frac{1}{2}}_{\nu_1,\nu_2} - \sqrt{1 - \frac{1}{2\nu_2}}\right]\left[\frac{1}{2\nu_2}F_{\nu_1,\nu_2} + \frac{1}{2\nu_1}\right]^{-\frac{1}{2}} \tag{21}$$

as a unit normal variable. Since the Fisher transformation is generally less accurate than the Wilson-Hilferty transformation, one would expect (20) to be generally more accurate than (21). While this is so, the comparison is not as disadvantageous to (21) as might be expected.

If one only of ν_1 and ν_2 is large (ν_2, say) then the approximation of taking F_{ν_1,ν_2} to be distributed as $\chi^2_{\nu_1}/\nu_1$ is a natural one to use. (It is clearly always possible to arrange that $\nu_2 > \nu_1$ in this case.) Scheffé and Tukey [51] have proposed a simple improvement on this which, in terms of the incomplete beta function ratio (Chapter 24, Equation (3)), is that if

$$I_p(n - r + 1,r) = \alpha,$$

then

$$n \doteqdot \tfrac{1}{4}\chi^2_{2r,\alpha}(1 + p)(1 - p)^{-1} + \tfrac{1}{2}(r - 1).$$

In terms of F-distributions, this can be stated as follows:

$$F_{\nu_1,\nu_2,\alpha} \doteqdot \left[\frac{\nu_1}{\chi^2_{\nu_1,\alpha}} + \frac{\nu_1}{\nu_2}\left(\frac{\frac{1}{2}\nu_1 - 1}{\chi^2_{\nu_1,\alpha}} - \frac{1}{2}\right)\right]^{-1}.$$

(At one stage, the derivation of this entails reversing the degrees of freedom of F.) McIntyre and Ward [42] have constructed a computer program using this approximation.

We may also note the following approximate result (Aroian [4]):

"For ν_1 and ν_2 large,

$$z\,\frac{\sqrt{2(\nu_1 + \nu_2 - 1)\nu_1\nu_2}}{\nu_1 + \nu_2}$$

is approximately distributed as t with $(\nu_1 + \nu_2 - 1)$ degrees of freedom."

Aroian also proposed the approximation

$$\Pr\left[z\,\frac{\sqrt{2(\nu_1 + \nu_2 - 1)}}{\nu_1 + \nu_2} > t_{\nu_1+\nu_2-1,1-\alpha}\right]$$
$$\doteqdot \alpha\sqrt{\frac{\nu_1 + \nu_2 - 1}{\nu_1 + \nu_2}}\exp\left[\tfrac{1}{6}(\nu_1^{-1} + \nu_2^{-1}) - \tfrac{5}{12}(\nu_1 + \nu_2 - 1)^{-1}\right].$$

The right hand side is always less than α.

The approximation is reported to be generally not as good as the Cochran-Fisher formula (18). However the stated approximate relation between F and t distributions might have some useful application for analytical purposes.

Nomograms for evaluation of incomplete beta function values can be used for calculation of values of cumulative F-distributions. There is a nomogram specifically designed for the F-distribution in Stammberger [54] which is reproduced in Figure 2.

From this nomogram one can evaluate any one of $\Pr[F_{\nu_1,\nu_2} \leq F]$, F, ν_1 and ν_2 given the values of the other three quantities, using two straight-edges (indicated by broken lines in Figure 2).

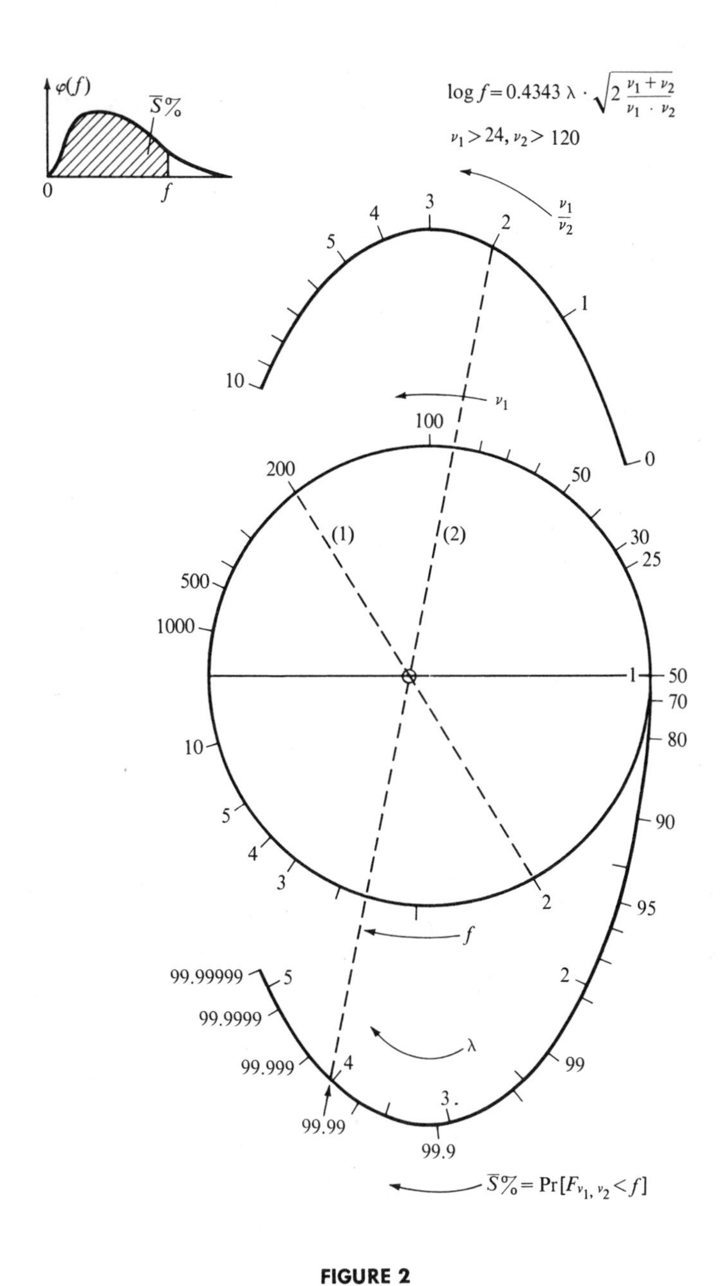

FIGURE 2

Nomogram for Cumulative F Distribution

5. Applications

The commonest application of the F-distribution in statistical work is in standard tests associated with the analysis of variance. Many of these are based on a general result of Kolodziejczyk [37] according to which the likelihood ratio test of a general linear hypothesis (H_0) about the parameters of a general linear model with normal residuals can be expressed in terms of a statistic which has an F-distribution when H_0 is valid.

The application of the F-distribution in testing equality of variances of two normal populations has been described in Section 1.

In some earlier tables of (upper) percentiles of F-distributions, the tables are stated to apply to the distribution of "greater divided by lesser mean square." While it is true that it is often convenient to take the ratio so that it exceeds 1, this affects the significance level of the test. If tables of upper $100\alpha\%$ percentiles are used ($F_{\nu_1,\nu_2,1-\alpha}$) then the *actual* significance level is not α, but 2α. This can be seen by noting that 'significance' is attained if the observed ratio is *either* greater than $F_{\nu_1,\nu_2,1-\alpha}$ or less than $(F_{\nu_2,\nu_1,1-\alpha})^{-1} = F_{\nu_1,\nu_2,\alpha}$.

The F-distribution is also used in the calculation of power functions of these tests, and of confidence limits for the ratios of variances of normal populations.

The relation between F-distributions and binomial distributions has already been noted in Section 3 (also Section 7). A consequence of this is that the approximate confidence limits $(\underline{p},\bar{p})$ for a binomial proportion obtained by solving the equations

$$\sum_{j=r}^{n} \binom{n}{j} \underline{p}^j(1-\underline{p})^{n-j} = \tfrac{1}{2}\alpha$$

$$\sum_{j=0}^{r} \binom{n}{j} \bar{p}^j(1-\bar{p})^{n-j} = \tfrac{1}{2}\alpha$$

can be expressed in terms of percentile points of an F-distribution:

$$\underline{p} = r[(n-r+1)F_{2(n-r+1),2r,\frac{1}{2}\alpha} + r]^{-1}$$

$$\bar{p} = (r+1)F_{2(r+1),2(n-r),\frac{1}{2}\alpha}[n-r+(r+1)F_{2(r+1),2(n-r),\frac{1}{2}\alpha}]^{-1}.$$

Box [12] expresses certain approximations to the distributions (under multinormality conditions) of multivariate test statistics in terms of F-distributions. These are just convenient ways of representing the results of fitting Pearson Type VI distributions. This family of distributions will be discussed in Section 6.

6. Pearson Type VI Distributions

We have already remarked that any F-distribution is a special form of Pearson Type VI distribution (3). Here we give an account of this general family, to which F-distributions belong.

The most general form of equation for the probability of a Type VI distribution can be written (Elderton and Johnson [21])

$$(22) \qquad p_X(x) = \frac{\Gamma(q_1)(a_2 - a_1)^{q_1-q_2-1}}{\Gamma(q_1 - q_2 - 1)\Gamma(q_2 + 1)} \frac{(x - a_2)^{q_2}}{(x - a_1)^{q_1}} \qquad (q_1 > q_2 > -1; x \geq a_2 > a_1)$$

(assuming that the sign of X has been chosen so that $\sqrt{\beta_1(X)} > 0$).

The graph of $p_X(x)$ against x has a single mode at

$$x = a_2 + q_2(q_1 - q_2)^{-1}(a_2 - a_1),$$

provided q_2 is positive. If q_2 is zero, the mode is at $x = a_2$, and if q_2 is negative, $p_X(x)$ tends to infinity as x tends to a_2. In the last two cases $p_X(x)$ decreases as x increases from a_2.

The rth moment of X about a_1 is

$$(23) \qquad E[(X - a_1)^r] = (a_2 - a_1)^r(q_1 - 1)^{(r)}/(q_1 - q_2 - 2)^{(r)}.$$

Note that if $r \geq q_1 - q_2 - 1$, the rth moment is infinite. The expected value and variance of the distribution are

$$(24) \qquad \begin{aligned} E[X] &= a_1 + (q_1 - 1)(a_2 - a_1)/(q_1 - q_2 - 2) \\ &= a_2 + (q_2 - 1)(a_2 - a_1)/(q_1 - q_2 - 2) \qquad (q_1 - q_2 > 2); \\ \text{var}(X) &= (a_2 - a_1)^2(q_1 - 1)(q_2 + 1)(q_1 + q_2 - 2)^{-2} \\ &\quad \times (q_1 - q_2 - 3)^{-1} \qquad (q_1 - q_2 > 3). \end{aligned}$$

The four parameters a_1, a_2, q_1 and q_2 can, of course, be expressed in terms of the first four moments of X. In fact q_2 and $-q_1$ are given by

$$(25) \qquad \tfrac{1}{2}(r - 2) \pm \tfrac{1}{2}r(r + 2)\sqrt{\frac{\beta_1}{\beta_1(r + 2)^2 + 16(r + 1)}}$$

where $r = -6(\beta_2 - \beta_1 - 1)/(2\beta_2 - 3\beta_1 - 6)$. Note that for Type VI curves, $2\beta_2 - 3\beta_1 - 6 > 0$, so that $r < 0$; and also

$$\frac{\beta_1(\beta_2 + 3)^2}{4(2\beta_2 - 3\beta_1 - 6)(4\beta_2 - 3\beta_1)} > 1.$$

There is a simple relation between Type VI and Type I distributions (Chapter 24): the distribution of $Y = (X - a_2)/(X - a_1)$ is of standard beta form, with parameters $q_2 + 1$ and $q_1 - q_2 - 1$.

If $X_1, X_2, \ldots, X_n$ are independent random variables, each having a distribution of form (22), the formal equations for the maximum likelihood estimators $\hat{a}_1$, $\hat{a}_2$, $\hat{q}_1$, and $\hat{q}_2$ can be written

(26.1)

$$n(\hat{q}_1 - \hat{q}_2 - 1)(\hat{a}_2 - \hat{a}_1)^{-1} = \hat{q}_1 \sum_{j=1}^{n} (x_j - \hat{a}_1)^{-1} = \hat{q}_2 \sum_{j=1}^{n} (x_j - \hat{a}_2)^{-1}$$

(26.2)

$$n[\psi(\hat{q}_1) - \psi(\hat{q}_1 - \hat{q}_2 - 1) + \log(\hat{a}_2 - \hat{a}_1)] = \sum_{j=1}^{n} \log(x_j - \hat{a}_1)$$

(26.3)

$$n[\psi(\hat{q}_2 + 1) - \psi(\hat{q}_1 - \hat{q}_2 - 1) + \log(\hat{a}_2 - \hat{a}_1)] = \sum_{j=1}^{n} \log(x_j - \hat{a}_2).$$

It is necessary to check that the solutions of (26) do satisfy the inequalities $\hat{a}_1 < \hat{a}_2 < \min(x_1, \ldots, x_n)$. It should also be noted that the standard formulas for asymptotic variances do not apply if $q_1 - q_2 \leq 2$. The standard formula for the asymptotic variance-covariance matrix of $\sqrt{n}\,\hat{a}_1$, $\sqrt{n}\,\hat{a}_2$, $\sqrt{n}\,\hat{q}_1$ and $\sqrt{n}\,\hat{q}_2$ is the inverse of the symmetric matrix:

$$\begin{pmatrix} \dfrac{(q_2+1)(q_1-q_2-1)}{(q_1+1)(a_2-a_1)^2} & \dfrac{-(q_1-q_2-1)}{(a_2-a_1)^2} & --- & --- \\[2ex] \dfrac{-(q_1-q_2-1)}{(a_2-a_1)^2} & \dfrac{(q_1-1)(q_1-q_2-1)}{(q_2-1)(a_2-a_1)^2} & --- & --- \\[2ex] \dfrac{q_2+1}{q_1(a_2-a_1)} & \dfrac{q_1-1}{q_2(a_2-a_1)} & \begin{matrix}-\psi'(q_1)+\\ \psi'(q_1-q_2-1)\end{matrix} & --- \\[2ex] \dfrac{-1}{a_2-a_1} & \dfrac{-1}{a_2-a_1} & -\psi'(q_1-q_2-1) & \begin{matrix}\psi'(q_2+1)+\\ \psi'(q_1-q_2-1)\end{matrix} \end{pmatrix}$$

7. Other Related Distributions

The relation between F-distributions and Type I (beta) distributions has already been described in Section 2 (and, more generally, between Type VI and Type I in Section 6). A less evident (though more specialized and less useful) relationship can be summarized as follows:

"The distribution of $\frac{1}{2}\sqrt{\nu}\,(F_{\nu,\nu}^{\frac{1}{2}} - F_{\nu,\nu}^{-\frac{1}{2}})$ is a t distribution with ν degrees of freedom."

This result has been reported by several different authors (see Aroian [7], Cacoullos [14]).

There is also a relationship (briefly mentioned in Chapter 3) between the binomial and F-distributions. This can be summarized by the equation

$$\Pr\left[F_{2(n-r+1),2r} > \frac{1-p}{p}\cdot\frac{r}{n-r+1}\right] = \Pr[Y \geq r]$$

(r an integer, $0 \leq r \leq n$) where Y is a binomial variable with parameters (n,p) (Bizley [9], Jowett [35]).

Non-central F-distributions are discussed in Chapter 30. Multivariate generalizations are discussed in Chapter 40.

There are several pseudo-F-distributions corresponding to replacement of either or both, of S_1 and S_2 in (2) by some other sample measure of dispersion such as sample range or mean deviation. (See, e.g., Newman [44].)

The works of David [20], Gayen [26], Horsnell [32], Swain [55], Tiku [57], and Zeigler [62] are the most comprehensive among the numerous investigations of F-distributions under non-normal conditions, i.e. when the variables $X_1, X_2, \ldots, X_n$ in the definitions of $S_t(t = 1,2)$ have non-normal distribution. Additional references will be indicated in Chapter 27, Section 7 in connection with the similar problem for t-distributions.

The truncated Type VI distribution with density function

$$p_X(x) = [\beta/\log(1+\beta)](1+\beta x)^{-1} \qquad (0 < x < 1; \beta > -1)$$

has been used to represent the distribution of references among different sources (Bradford [13], Leimkuhler [40]). (x denotes a proportion.) When used for this purpose it is called the *Bradford distribution*. Of course X would be more naturally represented as a discrete variable and the Bradford distribution may be regarded as an approximation to a Zipf or Yule distribution (Chapter 11).

In the analysis of variance one often has a situation wherein the ratios of each of a number of "mean squares" $M_1, M_2, \ldots, M_k$ to a residual mean square M_0 have to be considered. M_j can be regarded as being distributed as $\sigma^2(\chi^2_{\nu_j}/\nu_j)$ $(j = 0,1,\ldots,k)$ and the M's are mutually independent. If one of the ratios M_j/M_0 $(j \neq 0)$ is large it is helpful to compare it with the distribution of $\max_{1\le j\le k}(M_j/M_0)$. In the case when $\nu_1 = \nu_2 = \cdots = \nu_k = \nu$ (not necessarily equal to ν_0), this distribution is related to that of "Studentized" χ^2 — i.e. of the ratio of the minimum of k independent χ^2_ν variables to an independent $\chi^2_{\nu_0}$ variable.

Armitage and Krishnaiah [2] have given tables of the upper 1%, $2\frac{1}{2}$%, 5% and 10% points of this distribution, to 2 decimal places for

$$k = 1(1)12; \qquad \nu = 1(1)19$$
$$\nu_0 = 6(1)45 \qquad (\text{also } \nu_0 = 5 \text{ for } 5\% \text{ and } 10\% \text{ points}).$$

Tables of the upper 1% and 5% points of $(\max_{1\le j\le k} M_j)/(\min_{1\le j\le k} M_j)$ for $k = 2(1)12$ and $\nu_1 = \nu_2 = \cdots = \nu_k = \nu = 2(1)10, 12, 15, 20, 30, 60, \infty$ are contained in Pearson and Hartley [47].

REFERENCES

[1] Ashby, T. (1968). A modification to Paulson's approximation to the variance ratio distribution, *The Computer Journal*, **11,** 209–210.

[2] Armitage, J. V. and Krishnaiah, P. R. (1964). *Tables of the Studentized largest chi-square distribution and their applications*, Report ARL 64–188, Aerospace Research Laboratories, Wright-Patterson Air Force Base, Ohio.

[3] Aroian, L. A. (1941). A study of R. A. Fisher's z-distribution and the related F distribution, *Annals of Mathematical Statistics*, **12,** 429–448.

[4] Aroian, L. A. (1942). The relationship of Fisher's z-distribution to Student's t distribution, (Abstract), *Annals of Mathematical Statistics*, **13,** 451–452.

[5] Aroian, L. A. (1947). Note on the cumulants of Fisher's z-distribution, *Biometrika*, **34,** 359–360.

[6] Aroian, L. A. (1950). On the levels of significance of the incomplete beta function and the F-distributions, *Biometrika*, **37,** 219–223.

[7] Aroian, L. A. (1953). A certain type of integral, *American Mathematical Monthly*, **50,** 382–383.

[8] Bennett, G. W. and Cornish, E. A. (1964). A comparison of the simultaneous fiducial distributions derived from the multivariate normal distribution, *Bulletin of the International Statistical Institute*, **46,** 918.

[9] Bizley, M. T. L. (1950). A note on the variance-ratio distribution, *Journal of the Institute of Actuaries Students' Society*, **10,** 62–64.

[10] Bol'shev, L. N. (1960). On estimates of probabilities, *Teoriya Veroyatnostei i ee Primeneniya*, **5,** 453–457. (In Russian. English translation: pp. 411–415.)

[11] Bol'shev, L. N., Gladkov, B. V. and Shcheglova, M. V. (1961). Tables for the calculation of B and z-distribution functions, *Teoriya Veroyatnostei i ee Primeneniya*, **6,** 446–454. (English translation: pp. 410–419.)

[12] Box, G. E. P. (1949). A general distribution theory for a class of likelihood criteria, *Biometrika*, **36,** 317–346.

[13] Bradford, S. C. (1948). *Documentation*, London: Crosby Lockwood.

[14] Cacoullos, T. (1965). A relation between t and F distributions, *Journal of the American Statistical Association*, **60,** 528–531. (Correction, 1249).

[15] Carter, A. H. (1947). Approximation to percentage points of the z-distribution, *Biometrika*, **34,** 352–358.

[16] Cochran, W. G. (1940). Note on an approximate formula for the significance levels of z, *Annals of Mathematical Statistics*, **11,** 93–95.

[17] Colcord, C. G. and Deming, Lola S. (1936). The one-tenth percent level of 'z', *Sankhyā*, **2,** 423–424.

[18] Cornish, E. A. and Fisher, R. A. (1937). Moments and cumulants in the specification of distributions, *Review of the International Statistical Institute*, **5,** 307–322.

[19] Curtiss, J. H. (1953). Convergent sequences of probability distributions, *American Mathematical Monthly*, **50,** 100–107.

[20] David, F. N. (1949). The moments of the z and F distributions, *Biometrika*, **36,** 394–403.

[21] Elderton, W. P. and Johnson, N. L. (1969). *Systems of Frequency Curves*, London: Cambridge University Press.

[22] Fisher, R. A. (1924). On a distribution yielding the error functions of several well-known statistics, *Proceedings of the International Mathematical Congress, Toronto*, 805–813.

[23] Fisher, R. A. and Cornish, E. A. (1960). The percentile points of distributions having known cumulants, *Technometrics*, **2,** 209–225.

[24] Fisher, R. A. and Yates, F. (1953). *Statistical Tables for Biological, Agricultural, and Medical Research*, London: Oliver and Boyd. (4th edition)

[25] Fox, B. L. (1963). Generation of random samples from the Beta F distributions, *Technometrics*, **5,** 269–270.

[26] Gayen, A. K. (1950). The distribution of the variance ratio in random samples of any size drawn from non-normal universes, *Biometrika*, **37,** 236–255.

[27] Geisser, S. (1964). Estimation in the uniform covariance case, *Journal of the Royal Statistical Society, Series B*, **26,** 477–483.

[28] Golden, R. R., Weiss, D. J. and Dawis, R. V. (1968). An evaluation of Jaspen's approximation of the distribution functions of the F, t and chi-square statistics, *Educational and Psychological Measurement*, **28,** 163–165.

[29] Greenwood, J. and Hartley, H. O. (1961). *Guide to Tables in Mathematical Statistics*, Princeton: Princeton University Press.

[30] Hald, A. (1952). *Statistical Tables and Formulas*, New York: John Wiley & Sons.

[31] Hartley, H. O. (1950). The use of range in analysis of variance, *Biometrika*, **37,** 271–280.

[32] Horsnell, G. (1953). The effect of unequal group variances on the F-test for homogeneity of group means, *Biometrika*, **40,** 128–136.

[33] Jaspen, N. (1965). The calculation of probabilities corresponding to values of z, t, F, and chi-square, *Educational and Psychological Measurement*, **25,** 877–880.

[34] Johnson, N. L. (1959). On an extension of the connexion between Poisson and χ^2 distributions, *Biometrika*, **46,** 352–363.

[35] Jowett, G. H. (1963). The relationship between the binomial and F distributions, *The Statistician*, **13,** 55–57.

[36] Kelley, T. L. (1948). *The Kelley Statistical Tables* (*Revised*), Cambridge, Mass.: Harvard University Press.

[37] Kolodziejczyk, S. (1935). On an important class of statistical hypotheses, *Biometrika*, **27,** 161–190.

[38] Kullback, S. (1935). The distribution laws of the difference and quotient of variables independently distributed in Pearson Type III laws, *Annals of Mathematical Statistics*, **7,** 51–53.

[39] Laubscher, N. F. (1965). Interpolation in F-tables, *American Statistician*, **19,** No. 1, 28, 40.

[40] Leimkuhler, F. F. (1967). The Bradford distribution, *Journal of Documentation*, **23,** 197–207.

[41] Mantel, N. (1966). F-ratio probabilities from binomial tables, *Biometrics*, **22,** 404–407.

[42] McIntyre, G. A. and Ward, M. M. (1968). Estimates of percentile points of the F-distribution, *Australian Computation Journal*, **1,** 113–114.

[43] Merrington, M. and Thompson, C. M. (1943). Tables of percentage points of the inverted beta (F) distribution, *Biometrika*, **33,** 73–88.

[44] Newman, D. (1939). The distribution of range in samples from a normal population, expressed in terms of an independent estimate of standard deviation, *Biometrika*, **31,** 20–30.

[45] Owen, D. B. (1962). *Handbook of Statistical Tables*, Reading, Massachusetts: Addison-Wesley Publishing Company, Inc.

[46] Paulson, E. (1942). An approximate normalization of the analysis of variance distribution, *Annals of Mathematical Statistics*, **13,** 233–235.

[47] Pearson, E. S. and Hartley, H. O. (1958). *Biometrika Tables for Statisticians*, **1,** London: Cambridge University Press. (2nd edition)

[48] Prins, H. J. (1960). Transforms for finding probabilities and variates of a distribution in terms of a related distribution function, *Statistica Neerlandica*, **14,** 1–17.

[49] Rider, P. R. (1931). A note on small sample theory, *Journal of the American Statistical Association*, **26,** 172–174.

[50] Scheffé, H. (1942). On the ratio of variances of two normal populations, *Annals of Mathematical Statistics*, **13,** 371–388.

[51] Scheffé, H. and Tukey, J. W. (1944). A formula for sample sizes for population tolerance limits, *Annals of Mathematical Statistics*, **15,** 217.

[52] Smillie, K. W. and Anstey, T. H. (1964). A note on the calculation of probabilities in an F-distribution, *Communications of the Association for Computing Machinery*, **7,** 725.

[53] Snedecor, G. W. (1934). *Calculation and Interpretation of the Analysis of Variance*, Ames, Iowa: Collegiate Press.

[54] Stammberger, A. (1967). Über einige Nomogramme zür Statistik, *Wissenschaftliche Zeitschrift der Humboldt-Universität Berlin. Mathematisch-Naturwissenschaftliche Reihe*, **16,** 86–93.

[55] Swain, A. K. P. C. (1965). A lower bound to the probability of variance-ratio, *Annals of the Institute of Statistical Mathematics, Tokyo*, **17,** 81–84.

[56] Tiao, G. C. and Lochner, R. H. (1966). *Tables for the Comparison of the Spread of two Normal Distributions*, Technical Report No. 88, Department of Statistics, University of Wisconsin.

[57] Tiku, M. L. (1964). Approximating the general non-normal variance-ratio sampling distributions, *Biometrika*, **51,** 83–95.

[58] Vogler, L. E. and Norton, K. A. (1957). *Graphs and Tables of the Significance Levels ($F(r_1,r_2,p)$) of the Fisher-Snedecor Variance Ratio*, National Bureau of Standards Report No. 5069, U. S. Government Printing Office; Washington, D. C.

[59] Wishart, J. (1946). The variance ratio test in statistics, *Journal of the Institute of Actuaries Students' Society*, **6,** 172–184.

[60] Wishart, J. (1947). The cumulants of the z and of the logarithmic χ^2 and t distributions, *Biometrika*, **34,** 170–178, (Correction, 374).

[61] Wishart, J. (1957). An approximate formula for the cumulative z-distribution, *Annals of Mathematical Statistics*, **28,** 504–510.

[62] Zeigler, C. O. (1965). *The effect of the normality and homogeneity of variance assumptions upon the validity of the F table for interaction terms*, M. Sc. thesis, Department of Industrial Engineering, Texas Technological College, Lubbock, Texas.

[63] Zinger, A. (1964). On interpolation in tables of the *F*-distribution, *Applied Statistics*, **13,** 51–53.

27

t Distribution

1. Genesis and Historical Remarks

If $X_1, X_2, \ldots, X_n$ are independent random variables each having the same normal distribution, with expected value ξ and standard deviation σ, then, $\sqrt{n}(\overline{X} - \xi)/\sigma$ (with $\overline{X} = n^{-1} \sum_{j=1}^{n} X_j$) has a unit normal distribution. This statistic can be used in the construction of tests and confidence intervals relating to the value of ξ, provided σ be known. If σ be not known, it is reasonable to replace it by the estimator $S = [(n-1)^{-1} \sum_{j=1}^{n} (X_j - \overline{X})^2]^{\frac{1}{2}}$. This procedure was adopted for some time, without making allowance for differences between the distributions of $\sqrt{n}(\overline{X} - \xi)/\sigma$ and $\sqrt{n}(\overline{X} - \xi)/S$. It was realized that the two distributions are not identical but the determination of the actual distribution presented difficulties. In 1908, 'Student' [83] obtained the distribution of

$$Z = \sqrt{n}\,(\overline{X} - \xi)\left[\sum_{j=1}^{n} (X_j - \overline{X})^2\right]^{-\frac{1}{2}} = \sqrt{n-1}\,\{\sqrt{n}\,(\overline{X} - \xi)/S\}$$

and gave a short table of its cumulative distribution function.

Remembering the results on the joint distribution of $\overline{X}$ and S described in Chapter 13 it can be seen that Z is distributed as a ratio U/χ_{n-1}, the two variables U and χ_{n-1} being mutually independent. The divisor $\sqrt{n-1}$ in denominator was introduced by Fisher [25] in 1925 defining '*t with* ν *degrees of freedom*' as

$$t_\nu = U[\chi_\nu^2/\nu]^{-\frac{1}{2}}. \tag{1}$$

This quantity is usually called '*Student's t*' and the corresponding distribution is called '*Student's distribution.*' Occasionally they are called *Fisher's statistic* and *distribution* respectively, but these latter terms more commonly refer to the variance-ratio, and its distribution, discussed in Chapter 26.

Cacoullos [14] has shown that if X_0 and X_1 are independent χ^2's each with ν degrees of freedom, then $\frac{1}{2}\sqrt{\nu}(X_1 - X_0)(X_0X_1)^{-\frac{1}{2}}$ has a t_ν distribution (Chapter 17).

2. Properties

The probability density function of $t_\nu = U[\chi_\nu/\sqrt{\nu}\,]^{-1}$ is

$$p_{t_\nu}(t) = \frac{1}{\sqrt{\nu}\,B(\frac{1}{2},\frac{1}{2}\nu)}\left(1 + \frac{t^2}{\nu}\right)^{-\frac{1}{2}(\nu+1)}. \tag{2}$$

This is a special form of Pearson Type VII distribution. It is symmetrical about $t = 0$ and has a single mode at $t = 0$. It is easy to show that

$$\lim_{\nu\to\infty} p_{t_\nu}(t) = (\sqrt{2\pi})^{-1}e^{-\frac{1}{2}t^2}.$$

In fact, as $\nu \to \infty$, the distribution of t_ν tends to the unit normal distribution. (This fact is the basis of most of the methods of approximation described in Section 4.)

If $t_{\nu,\alpha}$ is defined by the equation

$$\Pr[t_\nu \leq t_{\nu,\alpha}] = \alpha \tag{3}$$

then (from the symmetry)

$$t_{\nu,0.5} = 0 = U_{0.5}$$

while, for $\alpha > 0.5$

$$t_{\nu,\alpha} > U_\alpha > 0$$

and for $\alpha < 0.5$

$$t_{\nu,\alpha} < -U_\alpha < 0$$

(of course $t_{\nu,1-\alpha} = -t_{\nu,\alpha}$).

Making the transformation $w = \nu(\nu + y^2)^{-1}$ we have, for $t \geq 0$

$$\begin{aligned}\Pr[t_\nu \leq t] &= \Pr[t_\nu \leq 0] + [\sqrt{\nu}\,B(\tfrac{1}{2},\tfrac{1}{2}\nu)]^{-1}\int_0^t (1 + y^2/\nu)^{-\frac{1}{2}(\nu+1)}\,dy \\ &= \tfrac{1}{2} + \tfrac{1}{2}[B(\tfrac{1}{2},\tfrac{1}{2}\nu)]^{-1}\int_{\nu/(\nu+t^2)}^1 w^{\frac{1}{2}(\nu+1)-\frac{3}{2}}(1 - w)^{-\frac{1}{2}}\,dw \\ &= 1 - \tfrac{1}{2}I_{\nu/(\nu+t^2)}(\tfrac{1}{2}\nu,\tfrac{1}{2}) = \tfrac{1}{2}[1 + I_{t^2/(\nu+t^2)}(\tfrac{1}{2},\tfrac{1}{2}\nu)].\end{aligned} \tag{4.1}$$

Alternative expressions (valid for all t) are

$$\Pr[t_1 \leq t] = \frac{1}{2} + \frac{1}{\pi}\tan^{-1} t \qquad \text{for } \nu = 1, \tag{4.2}$$

and, putting $\theta = \tan^{-1}(t/\sqrt{\nu})$,

$$\text{(4.3)} \quad \Pr[t_\nu \le t) = \frac{1}{2} + \frac{1}{\pi}\left[\theta + \left\{\cos\theta + \tfrac{2}{3}\cos^3\theta + \cdots + \frac{(2)(4)\cdots(\nu-3)}{(3)(5)\cdots(\nu-2)}\cos^{\nu-2}\theta\right\}\sin\theta\right]$$

for ν odd and greater than 1;

$$\text{(4.4)} \quad \Pr[t_\nu \le t] = \frac{1}{2} + \left\{1 + \tfrac{1}{2}\cos^2\theta + \frac{(1)(3)}{(2)(4)}\cos^4\theta + \cdots + \frac{(1)(3)\cdots(\nu-3)}{(2)(4)\cdots(\nu-2)}\cos^{\nu-2}\theta\right\}\sin\theta$$

for ν even

(see Zelen and Severo [103]).

Amos [1] has obtained several expressions for $\Pr[t_\nu \le t]$ in terms of hypergeometric functions. For example,

$$\text{(4.5)} \quad \Pr[t_\nu \le t] = \frac{1}{2} + \frac{t}{\sqrt{\pi\nu}}\frac{\Gamma(\frac{1}{2}(\nu+1))}{\Gamma(\frac{1}{2}\nu)} F\left(\frac{1}{2}(\nu+1); \frac{1}{2}; \frac{3}{2}; \frac{-t^2}{\nu}\right)$$

for $t^2 < \nu$

which is useful when both $|t|/\nu^{\frac{1}{2}}$ and ν are small.

All odd moments of t_ν about zero are zero. If r be even then the rth central moment is:

$$\text{(5)} \quad \mu_r(t_\nu) = \nu^{\frac{1}{2}r} \cdot \frac{\Gamma(\frac{1}{2}(r+1))\Gamma(\frac{1}{2}(\nu-r))}{\Gamma(\frac{1}{2})\Gamma(\frac{1}{2}\nu)}$$

$$= \nu^{\frac{1}{2}r} \cdot \frac{1\cdot 3\cdots(r-1)}{(\nu-r)(\nu-r+2)\cdots(\nu-2)}.$$

(Note that if $r \ge \nu$, the rth moment is infinite.)

Putting $r = 1$ we obtain a formula for the mean deviation:

$$\text{(6)} \quad E[|t_\nu|] = \sqrt{\nu}\,\frac{\Gamma(\frac{1}{2}(\nu-1))}{\sqrt{\pi}\,\Gamma(\frac{1}{2}\nu)}.$$

From (5)

$$\text{(7.1)} \quad \operatorname{var}(t_\nu) = \frac{\nu}{\nu-2} \qquad (\nu \ge 2);$$

$$\text{(7.2)} \quad \alpha_4(t_\nu) = \beta_2(t_\nu) = 3 + \frac{6}{\nu-4} \qquad (\nu \ge 4).$$

Also, $\alpha_3(t_\nu) = \sqrt{\beta_1(t_\nu)} = 0$ (Wishart [101]).

From (6) and (7.1)

$$\text{(8)} \quad \frac{\text{mean deviation}}{\text{standard deviation}} = \sqrt{\frac{2}{\pi}}\cdot\left[\sqrt{\tfrac{1}{2}\nu - 1}\cdot\frac{\Gamma(\frac{1}{2}(\nu-1))}{\Gamma(\frac{1}{2}\nu)}\right].$$

TABLE 1

Ratio (m.d./s.d.) for t_ν Distribution

ν	3	4	5	6	7	8	9	10	11	12
ratio	0.637	0.707	0.735	0.750	0.759	0.765	0.770	0.773	0.776	0.778

The multiplier of $\sqrt{2/\pi}$ tends rapidly to 1 as ν increases, as can be seen from Table 1.

The probability density function of t_ν has points of inflexion at

$$t_\nu = \pm\sqrt{\nu/(\nu + 2)}.$$

As ν increases the distribution of t_ν approximates quite rapidly to the unit normal distribution. Figure 1 compares the t_4 and unit normal probability density functions. Even for such a small number of degrees of freedom, the two functions are not markedly dissimilar. If the 'standardized t_4' — that is, $t_4/\sqrt{2}$ — distribution is used the agreement is even closer (Weir [95]).

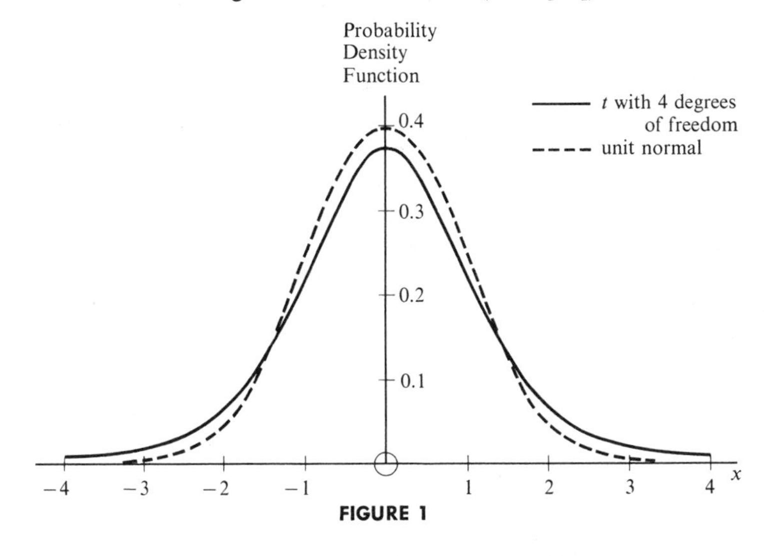

FIGURE 1

Comparison of Unit Normal and t_4 Density Functions

3. Tables and Nomograms

3.1 *Tables*

The cumulative t distribution and percentile points have been rather thoroughly tabulated. We give here a list, roughly in chronological order. There are, in addition, short tables in many text-books, which are mostly derived from tables in our list. (Note that $\nu = \infty$ corresponds to the unit normal distribution.)

The earliest published tables were provided (in 1908) by 'Student' [83]. These gave values of $\Pr[z_\nu \leq z]$ where $z_\nu = t_\nu/\sqrt{\nu + 1}$. Later (in 1925) the same author [84] gave the values of $\Pr[t_\nu \leq t]$ to 4 decimal places for $\nu = 1(1)20, \infty$ and $t = 0(0.1)6.0$; and also to 6 decimal places for $\nu = 3(1)11$ and $t = 6.0(0.5)10.0(1)12(2)16(4)28$.

In Biometrika Tables [62] there are tables of $\Pr[t_\nu \leq t]$, to 5 decimal places, for

$$\nu = 1(1)24, 30, 40, 60, 120, \infty$$

and

$$t = \begin{cases} 0.0(0.1)4.0(0.2)8.0 & (\nu \leq 20) \\ 0.00(0.05)2.0(0.1)4.0, 5 & (\nu \geq 20); \end{cases}$$

and of $t_{\nu,\alpha}$ to 3 decimal places for

$$\nu = 1(1)30, 40, 60, 120, \infty$$

and

$$\alpha = 0.6, 0.75, 0.9, 0.95, 0.975, 0.99, 0.995, 0.9975, 0.999, 0.9995.$$

(Also, $t_{\nu,\alpha}$ is given to at least 3 significant figures for

$$\nu = 1(1)10 \qquad \text{and} \qquad \alpha = 0.9999, 0.99999, 0.999995.)$$

Parts of these tables appeared earlier in Baldwin [6] and Hartley and Pearson [41]; parts are reproduced in Janko [47]. Rao *et al.* [69] give similar tables with the addition of $\nu = 80$ (and exclusion of $\nu = 120$) and $\alpha = 0.7, 0.8$.

These tables are notable for the extreme tail percentiles included. For more extensive sets of values of ν, however, some of the following tables are more useful.

Fisher and Yates [31] give values of $t_{\nu,\alpha}$ to 3 decimal places for

$$\nu = 1(1)30, 40, 60, 120$$

and

$$\alpha = 0.55(0.05)0.95, 0.975, 0.99, 0.995, 0.9995.$$

Veselá [88] gives values of $t_{\nu,\alpha}$ to 4 decimal places for $\nu = 30(1)120$ and $\alpha = 0.95, 0.975, 0.995$.

In Owen [59] there are values of $t_{\nu,\alpha}$ to 4 decimal places for

$$\nu = 1(1)100, 150, 200(100)1000, \infty$$

and

$$\alpha = 0.75, 0.90, 0.95, 0.975, 0.99, 0.995.$$

Also given are values of $t_{\nu,\alpha}$ to 5 decimal places for

$$\nu = 1(1)30(5)100(10)200, \infty$$

with

$$\alpha = 0.95,\ 0.975,\ 0.99,\ 0.995.$$

These tables are notable for the extensive series of values of ν.

Federighi [23] has concentrated on values of α very near to 1 (i.e. the extreme tail of the distribution). His tables give values of $t_{\nu,\alpha}$ to 3 decimal places for

$$\nu = 1(1)30(5)60(10)100,\ 200,\ 500,\ 1000,\ 2000,\ 10000,\ \infty$$

and

$$1 - \alpha = 0.25, 0.1, 0.05, 0.025, 0.01, 0.005, 0.0025, 10^{-3}, \tfrac{1}{2} \times 10^{-3}, \tfrac{1}{4} \times 10^{-3}, 1 \times 10^{-4}, \tfrac{1}{2} \times 10^{-4}, \tfrac{1}{4} \times 10^{-4}, 1 \times 10^{-5}, \tfrac{1}{2} \times 10^{-5}, \tfrac{1}{4} \times 10^{-5}, 1 \times 10^{-6}, \tfrac{1}{2} \times 10^{-6}, \tfrac{1}{4} \times 10^{-6}, 1 \times 10^{-7}.$$

Hald [40] gives the values of $t_{\nu,\alpha}$ to 3 decimal places for

$$\nu = 1(1)30(10)60(20)100,\ 200,\ 500,\ \infty$$

and

$$\alpha = 0.6(0.1)0.9,\ 0.95,\ 0.975,\ 0.99,\ 0.995,\ 0.999,\ 0.9995.$$

Values of the density function $p_{t_\nu}(t)$ are tabulated in Bracken and Schleifer [11], Smirnov [80] and Sukhatme [85]. In 1938, Sukhatme published values to 7 decimal places for

$$\nu = 1(1)10,\ 12,\ 15,\ 20,\ 24,\ 30$$

with

$$t = 0.05(0.1)7.25.$$

In addition there are values for $\nu = 60$ with $t = 0.05(0.1)6.35$.

The tables of Bracken and Schleifer [11] cover greater ranges of values of each of the arguments (note especially the fractional value $\nu = 1.5$):

$$\nu = 1,\ 1.5,\ 2(1)10,\ 12,\ 15,\ 20,\ 24,\ 30,\ 40,\ 60,\ 120,\ \infty$$

and

$$t = 0.00(0.01)8.00.$$

The collection [80] includes values of $p_{t_\nu}(t)$ and $\Pr[t_\nu \leq t]$ to 6 decimal places for

$$\nu = 1(1)12 \text{ with } t = 0.00(0.01)3.00(0.02)4.50(0.05)6.50$$

and for

$$\nu = 13(1)24 \text{ with } t = 0.00(0.01)2.50(0.02)3.50(0.05)6.50.$$

This collection further includes values of $\Pr(t_\nu \leq t)$ also to 6 decimal places for

$$\nu = 1(1)10 \text{ with } t = 6.5(0.1)9.0;$$

for

$$\nu = 25(1)35 \text{ with } t = 0.00(0.01)2.50(0.02)3.50(0.05)5.00$$

and for

$$10^3\nu^{-1} = 30(-2)0 \text{ with } t = 0.00(0.01)2.50(0.02)5.00.$$

There are also extensive tables of $t_{\nu,\alpha}$ to 4 decimal places for

$$\nu = 1(1)30(10)100,\ 120,\ 150(50)500(100)1000,\ 1500,\ 2000(1000)6000,\ 8000,\ 10000,\ \infty$$

with

$$\alpha = 0.6, 0.75, 0.9, 0.95, 0.975, 0.99, 0.995, 0.9975, 0.999, 0.9995.$$

These are the most extensive tables of $t_{\nu,\alpha}$ currently available. They also include values of the multiplying constant $K_\nu = (\pi\nu)^{-\frac{1}{2}}\Gamma(\frac{1}{2}(\nu + 1))/\Gamma(\frac{1}{2}\nu)$, and $\log K_\nu$, to ten significant figures for $\nu = 1(1)24$.

Cotterman and Knoop [17] have given tables from which $\Pr[t_\nu \geq T]$ can be evaluated to three places of decimals for $\nu = 1(1)15$. The quantities actually tabulated (to 5 decimal places) are boundary values $T_1(p)$, $T_2(p)$ such that, to 3 decimal places, $\Pr[t_\nu \geq T]$ is equal to p ($p = 0.000(0.001)0.500$) for

$$T_1(p) < T < T_2(p).$$

Laumann [54] gives $\Pr[t_\nu \leq t]$ to 7 decimal places for $t = 0(0.01)4.50$; $\nu = 20(2)40(10)100(20)200, 300, 500, 1000$.

3.2 *Nomograms*

The above list shows that the distribution of t_ν has been thoroughly tabulated. The available tables are more than sufficient for almost all applications. However, practical situations arise which call for quick evaluation of values of $t_{\nu,\alpha}$ or $\Pr[t_\nu \leq t]$. In such cases it is useful to have a reliable graphical method of determining the required value with sufficient accuracy.

James-Levy [46] gave a nomogram relating ν, t and $\Pr[t_\nu \leq t]$ which is reproduced in Figure 2. It is used by placing a straightedge joining given values of any two of these quantities. The intersection with the third line then gives the required value of the third quantity. For example, to find $t_{\nu,\alpha}$ the appropriate points on the ν and $\Pr[t_\nu \leq t]$ are joined, and the intersection with the t line gives $t_{\nu,\alpha}$. (A table in [46] indicates that with $\alpha = 0.950$–0.999 an accuracy of about 0.001 in $t_{\nu,\alpha}$ can usually be attained.)

Stammberger [81] has published a simple nomogram from which the value of any one of ν, $\Pr[t_\nu > t]$, and t given the values of the other two quantities, can be read off using a straight-edge. This nomogram is reproduced in Figure 3.

Babanin [5] has provided a nomogram (or abac) from which values can be read directly, without using a straightedge. However, the nomogram itself is not so simple as those of James-Levy and Stammberger.

4. Approximations

There has been an intense study of possible approximations to the t distribution. This has produced approximations of very high accuracy, though sometimes rather complicated. In the present context bear in mind the general

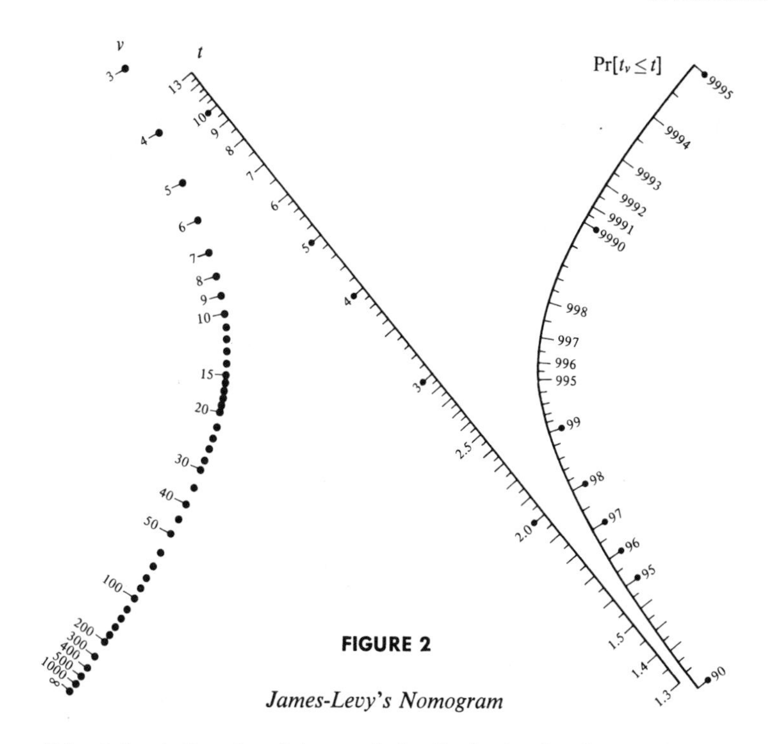

FIGURE 2

James-Levy's Nomogram

Notes: $\Pr[t_\nu \leq t]$ — is the value of the cumulative distribution function at point t.
ν — is the number of degrees of freedom.
Given any two of ν, t or $\Pr[t_\nu \leq t]$ we can determine the third.

rule that simplicity as well as accuracy, is an important factor in assessing the value of an approximation.

The simplest approximation is made by regarding t_ν as a unit normal variable. This is very crude unless ν is at least 30, and unsatisfactory for considerably larger values of ν if extreme tails (say $|t_\nu| > 4$) of the distribution are being considered. The simple modification of standardizing the t variable, and regarding $t_\nu\sqrt{1 - 2\nu^{-1}}$ as having a unit normal distribution (suggested by Weir [95]) effects a substantial improvement, but the approximation is still only moderately good (for analytical purposes) if ν is less than 20 (or if extreme tails are being considered).

As in Section 3 we will present a list, very roughly in chronological order.

Fisher [26] gave a direct expansion of the probability density, and hence $\Pr[t_\nu \leq t]$ as a series in ν^{-1}. For the probability density function, he gives the formula

$$
\begin{aligned}
(9) \qquad p_{t_\nu}(t) = Z(t)[1 &+ \tfrac{1}{4}(t^4 - 2t^2 - 1)\nu^{-1} \\
&+ \tfrac{1}{96}(3t^8 - 28t^6 + 30t^4 + 12t^2 + 3)\nu^{-2} \\
&+ \tfrac{1}{384}(t^{12} - 22t^{10} + 113t^8 - 92t^6 \\
&\qquad - 33t^4 - 6t^2 + 15)\nu^{-3} + \cdots].
\end{aligned}
$$

FIGURE 3

Stammberger's Nomogram

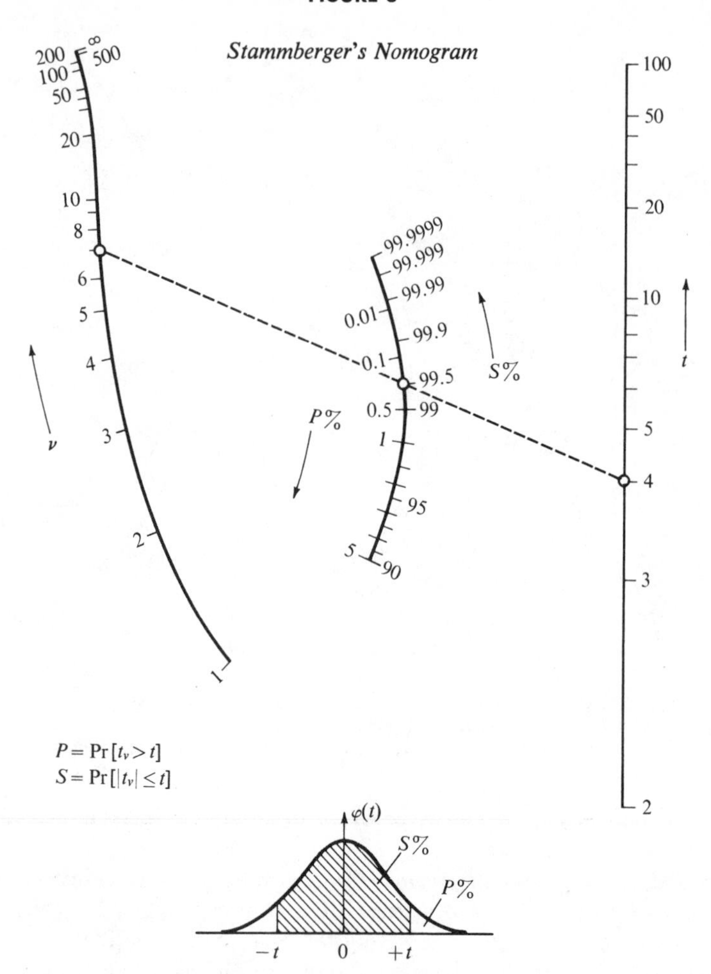

Integrating (9) we obtain

$$\text{(10)}\qquad \Pr[t_\nu \geq t] \doteqdot 1 - \Phi(t) + Z(t)[\tfrac{1}{4}t(t^2+1)\nu^{-1} + \tfrac{1}{96}t(3t^6 - 7t^4 - 5t^2 - 3)\nu^{-2} + \tfrac{1}{384}t(t^{10} - 11t^8 + 14t^6 + 6t^4 - 3t^2 - 15)\nu^{-3}].$$

This formula has a maximum absolute error of 0.000005.

Much later, Fisher and Cornish [29] inverted this series, obtaining the approximation

$$\text{(11)}\qquad t_{\nu,\alpha} \doteqdot U_\alpha + \tfrac{1}{4}U_\alpha(U_\alpha^2+1)\nu^{-1} + \tfrac{1}{96}U_\alpha(5U_\alpha^4 + 16U_\alpha^2 + 3)\nu^{-2} + \tfrac{1}{384}U_\alpha(3U_\alpha^6 + 19U_\alpha^4 + 17U_\alpha^2 - 15)\nu^{-3}.$$

Later still, Dickey [20] obtained an asymptotic (divergent) series approximation for $t < \sqrt{\nu}$ in terms of Appell polynomials $A_r(x)$ defined by the identity

$$e^{-x}(1 - ux)^{-1/u} = \sum_{r=0}^{\infty} A_r(x)u^r \qquad (|tx| < 1).$$

Putting $u = z(\nu + 1)^{-1}$; $x = \frac{1}{2}(1 + \nu^{-1})t^2$, we have

$$\left(1 + \frac{t^2}{\nu}\right)^{-\frac{1}{2}(\nu+1)} = e^{-\frac{1}{2}(1+\nu^{-1})t^2} \sum_{r=0}^{\infty} A_r(-\tfrac{1}{2}(1 + \nu^{-1})t^2)\left(\frac{2}{\nu + 1}\right)^r \qquad (t^2 < \nu).$$

Dickey gives the following table of coefficients $B_{r,j}$ in $A_r(x) = x^r \sum_{j=0}^{r} B_{r,j}x^j$.

TABLE 2

Values of Coefficients $B_{r,j}$ in Appell Polynomials

$$A_r(x) = x^r \sum_{j=0}^{r} B_{r,j}x^j$$

	j						
r	0	1	2	3	4	5	6
0	1						
1	2	$\frac{1}{2}$					
2	0	$\frac{1}{3}$	$\frac{1}{8}$				
3	0	$\frac{1}{4}$	$\frac{1}{6}$	$\frac{1}{48}$			
4	0	$\frac{1}{5}$	$\frac{13}{72}$	$\frac{1}{24}$	$\frac{1}{384}$		
5	0	$\frac{1}{6}$	$\frac{11}{60}$	$\frac{17}{288}$	$\frac{1}{144}$	$\frac{1}{3840}$	
6	0	$\frac{1}{7}$	$\frac{29}{160}$	$\frac{59}{810}$	$\frac{7}{576}$	$\frac{1}{1152}$	$\frac{1}{46080}$

Hendricks [42] also approximated directly to the probability density function, obtaining

$$(12) \quad p_{t_\nu}(t) \doteqdot 2\nu c_\nu \sqrt{\frac{\nu + 1}{\pi}} [t^2 + 2\nu]^{-\frac{3}{2}} \exp\left[-(\nu + 1)c_\nu^2 t^2(t^2 + 2\nu)^{-1}\right]$$

where $\qquad c_\nu = 1 - \frac{3}{4}(\nu + 1)^{-1} - \frac{7}{32}(\nu + 1)^{-2}.$

This gives quite good results in the 'center' of the distribution ($|t| < 2$), but not in the tails.

Formula (12) is equivalent to the approximation "$\sqrt{2(\nu + 1)}\, c_\nu t_\nu[t_\nu^2 + 2\nu]^{-\frac{1}{2}}$ has a unit normal distribution". In practice $\sqrt{2(\nu + 1)}\, c_\nu$ may be replaced by $\sqrt{2\nu - 1}$, unless ν is small.

Some numerical comparisons are shown in Table 3.

Another approximation, of a similar nature, has been obtained by Elfving [22]. This is

$$\text{(13)}\quad \Pr[t_\nu \leq t] \doteqdot \Phi(\sigma t) + \tfrac{5}{96}(\sigma t^5 \nu^{-2}(1 + \tfrac{1}{2}t^2\nu^{-1})^{-\frac{1}{2}(\nu+4)} - Z(\sigma t/\sqrt{2})$$

where

$$\sigma = \left[\frac{\nu - \frac{1}{2}}{\nu + \frac{1}{2}t^2}\right]^{\frac{1}{2}}.$$

The error can be shown to be less than $\frac{1}{2}\nu^{-2}$ times the true value of $\Pr[t_\nu \leq t]$, for all values of t.

In 1938 Hotelling and Frankel [44] sought to find a function of t_ν which should have a distribution which is well approximated by the unit normal distribution. The leading terms of their series (which is, in fact, a Cornish-

TABLE 3

	α	Values of $t_{\nu,\alpha}$	
		Exact Value	Hendricks' Approximation
$\nu = 9$	0.55	0.129	0.129
	0.65	0.398	0.398
	0.75	0.703	0.703
	0.85	1.100	1.104
	0.95	1.833	1.844
	0.975	2.262	2.290
	0.99	2.821	2.869
	0.995	3.250	3.389
$\nu = 29$	0.55	0.127	0.127
	0.65	0.389	0.389
	0.75	0.683	0.683
	0.85	1.055	1.055
	0.95	1.699	1.700
	0.975	2.045	2.047
	0.99	2.462	2.466
	0.995	2.756	2.764

TABLE 4

Approximations—Values of U Using Expansion (14)

α	0.95		0.975		0.995	
	$\nu = 10$	$\nu = 30$	$\nu = 10$	$\nu = 30$	$\nu = 10$	$\nu = 30$
$t_{\nu,\alpha}$	1.812	1.697	2.228	2.042	3.169	2.750
x_1	1.618	1.642	1.896	1.954	2.294	2.554
x_2	1.650	1.645	1.980	1.960	2.754	2.579
x_3	1.643	1.645	1.953	1.960	2.446	2.575
x_4						
U_α	1.645		1.960		2.576	

α	0.9995		0.99995		
	$\nu = 10$	$\nu = 30$	$\nu = 10$	$\nu = 30$	$\nu = 100$
$t_{\nu,\alpha}$	4.587	3.646	6.22	4.482	4.052
x_1	2.059	3.212	.05	3.69	3.88
x_2	4.981	3.313	12.86	3.98	3.89
x_3	0.896	3.283	−20.44	3.85	3.89
x_4	7.163	3.293	75.66	3.91	3.89
U_α	3.291		3.891		

Fisher form of expansion) are

$$(14) \qquad U = t\{1 - \tfrac{1}{4}(t^2 + 1)\nu^{-1} + \tfrac{1}{96}(13t^4 + 8t^2 + 3)\nu^{-2} - \tfrac{1}{384}(35t^6 + 19t^4 + t^2 - 15)\nu^{-3} + \tfrac{1}{92160}(6271t^8 + 3224t^6 - 102t^4 - 1680t^2 - 945)\nu^{-4}\}.$$

The successive terms rapidly become more complicated. Table 4 (taken from [44]) gives values of U corresponding to $t_{\nu,\alpha}$ for various values of ν and α using the first two (x_1), three (x_2) and four (x_3) and five (x_4) terms of (14). The correct values of U_α are also shown. For extreme tails poor results are obtained with $\nu = 10$, but even for extreme tails ($\alpha = 0.99995$) quite good results are obtained with $\nu \geq 30$, if five terms are used.

Among other investigations of expansions of Cornish-Fisher type we note the work of Peiser [63] in 1943, and Goldberg and Levine [38] in 1946. Peiser used the simple formula

$$t_{\nu,\alpha} \doteqdot U_\alpha + \tfrac{1}{4}(U_\alpha^3 + U_\alpha)\nu^{-1}. \tag{15}$$

Table 5 (taken from [63]) shows that this gives useful results for $\nu \geq 30$.

TABLE 5

Approximation to $t_{\nu,\alpha}$ Using (15)

ν		α = 0.9875	0.975	0.95	0.875	0.75
10	(15)	2.579	2.197	1.797	1.212	0.700
	Exact	2.634	2.228	1.813	1.221	0.700
30	(15)	2.354	2.039	1.696	1.171	0.683
	Exact	2.360	2.042	1.697	1.173	0.683
60	(15)	2.298	2.000	1.670	1.161	0.679
	Exact	2.299	2.000	1.671	1.162	0.679
120	(15)	2.270	1.980	1.658	1.156	0.677
	Exact	2.270	1.980	1.658	1.156	0.677

Goldberg and Levine included one further term in the expansion, giving

$$t_{\nu,\alpha} \doteqdot U_\alpha + \tfrac{1}{4}(U_\alpha^3 + U_\alpha)\nu^{-1} + \tfrac{1}{96}(5U_\alpha^5 + 16U_\alpha^3 + 3U_\alpha)\nu^{-2}. \tag{16}$$

Table 6, taken from [38] compares exact values with approximate values calculated from (16). (The original tables also give values obtained using only the first two terms, as in Peiser's work.)

Inclusion of the third term considerably improves the approximation, which is now reasonably good for ν as small as 20.

Simaika [79] improved the approximation $U \doteqdot t\sqrt{1 - 2\nu^{-1}}$ by introducing higher powers of U.

The approximation "$\sinh^{-1}$ $(t_\nu\sqrt{\tfrac{3}{2}\nu^{-1}})$ has a unit normal distribution" (Anscombe [2]) is a special case of a transformation of noncentral t (Chapter 31). It is not much used for central t.

Chu [15] obtained the following inequalities, which aid in assessing the accuracy of simple normal approximations to the distribution of t_ν:

$$\frac{\nu}{\nu+1}\left[\Phi\left(B\sqrt{\frac{\nu+1}{\nu}}\right) - \Phi\left(A\sqrt{\frac{\nu+1}{\nu}}\right)\right] \leq \Pr[A < t_\nu < B] \tag{17}$$

$$\leq \sqrt{\frac{7\nu-3}{7\nu-14}}\left[\Phi\left(B\sqrt{\frac{\nu-2}{\nu}}\right) - \Phi\left(A\sqrt{\frac{\nu-2}{\nu}}\right)\right].$$

He showed that for ν large, the ratio of absolute error to correct value of $\Pr[A < t_\nu < B]$ using the unit normal approximation to t_ν, is less than ν^{-1}.

TABLE 6

Comparative Table of Approximate and Exact Values of the Percentage Points of the t-Distribution

Probability Integral (α)	Degrees of Freedom (ν)	Approximate Percentage Point (from (16))	Exact Percentage Point
.9975	1	21.8892	127.32
	2	9.1354	14.089
	10	3.5587	3.5814
	20	3.1507	3.1534
	40	2.9708	2.9712
	60	2.9145	2.9146
	120	2.8599	2.8599
.9950	1	16.3271	63.657
	2	7.2428	9.9248
	10	3.1558	3.1693
	20	2.8437	2.8453
	40	2.7043	2.7045
	60	2.6602	2.6603
	120	2.6174	2.6174
.9750	1	7.1547	12.706
	2	3.8517	4.3027
	10	2.2254	2.2281
	20	2.0856	2.0860
	40	2.0210	2.0211
	60	2.0003	2.0003
	120	1.9799	1.9799
.9500	1	4.5888	6.3138
	2	2.7618	2.9200
	10	1.8114	1.8125
	20	1.7246	1.7247
	40	1.6838	1.6839
	60	1.6706	1.6707
	120	1.6577	1.6577
.7500	1	.9993	1.0000
	2	.8170	.8165
	10	.6998	.6998
	20	.6870	.6870
	40	.6807	.6807
	60	.6786	.6786
	120	.6765	.6766

Wallace [89], developing the methods used by Chu [15], has obtained *bounds* for the cumulative distribution function of t_ν. These are most easily expressed in terms of bounds on the corresponding normal deviate $u(t)$, defined by the equation

$$\Phi(u(t)) = \Pr[t_\nu \leq t]. \tag{18}$$

Wallace summarized his results as follows:

(19.1) $$u(t) \leq [\nu \log (1 + t^2\nu^{-1})]^{\frac{1}{2}}$$

(19.2) $$u(t) \geq (1 - \tfrac{1}{2}\nu^{-1})^{\frac{1}{2}}[\nu \log (1 + t^2\nu^{-1})]^{\frac{1}{2}}, \qquad \text{for } \nu > \tfrac{1}{2}$$

(19.3) $$u(t) \geq [\nu \log (1 + t^2\nu^{-1})]^{\frac{1}{2}} - 0.368\nu^{-\frac{1}{2}}, \qquad \text{for } \nu \geq \tfrac{1}{2}.$$

From (19.1) and (19.2) it can be seen that $[\nu \log (1 + t^2\nu^{-1})]^{\frac{1}{2}}$ differs from $u(t)$ by an amount not exceeding $25\nu^{-1}\%$; (19.3) shows that the absolute error does not exceed $0.368\nu^{-\frac{1}{2}}$. Usually (19.2) gives a better (i.e. greater) lower bound than (19.3).

Wallace further obtains two good approximations (though not giving precise *bounds*, in these cases):

(20) $$\left(\frac{8\nu + 1}{8\nu + 3}\right) [\nu \log (1 + t^2\nu^{-1}]^{\frac{1}{2}},$$

(21) $$\left[1 - \frac{2}{8x + 3}\{1 - e^{-s^2}\}^{\frac{1}{2}}\right][\nu \log (1 + t^2\nu^{-1})]^{\frac{1}{2}}$$

with $$s = 0.183(8\nu + 3)\nu^{-1}[\log (1 + t^2\nu^{-1})]^{-\frac{1}{2}}.$$

He states that (21) is within 0.02 of the true value of $u(t)$ for a wide range of values of t.

Wallace compares the values given by (19.1), (19.2), (19.3), (20) and (21), and by the formula corresponding to the Paulson approximation to the F-distribution (Chapter 26) (putting $\nu_1 = 1$); namely

(22) $$\Pr[|t_\nu| \leq t] \doteqdot \Pr\left[U \leq \frac{1}{3\sqrt{2}}[(9 - 2\nu^{-1})t^{\frac{2}{3}} - 7][\nu^{-1}t^{\frac{4}{3}} + 1]^{-\frac{1}{2}}\right].$$

His results (from [89]) are shown in Table 7.

The accuracy of (21) is noteworthy, though the formula is rather complicated.

Recently, Peizer and Pratt [64] have proposed other formulas of this kind, namely

(23.1) $$u(t) \doteqdot (\nu - \tfrac{2}{3})\left[\frac{\log (1 + t^2\nu^{-1})}{\nu - \frac{5}{6}}\right]^{\frac{1}{2}}$$

and

(23.2) $$u(t) \doteqdot (\nu - \tfrac{2}{3} + \tfrac{1}{10}\nu^{-1})\left[\frac{\log (1 + t^2\nu^{-1})}{\nu - \frac{5}{6}}\right]^{-1}.$$

Cornish [a]* reported Hill's [b]† Cornish-Fisher type expansions in terms of $u = [a_\nu \log (1 + t^2\nu^{-1})]^{\frac{1}{2}}$, where $a_\nu = \nu - \frac{1}{2}$,

*[a] Cornish, E. A. (1969). Fisher Memorial Lecture. (37th. Session I.S.I., London.)
†[b] Hill, G. W. (1969). *Progress results on asymptotic approximations for Student's t and chi-squared*, Personal communication.

TABLE 7

Bounds on the Equivalent Normal Deviate $u(t)$ for t_ν

ν	t	Exact Value	$u(t)$ (Bounds) (19.1)	(19.2)	(19.3)	Approximations (20)	(21)	Paulson (22)
1	0.3	.235	.294	.208	<0	.241	.241	.257
	1	.674	.832	.589	.465	.680	.681	.674
	2	1.047	1.269	.897	.901	1.038	1.048	1.031
	4	1.419	1.683	1.190	1.315	1.377	1.416	1.349
	8	1.756	2.043	1.445	1.675	1.672	1.750	1.576
	12	1.935	2.231	1.577	1.863	1.825	1.927	1.670
	10^2	2.729	3.035	2.146	2.667	2.177	2.704	1.896
	10^5	4.514	4.799	3.393	4.431	3.926	4.447	1.964
3	1	.858	.929	.848	.717	.860	.860	.855
	2	1.478	1.594	1.455	1.382	1.476	1.478	1.477
	4	2.197	2.353	2.148	2.141	2.179	2.197	2.160
	8	2.872	3.053	2.787	2.840	2.826	2.879	2.705
	12	3.228	3.417	3.119	3.204	3.164	3.237	2.953
	$\sqrt{3} \times 10^2$	5.057	5.256	4.797	5.044	4.866	5.058	3.493
10	1	.952	.976	.952	.860	.953	.953	.948
	2	1.790	1.834	1.788	1.718	1.790	1.790	1.805
	4	3.021	3.091	3.013	2.975	3.017	3.020	3.014
	8	4.382	4.474	4.361	4.357	4.366	4.384	4.279
	12	5.128	5.229	5.097	5.113	5.103	5.133	4.902
100	100	21.447	21.483	21.429	21.446	21.429	21.450	18.541

(23.3)

$$u(t) \doteqdot u + \tfrac{1}{48}(u^3 + 3u)a_\nu^{-2} - \tfrac{1}{23040}(4u^7 + 33u^5 + 240u^3 + 855u)a_\nu^{-4},$$

and, inversely

$$t_{\nu,\alpha} = [\nu(\exp[u^2a_\nu^{-1}] - 1)]^{\frac{1}{2}} \tag{23.4}$$

where

$$u \doteqdot U_\alpha - \tfrac{1}{48}(U_\alpha^3 + 3U_\alpha)a_\nu^{-2} + \tfrac{1}{23040}(4U_\alpha^7 + 63U_\alpha^5 + 360U_\alpha^3 + 945U_\alpha)a_\nu^{-4}.$$

Approximations (Hill [c]‡) based on (23.3) and (23.4) give $\Phi(u(t))$ to 10 decimal places and $t_{\nu,\alpha}$ to $(10 - U_\alpha)$ significant digits for $\nu > 13$.

Among formulas developed empirically, we mention results of four recent investigations, reported by Cucconi [18], Gardiner and Bombay [32], Moran [58] and Kramer [51].

‡[c] Hill, G. W. (1970). Algorithms '*Student*' and '*t quantile*', submitted to Comm. ACM.

Gardiner and Bombay gave formulas of the form

$$t_{\nu,\alpha} \doteqdot (a\nu + b + c\nu^{-1})(\nu + d + e\nu^{-1})^{-1}$$

for various percentiles. Values of a, b, c, d, e for $\alpha = 0.95, 0.975, 0.995$ are shown in Table 8 (note that $a = U_\alpha$).

TABLE 8

α	a	b	c	d	e
0.95	1.6449	3.5283	0.85602	1.2209	−1.5162
0.975	1.9600	0.60033	0.95910	−0.90259	0.11588
0.995	2.5758	−0.82847	1.8745	−2.2311	1.5631

The corresponding values of U_α are correct to 4 decimal places for $\nu > 3$.

These results are rather better, for smaller values of ν, than those obtained with Cucconi's [18] formulas:

(24.1) $$t_{\nu,0.975} \doteqdot 1.9600\nu(\nu^2 - 2.143\nu + 1.696)^{-\frac{1}{2}} \qquad (\nu > 1).$$

(24.2) $$t_{\nu,0.995} \doteqdot 2.5758\nu(\nu^2 - 3.185\nu + 4.212)^{-\frac{1}{2}} \qquad (\nu > 2).$$

Moran [58] sought to find a normalizing transformation. His investigations, which were confined to comparisons at 2.5%, 0.5% and 0.05% percentiles, led to the following approximations:

At 2.5% *point:*

(25.1) $$U_{0.975} = t_{\nu,0.975}(1 - 0.0550t_{\nu,0.975}\nu^{-1})$$

(25.2) $$U_{0.975} = t_{\nu,0.975}(1 - 0.6049t_{\nu,0.975}\nu^{-1} + 0.2783t^2_{\nu,0.975}\nu^{-2}).$$

At 0.5%:

(26) $$U_{0.995} = t_{\nu,0.995}[1 + 0.613t_{\nu,0.995}\nu^{-1}]^{-1} - 0.8\nu^{-1}.$$

At 0.05%:

(27) $$U_{0.9995} = t_{\nu,0.9995}[1 + 0.87t_{\nu,0.9995}\nu^{-1}]^{-1}.$$

The formula

(28) $$U_\alpha = t_{\nu,\alpha}[1 + 0.613t_{\nu,\alpha}\nu^{-1}]^{-1}$$

was found to give fairly good results both for $\alpha = 0.975$ and $\alpha = 0.995$. Values given by (25.2), (26) and (28) for $\alpha = 0.975, 0.995$ are shown in Table 9.

It can be seen that (26) does give a substantial improvement over (25.2) for $\alpha = 0.995$. Formula (28) gives reasonably good results for $\nu \geq 10$.

Kramer's [51] approximations are based on unpublished results obtained by Ray [71]. He states that the following formulas have errors less than 0.001 for $3 \leq \nu \leq 120$:

TABLE 9

	$\alpha = 0.975$ $U_\alpha = 1.9600$		$\alpha = 0.995$ $U_\alpha = 2.5758$		
	Value from		Value from		
ν	(25.2)	(28)	(25.2)	(28)	(26)
3	2.1369	1.9284	5.1238	2.6628	2.3961
4	1.9830	1.9477	3.0961	2.6994	2.4994
5	1.9603	1.9546	2.7950	2.6983	2.5383
6	1.9565	1.9575	2.7156	2.6889	2.5556
8	1.9573	1.9597	2.6684	2.6691	2.5691
10	1.9586	1.9604	2.6503	2.6537	2.5737
15	1.9603	1.9607	2.6282	2.6300	2.5767
20	1.9607	1.9606	2.6165	2.6171	2.5771
30	1.9608	1.9605	2.6039	2.6037	2.5770
40	1.9607	1.9604	2.5973	2.5969	2.5769

For $0 \leq t \leq 1$:

(29.1) $$\Pr[0 < t_\nu < t] \doteqdot 0.399622t - 0.068492t^3 + 0.010272t^5 - 0.111604(t/\nu) - 0.009310(t^3/\nu) + 0.02865(t/\nu^2).$$

For $1 \leq t \leq 2$:

(29.2) $$\Pr[0 < t_\nu < t] \doteqdot -0.060820 + 0.585243t - 0.208977t^3 + 0.025489t^5 + 0.082228(1/\nu) - 0.276747(t/\nu) + 0.080726(t^2/\nu) + 0.011192(t/\nu^2).$$

For $t > 2$:

(29.3) $$\Pr[0 < t_\nu < t] \doteqdot 0.503226 - 0.044928(1/\nu) + 0.112057(1/\nu^2) + 1.949790(\nu t^2)^{-1} - 5.917356(\nu^2 t^2)^{-1} - 7.549051(\nu t^3)^{-1} + 11.311627(\nu^2 t^3)^{-1} - 0.399205t^{-4} + 5.487170(\nu t^4)^{-1}.$$

Similar accuracy for $\nu = 1$ is obtained with the formulas:

For $0 \leq t \leq \frac{1}{2}$:

(30.1) $$\Pr[0 < t_1 < t] \doteqdot \pi^{-1}(t - \tfrac{1}{3}t^3 + \tfrac{1}{5}t^5 - \tfrac{1}{7}t^7).$$

For $\frac{1}{2} < t < \frac{3}{2}$:

$$\text{(30.2)} \qquad \Pr[0 < t_1 < t] \doteq \tfrac{1}{4} + \pi^{-1}[\tfrac{1}{2}(t-1) - \tfrac{1}{4}(t-1)^2 + \tfrac{1}{12}(t-1)^3 - \tfrac{1}{40}(t-1)^5].$$

For $t \geq \frac{3}{2}$:

$$\text{(30.3)} \qquad \Pr[0 < t_1 < t] \doteq \tfrac{1}{2} - \pi^{-1}[t^{-1} - \tfrac{1}{3}t^{-3} + \tfrac{1}{5}t^{-5} - \tfrac{1}{7}t^{-7}].$$

For $\nu = 2$, the formula

$$\text{(31)} \qquad \Pr[0 < t_2 < t] = t(8 + 4t^2)^{-\frac{1}{2}}$$

is exact for all $t \geq 0$.

The above formulas ((29)–(30)) are especially suited for use in electronic computers.

For calculating extreme tail probabilities of the t_ν distribution, Pinkham and Wilk [67] suggested using the expansion

$$\text{(32.1)} \qquad \int_t^\infty (1 + y^2\nu^{-1})^{-\frac{1}{2}(\nu+1)}\,dy = \sum_{j=1}^m w_j + R_m(t) \qquad (m < \tfrac{1}{2}(\nu+1))$$

where
$$w_1 = \nu(\nu-1)^{-1}\cdot(1 + t^2\nu^{-1})^{-\frac{1}{2}(\nu-1)}t^{-1}$$

$$w_{j+1} = w_j(1 + \nu t^{-2})\frac{2j-1}{2j+1-\nu} \qquad (j = 1,2,\ldots,m-1)$$

and the remainder term $R_m(t)$ does not exceed (in absolute value), w_m.

Table 10 (taken from [67]) shows how good approximations can be obtained with this formula, using only three terms in the expansion (i.e. $m = 3$). An expansion of the integral (32.1) in terms of $w = [1 + t^2\nu^{-1}]^{-\frac{1}{2}}$ was used by Hill [c]* for $t^2 > n \geq 1$:

(32.2)
$$\int_t^\infty (1 + y^2\nu^{-1})^{-\frac{1}{2}(\nu+1)}\,dy = \nu^{\frac{1}{2}}\int_0^w x^{\nu-1}(1-x^2)^{-\frac{1}{2}}\,dx$$
$$= \nu^{\frac{1}{2}}w^\nu\left(\frac{1}{\nu} + \frac{w^2}{2(\nu+2)} + \frac{1.3w^4}{2.4(\nu+4)} + \cdots\right).$$

The inverse of this series was used to express $t^2\nu^{-1}$ in terms of $z = [\nu^{\frac{1}{2}}c_\nu\alpha]^{2/\nu}$, where c_ν is the normalizing constant of the probability integral, yielding the formula

(32.3)
$$t_{\nu,\alpha}^2\nu^{-1} \doteq z^{-1} + \frac{\nu+1}{\nu+2}\left(-1 + \frac{z}{2(\nu+4)} + \frac{z}{3(\nu+2)[(\nu+6)(\nu z)^{-1} - 1]}\right)$$

which is exact for $\nu = 2$, and for larger ν correct to over 6 digits for $z < \nu^{-1}$.

*[c] Hill, G. W. (1970). Algorithms '*Student*' and '*t quantile*,' submitted to Comm. ACM.

TABLE 10

Exact and Approximate Tail Areas for the t-Distribution with ν Degrees of Freedom

Exact (*) Tail Area	Approximation (32.1) with $m = 3$ $\nu = 7$	$\nu = 15$	$\nu = 40$
.001	.000 816	.001 06	.001 02
.000 05	.000 042 8	.000 051 5	.000 050 3
.000 01	.000 008 66	.000 010 2	.000 010 05
.000 001	.000 000 873	.000 001 02	.000 001 003
.000 000 1	.000 000 087 7	.000 000 102	.000 000 100 1

*These tail areas are exact to the extent that Federighi's [24] tabled quantiles are exact.

In Zelen and Severo [103] the following approximations are stated:

For $\nu \le 5$, but t large:

$$\Pr[t_\nu \le t] \doteqdot 1 - a_\nu t^{-\nu} - b_\nu t^{-(\nu+1)} \tag{33}$$

with

$a_1 = 0.3183;\ a_2 = 0.4991;\ a_3 = 1.1094;\quad a_4 = 3.0941;\ a_5 = 9.948$
$b_1 = 0.0000;\ b_2 = 0.0518;\ b_3 = -0.0460;\ b_4 = -2.756;\ b_5 = -14.05.$

For ν large

$$\Pr[t_\nu \le t] \doteqdot \Phi(t(1 - \tfrac{1}{4}\nu^{-1})(1 + \tfrac{1}{2}t^2\nu^{-1})^{-\frac{1}{2}}). \tag{34}$$

Gentleman and Jenkins [36] have published an approximation, suitable for use on electronic computers, of form

$$\Pr[|t_\nu| < t] = 1 - \left(1 + \sum_{j=1}^{5} c_j t^j\right)^{-8}$$

where each c_j is the ratio of a quintic to a quadratic polynomial in ν^{-1}. This gives 5 decimal place accuracy for $\nu > 10$.

5. Applications

The major applications of the t distribution, construction of tests and confidence intervals relating to the expected values of normal distributions, have been discussed in Section 1.

In particular, if $X_1, X_2, \ldots, X_n$ are independent random variables, each having a normal distribution with expected value ξ and standard deviation σ, then the distribution of $\sqrt{n}(\bar{X} - \xi)/S'$ (where $\bar{X} = n^{-1} \sum_{j=1}^{n} X_j$ and $S'^2 = (n-1)^{-1} \sum_{j=1}^{n} (X_j - \bar{X})^2$) is a t distribution with $(n - 1)$ degrees of freedom. Since

$$\Pr[|\sqrt{n}(\bar{X} - \bar{\xi})/S'| < t_{n-1,1-\frac{1}{2}\alpha}] = 1 - \alpha$$

it follows that

$$\Pr[\bar{X} - (t_{n-1,1-\frac{1}{2}\alpha}/\sqrt{n})S' < \xi < \bar{X} + (t_{n-1,1-\frac{1}{2}\alpha}/\sqrt{n})S'] = 1 - \alpha$$

and so $\bar{X} \pm (t_{n-1,1-\frac{1}{2}\alpha}/\sqrt{n})S'$ is a confidence interval for ξ with confidence coefficient $100(1 - \alpha)\%$. For practical purposes, it is convenient to have a table of multipliers $b_{n,\alpha} = t_{n-1,1-\frac{1}{2}\alpha}/\sqrt{n}$ so that the limits of the interval are $\bar{X} \pm b_{n,\alpha}S'$. Table 11 contains a few values of $b_{n,\alpha}$.

TABLE 11

Values of $b_{n,\alpha}$

		n										
		4	5	6	7	8	9	10	11	12	15	20
	0.90	1.18	0.95	0.82	0.73	0.67	0.62	0.58	0.55	0.52	0.45	0.40
α	0.95	1.59	1.24	1.05	0.93	0.84	0.77	0.72	0.67	0.64	0.55	0.48
	0.99	2.92	2.06	1.65	1.40	1.24	1.12	1.03	0.96	0.90	0.77	0.66

In analysis of variance tests, when one of the sums of squares being compared has 1 degree of freedom, the appropriate null hypothesis distribution is F with $1, \nu$ degrees of freedom, which is identical with the distribution of t_ν^2. Confidence limits for a single specified linear function of parameters in a general linear model (Chapter 26) corresponding to a single degree of freedom are constructed in a similar way to that just described.

6. Pearson Type VII Distributions

The general Type VII distribution has a probability density function which can be expressed in the form

$$(35) \qquad p_X(x) = \frac{\Gamma(m)}{\sqrt{\pi}\,\Gamma(m - \frac{1}{2})} \frac{c^{2m-1}}{[c^2 + (x - \xi)^2]^m} \qquad (m > 0;\ c > 0)$$

This depends on the three parameters m, c and ξ. The t_ν distribution is obtained by putting $m = \frac{1}{2}(\nu + 1)$; $c = \sqrt{\nu}$ and $\xi = 0$. Thus if X has the distribution (35), then $\sqrt{2m - 1}\,(X - \xi)/\sigma$ is distributed as t_{2m-1}. The shape of the curve represented by (35) is therefore the same as the shape of the curve corresponding to the t_{2m-1} distribution. This has been described in Section 2, and will not be discussed here. The present section will be devoted to discussion of the problem of estimation of the parameters m, c and ξ given observed values of n independent random variables $X_1, X_2, \ldots, X_n$, each having the distribution (35). This problem was discussed by Fisher [24] in 1922 as one of the earliest illustrations of the use of the method of maximum likelihood. Some further formulas were published by Sichel [77] in 1949, who also applied his method of *frequency-moments* to the problem. Our discussion is based on these two papers.

The equations satisfied by the maximum likelihood estimators $\hat{m}$, $\hat{c}$, $\hat{\xi}$ can be written in the form:

$$n^{-1}\sum_{j=1}^{n}\log\left[1+\left(\frac{x_j-\hat{\xi}}{\hat{c}}\right)^2\right] = \psi(\hat{m}) - \psi(\hat{m}-\tfrac{1}{2}), \tag{36.1}$$

$$n^{-1}\sum_{j=1}^{n}\left[1+\left(\frac{x_j-\hat{\xi}}{\hat{c}}\right)^2\right]^{-1} = 1 - \frac{1}{2\hat{m}}, \tag{36.2}$$

$$\sum_{j=1}^{n}\left(\frac{x_j-\hat{\xi}}{\hat{c}}\right)\left[1+\left(\frac{x_j-\hat{\xi}}{\hat{c}}\right)^2\right]^{-1} = 0. \tag{36.3}$$

For large values of n the standard formulas give the following approximations:

$$n\operatorname{var}(\hat{m}) \doteqdot \left[\psi^{(1)}(m-\tfrac{1}{2}) - \psi^{(1)}(m) - \frac{m+1}{m^2(2m-1)}\right]^{-1}; \tag{37.1}$$

$$n\operatorname{var}(\hat{c}) \doteqdot \frac{[\psi^{(1)}(m-\tfrac{1}{2}) - \psi^{(1)}(m)]c^2}{\left(\dfrac{2m-1}{m+1}\right)[\psi^{(1)}(m-\tfrac{1}{2}) - \psi^{(1)}(m)] - \dfrac{1}{m^2}} \tag{37.2}$$

$$\doteqdot \frac{(m+1)c^2}{2m-1}\cdot\frac{1}{1-\dfrac{m+1}{m^2(2m-1)}[\psi^{(1)}(m-\tfrac{1}{2}) - \psi^{(1)}(m)]^{-1}};$$

$$n\operatorname{var}(\hat{\xi}) \doteqdot \frac{(m+1)c^2}{m(2m-1)}; \tag{37.3}$$

$$\operatorname{corr}(\hat{m},\hat{c}) \doteqdot \frac{\sqrt{m+1}}{m\sqrt{(2m-1)[\psi^{(1)}(m-\tfrac{1}{2}) - \psi^{(1)}(m)]}}; \tag{37.4}$$

$$\operatorname{corr}(\hat{m},\hat{\xi}) \doteqdot \operatorname{corr}(\hat{c},\hat{\xi}) \doteqdot 0. \tag{37.5}$$

Formula (37.3) also gives the approximate variance of the maximum likelihood estimator of ξ when the values of either, or both of m and c are known. The asymptotic formulas for the variances of (and correlation between) the maximum likelihood estimators of m and c are the same, whether the value of ξ is known or unknown.

If c is known then

$$n\cdot\operatorname{var}(\text{maximum likelihood estimator of } m) \doteqdot [\psi^{(1)}(m-\tfrac{1}{2}) - \psi^{(1)}(m)]^{-1}. \tag{38}$$

(The estimators of m and ξ are obtained by solving equations (36.1) and (36.3) with $\hat{c}$ replaced by c.) If m is known, then

$$n \cdot \text{var(maximum likelihood estimator of } c) \doteqdot \frac{(m+1)c^2}{2m-1}. \tag{39}$$

Formulas (38) and (39) are applicable, whether the value of ξ is known or not.

The parameters may also be estimated by equating sample and population values of the first, second and fourth moments. In terms of the population moments

$$\begin{aligned} m &= \tfrac{1}{2}(5\beta_2 - 9)/(\beta_2 - 3) \\ c^2 &= 2\mu_2\beta_2/(\beta_2 - 3) \\ \xi &= \mu_1'. \end{aligned} \tag{40}$$

Denoting by $\tilde{m}$, $\tilde{c}$, $\tilde{\xi}$ the estimators obtained by replacing population values by sample values on the right hand side of (40), we have the following approximate formulas (for n large):

(41.1)

$$n\,\text{var}(\tilde{m}) \doteqdot \tfrac{2}{3}(m-1)(2m-5)(2m-3)^2(2m^2-5m+12)(2m-7)^{-1} \times (2m-9)^{-1}$$

(41.2)

$$n\,\text{var}(\tilde{c}) \doteqdot \tfrac{1}{3}c^2(m-1)(2m-3)(8m^3-48m^2+108m-83) \times (2m-5)^{-1}(2m-7)^{-1}(2m-9)^{-1}$$

$$n\,\text{var}(\tilde{\xi}) = c^2(2m-3)^{-1}. \tag{41.3}$$

Note that:

formula (41.3) is exact and is also valid if either one or both of c and m have known values;
formulas (41.1) and (41.2) cannot be used unless m exceeds $4\frac{1}{2}$;
formulas (41.1) and (41.2) apply whether the value of ξ is known or not.

If either m or c is known, the other parameter (c or m respectively) can be estimated by equating

$$\text{var}(X) = c^2(2m-3)^{-1} \tag{42}$$

to the sample variance. For n large we have the following approximate formulas for the variance of estimators of m, c obtained from (40):

$$n \times \text{(variance of estimator of } m) \doteqdot (2m-3)^2(m-1)(2m-5)^{-1} \tag{43.1}$$

$$n \times \text{(variance of estimator of } c) \doteqdot c^2(m-1)(2m-5)^{-1}. \tag{43.2}$$

The parameter ξ may be estimated by the median. The variance of this estimator is approximately

$$\frac{c^2\pi}{4n}\left[\frac{\Gamma(m-\frac{1}{2})}{\Gamma(m)}\right]^2. \tag{44}$$

For $m < 2.8$, the median has a smaller asymptotic variance than the arithmetic mean. (The latter has infinite variance for $m \leq 1.5$.) The ratio of asymptotic variances of mean to median decreases as m increases, tending to 0.637 (the value for normal distributions) as $m \to \infty$.

7. Related Distributions

We have already noted that the distribution of t_ν^2 is identical with that of F with $1, \nu$ degrees of freedom. On account of the symmetry, about zero, of the t_ν distribution

$$\Pr[F_{1,\nu} < K] = \Pr[t_\nu^2 < K] = \Pr[|t_\nu| < \sqrt{K}]$$

and so

$$\sqrt{F_{1,\nu,\alpha}} = t_{\nu,\frac{1}{2}(1+\alpha)}.$$

Other relations between the t and F distributions have been described in Chapter 26.

There are a number of *pseudo-t* distributions obtained by replacing of $\chi_\nu/\sqrt{\nu}$ in the denominator of

$$t_\nu = \frac{U}{\chi_\nu/\sqrt{\nu}}$$

by other distributions (still independent of the distribution of U). These correspond to replacing S' in $\sqrt{n}(\bar{X} - \xi)/S'$ by other sample measures of dispersion; in particular by the range $(\max(X_1, \ldots, X_n) - \min(X_1, \ldots, X_n))$ or the mean deviation $(n^{-1}\sum_{j=1}^{n}|X_j - \bar{X}|)$. These latter have the common feature that they are distributed as σT where T is the variable corresponding to the case $\sigma = 1$. Hence the ratio is distributed as U/T and so does not depend on the value of σ. (See, for example, Pillai [66].)

The use of statistics of this kind has been described in Chapter 13. If the distribution of the denominator is approximated by a $c\chi_\nu/\sqrt{\nu}$ distribution, the ratio is approximated by a $c^{-1}t_\nu$ distribution. Usually ν is fractional, and it is necessarily to interpolate if standard tables are used (in which ν is usually given for integer values only). Alternatively, approximate formulas may be used.

The distribution of $(X_j - \bar{X})/S'$ where X_j is randomly chosen from $X_1, X_2, \ldots, X_n$ was found by Thompson [86] to be related to the t-distribution. In fact (taking $j = n$ for convenience) and putting $E(X_j) = 0$, and $\text{var}(X_j) = 1$ (which does not affect the distribution),

(45)

$$\sum_{j=1}^{n} (X_j - \overline{X})^2 = X_1^2 + \tfrac{1}{2}(X_2 - X_1)^2 + \tfrac{2}{3}(X_3 - \tfrac{1}{2}(X_1 + X_2))^2 + \cdots + \frac{n-1}{n}\left[X_n - \frac{1}{n-1}(X_1 + \cdots + X_{n-1})\right]^2$$

$$= Y + \frac{n}{n-1}(X_n - \overline{X})^2,$$

with Y and $(X_n - \overline{X})^2$ mutually independent, and Y is distributed as χ^2_{n-2}.

Hence $(X_n - \overline{X})/S'$ is distributed as $\dfrac{n-1}{\sqrt{n}} U(\chi^2_{n-2} + U^2)^{-\frac{1}{2}}$, with U and χ^2_{n-2} mutually independent. The extremes of the range of variation are $\pm \dfrac{n-1}{\sqrt{n}}$. Also, putting $(X_n - \overline{X})/S' = \tau$ we find that τ is distributed as

$$\frac{n-1}{\sqrt{n}} \sqrt{n-2}\, t_{n-2}(1 + (n-2)t^2_{n-2})^{-\frac{1}{2}}. \tag{46}$$

The distribution of $n(\overline{X} - \xi)/S'$ when $X_1, X_2, \ldots, X_n$ have identical, but not normal, distributions has been studied for the following cases:

Rectangular: (Hotelling and Frankel [44], Perlo [65], Rider [72], [73], Rietz [74], Siddiqui [78]).

Exponential: (Geary [35]).

Cauchy and Squared Hyperbolic Secant: (Bradley [12], Hotelling [43]).

Edgeworth Series: (Bartlett [8], Ghurye [37], Gayen [33], [34], Tiku [87], Zackrisson [102]).

Compound (Mixture) Normal: (Hyrenius [45], Quensel [68]).

Various Truncated Distributions: (Rectangular, Laplace, χ^2, Beta) (Watanabe [92]).

Laderman [53] gave a general formula for the distribution of t, for any continuous parent population, for sample size two. In a number of cases exact distributions have been obtained only for small sample sizes ($(\nu + 1)$ usually up to 4 or 5). For the parent rectangular distribution, Siddiqui has given bounds for the probability integral of t_ν and obtained numerical values for sample size up to 7. Hotelling by a geometrical method has shown that the ratio of $\Pr[t_\nu > t]$ to its normal theory value tends to

$$\frac{\{\pi(\nu + 1)\}^{\frac{1}{2}(\nu+1)}}{2^{\nu+1}\Gamma(\frac{1}{2}(\nu + 3))}$$

as t tends to infinity and obtained a similar result for a U-shaped Pearson Type II parent distribution.

Of most general interest are the results for Edgeworth Series. The results are given for all n, and indicate the variation from the Student t-distribution that one might expect to be associated with given non-normal values of the moment-ratios. Ghurye [37] considered a series only up to the term in $\sqrt{\beta_1}$.

Gayen [33], [34] included further terms. The results of Tiku [87] appear to correspond to inclusion of still further terms in the expansion for the distribution in the parent population. Gayen shows

$$\Pr[t_\nu > t] = (\text{value for normal population}) + \sqrt{\beta_1}P_{\sqrt{\beta_1}}(t) - (\beta_2 - 3)P_{\beta_2}(t) + \beta_1 P_{\beta_1}(t)$$

and gives a table of the polynomials $P(\cdot)$. We reproduce his table as Table 12 on pages 122–124. (See also Chung [16].)

We may also note a formula obtained by Bradley ([12], p. 21) for a Cauchy parent population (for which we cannot use moment-ratios as an index of non-normality). This expresses the value of $\Pr[t_\nu \leq t]$, for positive t, as a series in t^{-2} (up to and including the t^{-6} term).*

Ratcliffe [70] has presented results of an empirical investigation of the distribution of t for five markedly non-normal parent distributions (including rectangular, exponential, Laplace, gamma and a U-shaped distribution). He studied, in particular, the reduction in the effect of non-normality with increasing sample size. He concluded that a sample size of 80 or more should eliminate effects of non-normality (mainly skewness) for most practical purposes. For symmetrical population distributions the necessary sample size is much less.

Efron [21] has given a theoretical discussion of the distribution of t under general symmetry conditions.

The distributions of linear functions of the form $a_1 t_{\nu_1} - a_2 t_{\nu_2}$, with a_1, a_2 both positive, and t_{ν_1}, t_{ν_2} both independent variables, have been studied in connection with tests of the hypothesis that expected values of two normal distributions are equal, when it cannot be assumed that the standard deviations of the two distributions are equal. Tests of this kind were proposed by Behrens [9] in 1929 and studied later by Fisher [28]; they are said to relate to the 'Behrens-Fisher problem'.

Suppose that $X_{j1}, X_{j2}, \ldots, X_{jn_j}$ are independent random variables each normally distributed with expected value ξ_j and standard deviation σ_j ($j = 1,2$). Then, for $j = 1, 2$, $\sqrt{n_j}(\overline{X}_j - \xi_j)/S'_j$ is distributed as t_{n_j-1} (using a notation similar to that employed in Section 1 of this chapter). According to one form of the *fiducial* argument (Chapter 13, Section 7.2) from

$$\text{“}\sqrt{n_j}(\overline{X}_j - \xi_j)/S'_j \text{ distributed as } t_{n_j-1}\text{”}$$

one can deduce

$$\text{“}\xi_j \text{ (fiducially) distributed as } \overline{X}_j - (S'_j/\sqrt{n_j})t_{n_j-1}\text{”}$$

(regarding X_j and S'_j as fixed). Formally one can proceed to say

$$\text{“}\xi_2 - \xi_1 \text{ is (fiducially) distributed as } \overline{X}_2 - \overline{X}_1 + [(S'_1/\sqrt{n_1})t_{n_1-1} - (S'_2/\sqrt{n_2})t_{n_2-1}]\text{”}$$

*In all of the above cases the X's are assumed to be mutually independent. Weibull [93] considers the case where they are normal but are serially correlated.

The Behrens-Fisher test procedure rejects the hypothesis $\xi_1 = \xi_2$ at 'level of significance α' if (according to the 'fiducial' distribution)

$$\Pr[\xi_2 - \xi_1 < 0] < \tfrac{1}{2}\alpha \qquad \text{or} \qquad \Pr[\xi_2 - \xi_1 > 0] < \tfrac{1}{2}\alpha.$$

It is found convenient to use the quantity

$$\frac{\overline{X}_1 - \overline{X}_2}{\sqrt{\dfrac{S_1'^2}{n_1} + \dfrac{S_2'^2}{n_2}}} + t_{n_1-1}\cos\theta - t_{n_2-1}\sin\theta$$

with
$$\cos\theta = \frac{S_1'}{\sqrt{n_1}}\left(\frac{S_1'^2}{n_1} + \frac{S_2'^2}{n_2}\right)^{-\frac{1}{2}}.$$

Tables of percentile points of the distribution of $t_{\nu_1}\cos\theta - t_{\nu_2}\sin\theta$ are available as follows:

Sukhatme [85] gave upper $2\frac{1}{2}\%$ values to 3 decimal places for

$$\nu_1, \nu_2 = 6, 8, 12, 24, \infty \qquad \text{and} \qquad \theta = 0°(15°)90°.$$

Fisher and Yates [31] included these values together with upper $\frac{1}{2}\%$ values (also to 3 decimal places) and also upper 5%, $2\frac{1}{2}\%$, 1%, $\frac{1}{2}\%$, $\frac{1}{4}\%$, and 0.1% values for $\nu_1 = 10, 12, 15, 20, 30, 60$, with $\nu_2 = \infty$.

Weir [97] gave upper 0.1% points to 3 decimal places for

$$\nu_1, \nu_2 = 6, 8, 12, 24, \infty$$
$$\theta = 0°(15°)90°.$$

Ruben [75] has shown that $t_{\nu_1}\cos\theta - t_{\nu_2}\sin\theta$ has the same distribution as the ratio of a $t_{\nu_1+\nu_2}$ variable to an independent variable

$$\phi(X) = \left[\frac{\nu_1 + \nu_2}{\nu_1 X^{-1}\cos\theta - \nu_2(1 - X)^{-1}\sin\theta}\right]^{\frac{1}{2}}$$

where X has a standard beta distribution (Chapter 24) with parameters $\frac{1}{2}\nu_1, \frac{1}{2}\nu_2$.

Patil [61] has suggested that the distribution of

$$t_{\nu_1}\cos\theta - t_{\nu_2}\sin\theta$$

be approximated by that of ct_f with

$$f = 4 + \left[\frac{\nu_1\cos^2\theta}{\nu_1 - 2} + \frac{\nu_2\sin^2\theta}{\nu_2 - 2}\right]^2\left[\frac{\nu_1^2\cos^4\theta}{(\nu_1 - 2)^2(\nu_1 - 4)} + \frac{\nu_2^2\sin^4\theta}{(\nu_2 - 2)^2(\nu_2 - 4)}\right]^{-1}$$

$$c = \left[\frac{f-2}{f}\left\{\frac{\nu_1\cos^2\theta}{\nu_2 - 2} + \frac{\nu_2\sin^2\theta}{\nu_2 - 2}\right\}\right]^{\frac{1}{2}}.$$

(These values are chosen to make the first four moments of the two distribution agree.)

The approximation is exact when $\cos\theta = 1$ or $\sin\theta = 1$. It gives good results, even for ν_1, ν_2 as small as 7, in the central part of the distribution ($|t_{\nu_1}\cos\theta - t_{\nu_2}\sin\theta| < 5$).

Weir [96] has also provided the following approximation for the upper percentage point of a related statistic. He found that

$$\frac{\bar{X}_1 - \bar{X}_2}{\sqrt{\dfrac{(n_1-1)S_1^2}{(n_1-3)n_1} + \dfrac{(n_2-1)S_2^2}{(n_2-3)n_2}}}$$

has upper $2\frac{1}{2}\%$ points between 1.96 and 2, provided $n_1 \geq 6$ and $n_2 \geq 6$.

Welch [98] suggested approximating the distribution of $n_1^{-1}S_1^2 + n_2^{-1}S_2^2$ by that of $c\chi_\nu^2$ with

$$cv = E[n_1^{-1}S_1'^2 + n_2^{-1}S_2'^2] = n_1^{-1}\sigma_1^2 + n_2^{-1}\sigma_2^2, \tag{47.1}$$

(47.2)

$$2c^2\nu = \text{var}[n_1^{-1}S_1'^2 + n_2^{-1}S_2'^2] = 2n_1^{-2}(n_1-1)^{-1}\sigma_1^4 + 2n_2^{-2}(n_2-1)^{-1}\sigma_2^4$$

that is

$$c = \frac{n_1^{-2}(n_1-1)^{-1}\sigma_1^4 + n_2^{-2}(n_2-1)\sigma_2^4}{n_1^{-1}\sigma_1^2 + n_2^{-1}\sigma_2^2}$$

$$\nu = \frac{(n_1^{-1}\sigma_1^2 + n_2^{-1}\sigma_2^2)^2}{n_1^{-2}(n_1-1)\sigma_1^4 + n_2^{-2}(n_2-1)\sigma_2^4}$$

Then $(\bar{X}_1 - \bar{X}_2)(n_1^{-1}S_1^2 + n_2^{-1}S_2^2)^{-\frac{1}{2}}$ is approximately distributed as

$$\frac{U\sqrt{n_1^{-1}\sigma_1^2 + n_2^{-1}\sigma_2^2}}{\sqrt{c}\,\nu(\chi_\nu/\sqrt{\nu})} = t\nu,$$

(since $cv = n_1^{-1}\delta_1^2 + n_2^{-1}\delta_2^2$).

Further work has been done on this problem by Aspin [3] who has provided some tables [4] from which exact probabilities can be obtained (also Welch [99]).

Hajek [39] has shown that if $\sum_{j=1}^{k}\lambda_j = 1$, and

$$T = U\left[\sum_{j=1}^{k}\lambda_j\chi_{\nu_j}^2\nu_j^{-1}\right]^{-\frac{1}{2}},$$

U being a unit normal variable, and if U and the χ^2's are mutually independent then, for $t' \leq 0 \leq t''$, $\Pr[t' \leq T \leq t'']$ lies between $\Pr[t' \leq t_\nu \leq t'']$ and $\Pr[t' \leq t_m \leq t'']$ (calculated on normal theory), where $m = \sum_{j=1}^{k}\nu_j$; and ν is any integer not exceeding $\min(\nu_j/\lambda_j)$.

TABLE 12

Corrective Factors for Distribution of t_ν in Non-normal (Edgeworth) Populations

t	Normal theory	$P_{\sqrt{\beta_1}}(t_0)$	$P_{\beta_2}(t_0)$	$P_{\beta_1}(t_0)$
		$\nu = 1$		
0.0	0.5000	0.0470	0.0000	0.0000
0.5	0.3524	0.0589	−0.0064	−0.0066
1.0	0.2500	0.0665	0.0000	0.0044
1.5	0.1872	0.0622	0.0047	0.0147
2.0	0.1476	0.0547	0.0064	0.0188
2.5	0.1211	0.0476	0.0066	0.0195
3.0	0.1024	0.0416	0.0064	0.0188
3.5	0.0886	0.0368	0.0059	0.0176
4.0	0.0780	0.0329	0.0055	0.0163
		$\nu = 2$		
0.0	0.5000	0.0384	0.0000	0.0000
0.5	0.3333	0.0495	−0.0069	−0.0066
1.0	0.2113	0.0597	−0.0027	0.0009
1.5	0.1362	0.0563	0.0025	0.0118
2.0	0.0918	0.0469	0.0047	0.0172
2.5	0.0648	0.0375	0.0051	0.0179
3.0	0.0477	0.0298	0.0047	0.0165
3.5	0.0364	0.0239	0.0041	0.0145
4.0	0.0286	0.0194	0.0035	0.0125
		$\nu = 3$		
0.0	0.5000	0.0332	0.0000	0.0000
0.5	0.3257	0.0431	−0.0062	−0.0056
1.0	0.1955	0.0540	−0.0034	−0.0002
1.5	0.1153	0.0513	0.0013	0.0098
2.0	0.0697	0.0413	0.0035	0.0152
2.5	0.0439	0.0310	0.0039	0.0157
3.0	0.0288	0.0229	0.0034	0.0139
3.5	0.0197	0.0169	0.0028	0.0114
4.0	0.0137	0.0126	0.0023	0.0093

(Table 12 continued)

t	Normal theory	$P_{\sqrt{\beta_1}}(t_0)$	$P_{\beta_2}(t_0)$	$P_{\beta_1}(t_0)$
		$\nu = 4$		
0.0	0.5000	0.0297	0.0000	0.0000
0.5	0.3217	0.0387	−0.0055	−0.0047
1.0	0.1870	0.0495	−0.0036	−0.0005
1.5	0.1040	0.0473	0.0006	0.0084
2.0	0.0581	0.0372	0.0028	0.0135
2.5	0.0334	0.0266	0.0031	0.0139
3.0	0.0200	0.0184	0.0027	0.0119
3.5	0.0124	0.0127	0.0021	0.0095
4.0	0.0081	0.0088	0.0016	0.0072
		$\nu = 5$		
0.0	0.5000	0.0271	0.0000	0.0000
0.5	0.3192	0.0355	−0.0049	−0.0041
1.0	0.1816	0.0397	−0.0035	−0.0005
1.5	0.0970	0.0440	0.0002	0.0074
2.0	0.0510	0.0340	0.0022	0.0122
2.5	0.0272	0.0234	0.0025	0.0125
3.0	0.0150	0.0154	0.0021	0.0104
3.5	0.0086	0.0099	0.0016	0.0079
4.0	0.0052	0.0065	0.0011	0.0057
		$\nu = 6$		
0.0	0.5000	0.0251	0.0000	0.0000
0.5	0.3174	0.0329	−0.0044	−0.0035
1.0	0.1780	0.0430	−0.0033	−0.0005
1.5	0.0921	0.0413	0.0000	0.0066
2.0	0.0462	0.0315	0.0019	0.0111
2.5	0.0233	0.0210	0.0021	0.0113
3.0	0.0120	0.0132	0.0017	0.0092
3.5	0.0064	0.0081	0.0012	0.0067
4.0	0.0036	0.0050	0.0008	0.0047

(Table 12 continued)

t	Normal theory	$P_{\sqrt{\beta_1}}(t_0)$	$P_{\beta_2}(t_0)$	$P_{\beta_1}(t_0)$
		$\nu = 8$		
0.0	0.5000	0.0222	0.0000	0.0000
0.5	0.3153	0.0291	−0.0037	−0.0028
1.0	0.1733	0.0384	−0.0030	−0.0005
1.5	0.0860	0.0371	−0.0002	−0.0055
2.0	0.0403	0.0277	0.0014	0.0094
2.5	0.0185	0.0177	0.0016	0.0095
3.0	0.0085	0.0103	0.0013	0.0074
3.5	0.0040	0.0058	0.0008	0.0051
4.0	0.0020	0.0032	0.0005	0.0033
		$\nu = 12$		
0.0	0.5000	0.0184	0.0000	0.0000
0.5	0.3131	0.0243	−0.0027	−0.0019
1.0	0.1685	0.0325	−0.0023	−0.0002
1.5	0.0797	0.0315	−0.0004	0.0042
2.0	0.0343	0.0230	0.0009	0.0073
2.5	0.0140	0.0137	0.0011	0.0072
3.0	0.0055	0.0072	0.0008	0.0053
3.5	0.0022	0.0036	0.0005	0.0033
4.0	0.0009	0.0017	0.0003	0.0019
		$\nu = 24$		
0.0	0.5000	0.0133	0.0000	0.0000
0.5	0.3101	0.0176	−0.0015	−0.0010
1.0	0.1636	0.0238	−0.0014	−0.0001
1.5	0.0733	0.0232	−0.0003	0.0025
2.0	0.0285	0.0164	0.0004	0.0043
2.5	0.0098	0.0090	0.0005	0.0041
3.0	0.0031	0.0041	0.0003	0.0028
3.5	0.0009	0.0016	0.0002	0.0015
4.0	0.0003	0.0006	0.0001	0.0007
		$\nu = \infty$		
0.0	0.5000			
0.5	0.3085			
1.0	0.1587			
1.5	0.0668			
2.0	0.0228			
2.5	0.0062			
3.0	0.0013			
3.5	0.0002			
4.0	0.0000			

REFERENCES

[1] Amos, D. E. (1964). Representations of the central and non-central t distributions, *Biometrika*, **51,** 451–458.

[2] Anscombe, F. J. (1950). Table of the hyperbolic transformation $\sinh^{-1}\sqrt{x}$, *Journal of the Royal Statistical Society, Series A*, **113,** 228–229.

[3] Aspin, Alice A. (1948). An examination and further development of a formula arising in the problem of comparing two mean values, *Biometrika*, **35,** 88–96.

[4] Aspin, Alice A. (1949). Tables for use in comparisons whose accuracy involves two variances, separately estimated, *Biometrika*, **36,** 290–293.

[5] Babanin, B. V. (1952). Nomogram of basic statistical distributions and its application to some problems of sampling method, *Academy of Sciences, USSR, Institute of Mechanics, Engineering Transactions*, **11,** 169–180. (In Russian)

[6] Baldwin, Elizabeth M. (1946). Table of percentage points of the t-distribution. *Biometrika*, **33,** 362.

[7] Banerjee, S. K. (1957). A lower bound to the probability of Student's ratio, *Sankhyā, Series B*, **19,** 391–394.

[8] Bartlett, M. S. (1935). The effect of non-normality on the t distribution, *Proceedings of the Cambridge Philosophical Society*, **31,** 223–231.

[9] Behrens, W. V. (1929). Ein Beitrag zur Fehlenberechnung bei wenigen Beobachtungen, *Landwirtschaftliche Jahrbucher*, **68,** 807–837.

[10] Bhattacharyya, A. (1952). On the uses of the t-distribution in multivariate analysis, *Sankhyā*, **12,** 89–104.

[11] Bracken, J. and Schleifer, Jr., A. (1964). *Tables for normal sampling with unknown variances: the Student distribution and economically optimum sampling plans*, Division of Research, Harvard University.

[12] Bradley, R. A. (1952). The distribution of the t and F statistics for a class of non-normal populations, *Virginia Journal of Science*, **3,** 1–32.

[13] Buehler, R. J. and Feddersen, A. P. (1963). Note on a conditional property of Student's t, *Annals of Mathematical Statistics*, **34,** 1098–1100.

[14] Cacoullos, T. (1965). A relation between t and F distributions, *Journal of the American Statistical Association*, **60,** 528–531; (Correction **60,** 1249).

[15] Chu, J. T. (1956). Errors in normal approximations to the t, τ, and similar types of distribution, *Annals of Mathematical Statistics*, **27,** 780–789.

[16] Chung, K. L. (1946). The approximate distribution of Student's statistic, *Annals of Mathematical Statistics*, **17,** 447–465.

[17] Cotterman, T. E. and Knoop, Patricia A. (1968). *Tables of limiting t values for probabilities to the nearest* .001 ($n = 2 - 16$), Report AMRL-TR-67-161, Aerospace Medical Research Laboratories, Wright-Patterson Air Force Base, Ohio.

[18] Cucconi, O. (1962). On a simple relation between the number of degrees of freedom and the critical value of Student's t, *Memorie Accademia Patavina*, **74,** 179–187.

[19] Deming, W. E. and Birge, R. T. (1934). On the statistical theory of errors, *Review of Modern Physics*, **6,** 119–161.

[20] Dickey, J. M. (1967). Expansions of t densities and related complete integrals, *Annals of Mathematical Statistics*, **38,** 503–510.

[21] Efron, B. (1968). *Student's t-test under non-normal conditions*, Technical Report No. 21, Harvard University Department of Statistics.

[22] Elfving, G. (1955). An expansion principle for distribution functions, with application to Student's statistic, *Annales Academiae Scientiarum Fennicae, Series A*, **204,** 1–8.

[23] Federighi, E. T. (1959). Extended tables of the percentage points of Student's t-distribution, *Journal of the American Statistical Association*, **54,** 683–688.

[24] Fisher, R. A. (1922). On the mathematical foundations of theoretical statistics, *Philosophical Transactions of the Royal Society of London, Series A*, **222,** 309–368.

[25] Fisher, R. A. (1925). Applications of "Student's" distribution, *Metron*, **5,** 90–104.

[26] Fisher, R. A. (1925). Expansion of "Student's" integral in powers of n^{-1}, *Metron*, **5,** 109–112.

[27] Fisher, R. A. (1935). The mathematical distributions used in the common tests of significance, *Econometrica*, **3,** 353–365.

[28] Fisher, R. A. (1941). The asymptotic approach to Behrens' integral with further table for the d test of significance, *Annals of Eugenics, London*, **11,** 141–172.

[29] Fisher, R. A. and Cornish, E. A. (1960). The percentile points of distributions having known cumulants, *Technometrics*, **2,** 209–225.

[30] Fisher, R. A. and Healy, M. J. R. (1956). New tables of Behrens' test of significance, *Journal of the Royal Statistical Society, Series B*, **18,** 212–216.

[31] Fisher, R. A. and Yates, F. (1966). *Statistical Tables for Biological, Agricultural and Medical Research*, Edinburgh: Oliver and Boyd.

[32] Gardiner, D. A. and Bombay, Barbara F. (1965). An approximation to Student's t, *Technometrics*, **7,** 71–72.

[33] Gayen, A. K. (1949). The distribution of 'Student's' t in random samples of any size drawn from non-normal universes, *Biometrika*, **36,** 353–369.

[34] Gayen, A. K. (1952). The inverse hyperbolic sine transformation on Student's t for non-normal samples, *Sankhyā*, **12,** 105–108.

[35] Geary, R. C. (1936). The distribution of "Student's" ratio for non-normal samples, *Journal of the Royal Statistical Society, Series B*, **3,** 178–184.

[36] Gentleman, W. M. and Jenkins, M. A. (1968). An approximation for Student's t distribution, *Biometrika*, **55,** 571–572.

[37] Ghurye, S. G. (1959). On the use of Student's t-test in an asymmetrical population, *Biometrika*, **46,** 426–430.

[38] Goldberg, H. and Levine, H. (1946). Approximate formulas for the percentage points and normalization of t and χ^2, *Annals of Mathematical Statistics*, **17,** 216–225.

[39] Hajek, J. (1962). Inequalities for the generalized Student's distribution and their applications, *Selected Translations in Mathematical Statistics and Probability*, **2,** 63–74, American Mathematical Society: Providence, R. I.

[40] Hald, A. (1952). *Statistical Tables and Formulas*, New York and London: John Wiley & Sons, Inc.

[41] Hartley, H. O. and Pearson, E. S. (1950). Table of the probability integral of the t-distribution, *Biometrika*, **37,** 168–172.

[42] Hendricks, W. A. (1936). An approximation to "Student's" distribution, *Annals of Mathematical Statistics*, **7,** 210–221.

[43] Hotelling, H. (1961). The behavior of some standard statistical test under non-standard conditions, *Proceedings of the 4th Berkeley Symposium on Mathematical Statistics and Probability*, **1,** 319–359.

[44] Hotelling, H. and Frankel, L. R. (1938). The transformation of statistics to simplify their distribution, *Annals of Mathematical Statistics*, **9,** 87–96.

[45] Hyrenius, H. (1950). Distribution of "Student"-Fisher's t in samples from compound normal functions, *Biometrika*, **37,** 429–442.

[46] James-Levy, G. E. (1956). A nomogram for the integral law of Student's distribution, *Teoriya Veroyatnostei i ee Primeneniya*, **1,** 271–274 (In Russian). (pp. 246–248 in English translation.)

[47] Janko, J. (1958). *Statistické Tabulky*, Praha.

[48] Jeffreys, H. (1948). *Theory of Probability*, 2nd edition, Oxford: Clarendon Press.

[49] Kitagawa, T. (1954–56). Some contributions to the design of sample surveys, *Sankhyā*, **14,** 317–62; **17,** 1–36.

[50] Kotlarski, I. (1964). On bivariate random vectors where the quotient of their coordinates follows the Student's distribution, *Zeszyty Naukowe Politechniki Warszawskiej*, **99,** 207–220. (In Polish)

[51] Kramer, C. Y. (1966). Approximation to the cumulative t-distribution, *Technometrics*, **8,** 358–359.

[52] Krishnamoorthy, A. S. (1951). On the orthogonal polynomials associated with Student's distribution, *Sankhyā*, **11,** 37–44.

[53] Laderman, J. (1939). The distribution of "Student's" ratio for samples of two items drawn from non-normal universes, *Annals of Mathematical Statistics*, **10,** 376–379.

[54] Laumann, R. (1967). *Tafeln der STUDENT-oder t-Verteilung*, Deutsch-Französisches Forschungsinstitut, Saint-Louis, Akt. N21/67.

[55] Mauldon, J. G. (1956). Characterizing properties of statistical distributions, *Quarterly Journal of Mathematics, Oxford*, **7,** 155–160.

[56] McMullen, L. (1939). "Student" as a man, *Biometrika*, **30,** 205.

[57] Molina, E. C. and Wilkinson, R. I. (1929). The frequency distribution of the unknown mean of a sampled universe, *Bell System Technical Journal*, **8,** 632–645.

[58] Moran, P. A. P. (1966). Accurate approximations for t-tests. (*In Research Papers in Statistics* (*Festschrift for J. Neyman*) (Ed. F. N. David), pp. 225–230.)

[59] Owen, D. B. (1962). *Handbook of Statistical Tables*, Reading, Mass.: Addison-Wesley Publishing Company, Inc.

[60] Owen, D. B. (1965). The power of Student's t-test, *Journal of the American Statistical Association*, **60,** 320–333.

[61] Patil, V. H. (1965). Approximation to the Behrens-Fisher distributions, *Biometrika*, **52,** 267–271.

[62] Pearson, E. S. and Hartley, H. O. (1958). *Biometrika Tables for Statisticians*, **1,** London: Cambridge University Press. (2nd edition)

[63] Peiser, A. M. (1943). Asymptotic formulas for significance levels of certain distributions, *Annals of Mathematical Statistics*, **14,** 56–62.

[64] Peizer, D. B. and Pratt, J. W. (1968). A normal approximation for binomial, *F*, Beta, and other common, related tail probabilities, I., *Journal of the American Statistical Association*, **63,** 1416–1456.

[65] Perlo, V. (1933). On the distribution of Student's ratio for samples of three drawn from a rectangular distribution, *Biometrika*, **25,** 203–204.

[66] Pillai, K. C. S. (1951). On the distribution of an analogue of Student's *t*, *Annals of Mathematical Statistics*, **22,** 469–472.

[67] Pinkham, R. S. and Wilk, M. B. (1954). Tail areas of the *t*-distribution from a Mills'-ratio-like expansion, *Annals of Mathematical Statistics*, **34,** 335–337.

[68] Quensel, C. E. (1943). An extension of the validity of "Student"-Fisher's law of distribution, *Skandinavisk Aktuarietidskrift*, **26,** 210–219.

[69] Rao, C. R., Mitra, S. K. and Matthai, A. (1966). *Formulae and Tables for Statistical Work*, Calcutta, India: Statistical Publishing Society.

[70] Ratcliffe, J. F. (1968). The effect on the *t*-distribution of non-normality in the sampled population, *Applied Statistics*, **17,** 71–73.

[71] Ray, J. P. (1961). Unpublished thesis, Virginia Polytechnic Institute.

[72] Rider, P. R. (1929). On the distribution of the ratio of mean to standard deviation in small samples from non-normal universes, *Biometrika*, **21,** 124–143.

[73] Rider, P. R. (1931). On small samples from certain non-normal universes, *Annals of Mathematical Statistics*, **2,** 48–62.

[74] Rietz, H. L. (1939). On the distribution of the "Student" ratio for small samples from certain non-normal populations, *Annals of Mathematical Statistics*, **10,** 265–274.

[75] Ruben, H. (1960). On the distribution of the weighted difference of two independent Student variables, *Journal of the Royal Statistical Society*, *Series B*, **22,** 188–194.

[76] Scheffé, H. (1943). On solutions of the Behrens-Fisher problem, based on the *t*-distribution, *Annals of Mathematical Statistics*, **14,** 35–44.

[77] Sichel, H. S. (1949). The method of frequency-moments and its application to Type VII populations, *Biometrika*, **36,** 404–425.

[78] Siddiqui, M. M. (1964). Distribution of Student's *t* in samples from a rectangular universe, *Review of the International Statistical Institute*, **32,** 242–250.

[79] Simaika, J. B. (1942). Interpolation for fresh probability levels between the standard table levels of a function, *Biometrika*, **32,** 263–276.

[80] Smirnov, N. V. (1961). *Tables for the Distribution and Density Functions of t-Distribution* (*"Student's" Distribution*), Oxford: Pergamon Press.

[81] Stammberger, A. (1967). Über einige Nomogramme zur Statistik, *Wissenschaftliche Zeitschrift der Humboldt-Universität Berlin, Mathematisch-Naturwissenschaftliche Reihe*, **16,** 86–93.

[82] Stone, M. (1963). The posterior *t* distribution, *Annals of Mathematical Statistics*, **34,** 568–573.

[83] "Student" (1908). On the probable error of the mean, *Biometrika*, **6,** 1–25.

[84] "Student" (1925). New tables for testing the significance of observations, *Metron*, **5,** 105–108, 114–120.

[85] Sukhatme, P. V. (1938). On Fisher and Behrens' test of significance for the difference in means of two normal samples, *Sankhyā*, **4,** 39–48.

[86] Thompson, W. R. (1935). On a criterion for the rejection of observations and the distribution of the ratio of deviation to sample standard deviation, *Annals of Mathematical Statistics*, **6,** 213–219.

[87] Tiku, M. L. (1963). Approximation to Student's t distribution in terms of Hermite and Laguerre polynomials, *Journal of the Indian Mathematical Society*, **27,** 91–102.

[88] Veselá, A. (1964). Kritische Werte der Studentschen t-Verteilung für die Freiheitsgrade zwischen 30 und 120, *Biometrische Zeitschrift*, **6,** 123–127.

[89] Wallace, D. L. (1959). Bounds on normal approximations to Student's and the Chi-square distributions, *Annals of Mathematical Statistics*, **30,** 1121–1130.

[90] Walsh, J. E. (1947). Concerning the effect of intraclass correlation on certain significance tests, *Annals of Mathematical Statistics*, **18,** 88–96.

[91] Wasow, W. (1956). On the asymptotic transformation of certain distributions into the normal distribution, *Proceedings of the Symposium on Applied Mathematics*, **6,** (*Numerical Analysis*), pp. 251–259, New York: McGraw-Hill, Inc.

[92] Watanabe, Y. (1960), (1962), (1963), (1966). The Student's distribution for a universe bounded at one or both sides, *Journal of Gakugei, Tokushima University*, **11,** 11–51, **12,** 5–50, **13,** 1–42, **14,** 1–53, and **15,** 1–35.

[93] Weibull, C. (1958). The distribution of the Student ratio in the case of serially correlated normal variables, *Skandinavisk Aktuarietidskrift*, **41,** 172–176.

[94] Weibull, M. (1950). The distribution of the t and z variables in the case of stratified samples with individuals taken from normal parent populations with varying means, *Skandinavisk Aktuarietidskrift*, **33,** 137–167.

[95] Weir, J. B. de V. (1960). Standardized t, *Nature, London*, **185,** 558.

[96] Weir, J. B. de V. (1960). Significance of the difference between two means when the populations variances may be unequal, *Nature, London*, **187,** 438.

[97] Weir, J. B. de V. (1966). Table of 0.1 percentage points of Behrens's d, *Biometrika*, **53,** 367–368.

[98] Welch, B. L. (1938). The significance of the difference between two means when the population variances are unequal, *Biometrika*, **29,** 350–362.

[99] Welch, B. L. (1949). Further note on Mrs. Aspin's tables and on certain approximations to the tabled function, *Biometrika*, **36,** 293–296. (See also [4].)

[100] Welch, B. L. (1958). "Student" and small sample theory, *Journal of the American Statistical Association*, **53,** 777–788.

[101] Wishart, J. (1947). The cumulants of the z and of the logarithmic χ^2 and t distributions, *Biometrika*, **34,** 170–178.

[102] Zackrisson, U. (1959). The distribution of "Student's" t in samples from individual non-normal populations, *Statistical Institute of the University of Göteborg, Publications*, **6,** 7–32.

[103] Zelen, M. and Severo, N. C. (1964). *Probability Functions*. No. 26 in *Handbook of Mathematical Functions* (Ed. M. Abramowitz and I. A. Stegun), National Bureau of Standards, Applied Mathematics Series, **55,** Washington, D. C.: U. S. Government Printing Office.

28

Noncentral χ^2 Distribution

1. Definition and Genesis

If $U_1, U_2, \ldots, U_\nu$ are independent unit normal variables, and $\delta_1, \delta_2, \ldots, \delta_\nu$ are constants then the distribution of

$$\sum_{j=1}^{\nu} (U_j + \delta_j)^2$$

depends on $\delta_1, \delta_2, \ldots, \delta_\nu$ only through the sum of their squares. It is called the *non-central* χ^2 *distribution with* ν *degrees of freedom and noncentrality parameter* $\lambda = \sum_{j=1}^{\nu} \delta_j^2$.

The symbol $\chi_\nu'^2(\lambda)$ denotes a variable with this distribution. It is derived from the symbol χ_ν^2, denoting a *central* χ^2 variable with ν degree of freedom (Chapter 17), which has the same distribution as $\sum_{j=1}^{\nu} U_j^2$. In fact when $\lambda = 0$, the 'noncentral' distribution becomes the central χ^2 distribution.

Whenever justified by the context, the symbols ν and λ may be omitted and the symbols χ'^2 used. (The prime is retained to denote 'noncentral'.)

Sometimes $\sqrt{\lambda}$, and sometimes $\frac{1}{2}\lambda$, are called the noncentrality parameter. We will not use this notation.

A simple way in which the noncentral χ^2 distribution arises is in the distribution of the sum of squares

$$S = \sum_{j=1}^{n} (X_j - \overline{X})^2 \quad \text{where } \overline{X} = n^{-1} \sum_{j=1}^{n} X_j$$

when $X_1, X_2, \ldots, X_n$ are independent normal variables with X_j distributed normally with expected value ξ_j and standard deviation σ (the same for all j), for $j = 1, 2, \ldots, n$. Clearly we can write

$$X_j = \xi_j + \sigma U'_j$$

with U'_j's independent unit normal variables. Then

$$S = \sigma^2 \sum_{j=1}^{n} (U'_j + \xi_j\sigma^{-1} - (\bar{U}' + \bar{\xi}\sigma^{-1}))^2$$

where

$$\bar{U}' = n^{-1} \sum_{j=1}^{n} U'_j, \qquad \bar{\xi} = n^{-1} \sum_{j=1}^{n} \xi_j.$$

Applying a transformation from $U'_1, \ldots, U'_n$ to $U_1, \ldots, U_{n-1}, \bar{U}'$ (Chapter 13, Section 3) such that

$$\sum_{j=1}^{n} (U'_j - \bar{U}')^2 = \sum_{j=1}^{n-1} U_j^2$$

($U_1, \ldots, U_{n-1}$ being independent unit normal variables) we see that

$$S = \sigma^2 \sum_{j=1}^{n-1} (U_j + \delta_j)^2$$

where the δ_j's are linear functions of the ξ_j's, and U_j's of the U'_j's. Putting $U'_j = 0$ for all j, it follows that $U_j = 0$ for all j and hence

$$\sum_{j=1}^{n-1} \delta_j^2 = \sum_{j=1}^{n} (\xi_j - \bar{\xi})^2/\sigma^2$$

and so S is distributed as σ^2 times a noncentral χ^2 with $(n - 1)$ degrees of freedom and noncentrality parameter

$$\sum_{j=1}^{n} (\xi_j - \bar{\xi})^2/\sigma^2,$$

that is, as

$$\sigma^2 \chi'^2_{n-1}\left(\sum_{j=1}^{n} (\xi_j - \bar{\xi})^2/\sigma^2\right).$$

2. Historical Notes

The distribution was obtained by Fisher [13] (p. 663) in 1928 as a limiting case of the distribution of the multiple correlation coefficient (Chapter 32). He gave upper 5% points of the distribution for certain values of ν and λ (Section 7). The distribution has been obtained in a number of different ways, described in outline in Section 3.

Patnaik [38] in 1949, emphasized the relevance of this distribution in approximate determination of the power of the χ^2 test, and also suggested approximations to the noncentral χ^2 distribution itself.

The noncentral χ^2 distribution can be regarded as a *generalized Rayleigh* distribution (Miller *et al.* [35], Park [37]) also called the *Rayleigh-Rice* or *Rice* distribution. In this form, it is used in mathematical physics, and especially in communication theory (see, e.g. Helstrom [22]).

3. Distribution

The cumulative distribution function of $\chi_\nu'^2(\lambda)$ is

$$\text{(1)} \quad \Pr[\chi_\nu'^2(\lambda) \leq x] = F(x;\nu,\lambda) = e^{-\frac{1}{2}\lambda} \sum_{j=0}^{\infty} \frac{(\frac{1}{2}\lambda)^j}{j!} \cdot \frac{1}{2^{\frac{1}{2}\nu+j}\Gamma(\frac{1}{2}\nu + j)} \int_0^x y^{\frac{1}{2}\nu+j-1} e^{-\frac{1}{2}y}\, dy \qquad (x > 0)$$

while $F(x;\nu,\lambda) = 0$ for $x < 0$.

It is possible to express $F(x;\nu,\lambda)$, for $x > 0$, in an easily remembered form as a weighted sum of central χ^2 probabilities with weights equal to the probabilities of a Poisson distribution with expected value $\frac{1}{2}\lambda$. This is

$$\text{(2)} \qquad F(x;\nu,\lambda) = \sum_{j=0}^{\infty} \left[\frac{(\frac{1}{2}\lambda)^j}{j!} e^{-\frac{1}{2}\lambda}\right] \cdot \Pr[\chi_{\nu+2j}^2 \leq x].$$

Thus $\chi_\nu'^2(\lambda)$ can be regarded as a mixture of central χ^2 variables. This interpretation is often useful in deriving the distribution of functions of random variables, some (or all) of which are noncentral χ^2's. (See for example the discussion of the noncentral F distribution, Chapter 30, Section 3.)

The probability density function can, similarly, be expressed as a mixture of central χ^2 probability density functions

$$\text{(3)} \qquad p_{\chi'^2}(x) = \sum_{j=0}^{\infty} \left[\frac{(\frac{1}{2}\lambda)^j}{j!} e^{-\frac{1}{2}\lambda}\right] p_{\chi_{\nu+2j}^2}(x) = \frac{\exp\{-\frac{1}{2}(x+\lambda)\}}{2^{\frac{1}{2}\nu}} \sum_{j=0}^{\infty} \frac{(x)^{\frac{1}{2}\nu+j-1}\lambda^j}{\Gamma(\frac{1}{2}\nu + j)2^{2j}j!}.$$

The distribution of $\chi_\nu'^2(\lambda)$ has been derived in quite a number of different ways. Fisher [13] gave an indirect derivation (by a limiting process). The first direct derivation was given by Tang [49] in 1938. Geometric derivations have been given by Patnaik [38], Ruben [45] and Guenther [19]. The last of these is a particularly simple derivation. It is also possible to derive the distribution by a process of induction, first obtaining the distribution of $\chi_1'^2(\lambda)$ and then using the relation

$$\chi_\nu'^2(\lambda) = \chi_1'^2(\lambda) + \chi_{\nu-1}^2,$$

the noncentral and central χ^2's in the right-hand side being mutually independent. (See, for example, Johnson and Leone [26] (p. 245) and Kerridge [28].) The moment-generating function may be used (Graybill [17]) or the characteristic function may be inverted by contour integration (McNolty [33]). Alternatively (van de Vaart [52]), we note that the moment generating function is

$$E\left[\exp\left\{\sum_{j=1}^{\nu} t(U_j + \delta_j)^2\right\}\right] = \prod_{j=1}^{\nu} E[\exp\{t(U_j + \delta_j)^2\}] = \prod_{j=1}^{\nu}\left[\frac{1}{\sqrt{2\pi}} \int_{-\infty}^{\infty} \exp\{-\tfrac{1}{2}u^2 + t(u + \delta_j)^2\}\, du\right]$$

$$= \prod_{j=1}^{\nu} [(1-2t)^{-\frac{1}{2}} \exp\{\delta_j^2 t(1-2t)^{-1}\}]$$

$$= (1-2t)^{-\frac{1}{2}\nu} e^{-\frac{1}{2}\lambda} \exp\{\tfrac{1}{2}\lambda(1-2t)^{-1}\}$$

$$= e^{-\frac{1}{2}\lambda} \sum_{j=0}^{\infty} (\tfrac{1}{2}\lambda)^j (1-2t)^{-\frac{1}{2}(\nu+2j)}/j!.$$

Noting that $(1-2t)^{-\frac{1}{2}(\nu+2j)}$ is the moment generating function of $\chi^2_{\nu+2j}$ we obtain the formula (3).

There are a number of different forms in which the cumulative distribution function and the probability density function may be presented. We have first presented those forms which seem to be the most generally useful. We now discuss some other forms.

If ν is even, the cumulative distribution function of $\chi_\nu'^2(\lambda)$ can be expressed in terms of elementary functions. Using the relation (Chapter 17, Section 4) between the integral of a χ^2_ν distribution (with ν even) and a sum of Poisson probabilities, it can be shown that

$$\Pr[\chi_\nu'^2(\lambda) \leq x] = \Pr[X_1 - X_2 \geq \tfrac{1}{2}\nu] \tag{4}$$

where X_1, X_2 are independent Poisson variables with expected values $\frac{1}{2}x$, $\frac{1}{2}\lambda$ respectively. (Fisher [13], Johnson [23].)

It follows that the probability density function of $\chi_\nu'^2(\lambda)$ also can be expressed in terms of elementary functions when ν is even. This remains true when ν is odd. To demonstrate this we first note the alternative form (valid whether ν is even or odd) for the probability density function,

$$p_{\chi'^2}(x) = \tfrac{1}{2}(x/\lambda)^{\frac{1}{4}(\nu-2)} I_{\frac{1}{2}(\nu-2)}(\sqrt{\lambda}\, x) \exp[-\tfrac{1}{2}(\lambda + x)] \qquad (x > 0) \tag{5}$$

where I_k is the modified Bessel function of the first kind and order k. (Fisher [13], (p. 663) Equation (B).) If ν is odd, the probability density function can be written in terms of elementary functions by using the formula

$$I_{m+\frac{1}{2}}(z) = \sqrt{\frac{2}{\pi}}\, z^{m+\frac{1}{2}} \left(\frac{1}{z}\frac{d}{dz}\right)^m \left(\frac{\sinh z}{z}\right) \qquad (m \text{ an integer}). \tag{6}$$

Tiku [50] has obtained an expression for $p_{\chi'^2}(x)$ in terms of the *generalized Laguerre polynomials:*

$$L_j^{(m)}(x) = \sum_{i=0}^{j} \frac{(-x)^i}{i!(j-i)!} \cdot \frac{\Gamma(j+m+1)}{\Gamma(i+m+1)} \qquad (m > -1) \tag{7}$$

(as defined in Tiku [50]; see also Chapter 1, Section 3). Tiku showed that

$$p_{\chi'^2}(x) = \tfrac{1}{2}e^{-\frac{1}{2}x}(\tfrac{1}{2}x)^{\frac{1}{2}\nu-1} \sum_{j=0}^{\infty} \frac{(-\frac{1}{2}\lambda)^j}{\Gamma(\frac{1}{2}\nu + j)} L_j^{(\frac{1}{2}\nu-1)}(\tfrac{1}{2}x) \qquad (x > 0). \tag{8}$$

Formula (1) for the cumulative distribution can be rearranged by expanding $e^{-\frac{1}{2}\lambda}$ in powers of $\frac{1}{2}\lambda$, collecting together like powers of $\frac{1}{2}\lambda$, and interchanging the order of summation. The resulting expression is compactly represented as

(9) $$F(x;\nu,\lambda) = \sum_{j=0}^{\infty} \frac{(\frac{1}{2}\lambda)^j}{j!} \Delta^j g_0$$

or $e^{\frac{1}{2}\lambda\Delta}g_0$, symbolically, where $g_j = \Pr[\chi^2_{\nu+2j} \leq x]$ and Δ is the forward difference operator ($\Delta g_m = g_{m+1} - g_m$). This formula was given by Bol'shev and Kuznetzov [8].

4. Moments

The characteristic function of $(U_j + \delta_j)^2$ (where U_j is a unit normal variable) is $E(e^{it(U_j+\delta_j)^2}) = (1 - 2it)^{-\frac{1}{2}} \exp [it\delta_j^2(1 - 2it)^{-1}]$. Hence the characteristic function of $\chi_\nu'^2(\lambda)$ is

(10) $$(1 - 2it)^{-\frac{1}{2}\nu} \exp [it\lambda(1 - 2it)^{-1}]$$

and the moment generating function is

(11) $$M_\nu(\theta,\lambda) = E[\exp (\theta\chi_\nu'^2(\lambda)] = (1 - 2\theta)^{-\frac{1}{2}\nu} \exp\left[\frac{\lambda\theta}{1 - 2\theta}\right].$$

Alternatively

$$M_\nu(\theta,\lambda) = e^{-\frac{1}{2}\lambda} \sum_{j=0}^{\infty} \frac{(\frac{1}{2}\lambda)^j}{j!} (1 - 2\theta)^{-\frac{1}{2}(\nu+2j)}.$$

The cumulant generating function is

(12) $$\begin{aligned}\kappa_\nu(\theta,\lambda) &= \log M_\nu(\theta,\lambda)\\ &= -\tfrac{1}{2}\nu \log (1 - 2\theta) + \lambda\theta(1 - 2\theta)^{-1}.\end{aligned}$$

Hence the rth cumulant is

$$\kappa_r = 2^{r-1}(r - 1)!(\nu + r\lambda).$$

In particular

(13) $$\begin{cases}\kappa_1 = \nu + \lambda & = E(\chi'^2)\\ \kappa_2 = 2(\nu + 2\lambda) & = \operatorname{var}(\chi'^2) = [\sigma(\chi'^2)]^2\\ \kappa_3 = 8(\nu + 3\lambda) & = \mu_3(\chi'^2)\\ \kappa_4 = 48(\nu + 4\lambda); & \end{cases}$$

and hence

$$\mu_4(\chi'^2) = \kappa_4 + 3\kappa_2^2 = 48(\nu + 4\lambda) + 12(\nu + 2\lambda)^2.$$

From these formulas the values of the moment ratios can be calculated. These are

(14) $$\begin{aligned}\alpha_3 &= \sqrt{\beta_1} = \sqrt{8}(\nu + 3\lambda)/(\nu + 2\lambda)^{3/2}\\ \alpha_4 &= \beta_2 = 3 + 12(\nu + 4\lambda)/(\nu + 2\lambda)^2.\end{aligned}$$

From these equations it follows that

$$\frac{\beta_2 - 3}{\beta_1} = \frac{3(\nu + 4\lambda)(\nu + 2\lambda)}{2(\nu + 3\lambda)^2} = \frac{3}{2}\left[1 - \frac{\lambda^2}{(\nu + 3\lambda)^2}\right]$$

whence

$$\frac{4}{3} \le \frac{\beta_2 - 3}{\beta_1} \le \frac{3}{2}. \tag{15}$$

The expressions for moments of $\chi_\nu'^2(\lambda)$ about zero are not so elegant as those for the central moments and cumulants (Park [37]).

The following formula for the rth moment about zero was given (private communication) by D. W. Boyd:

$$\mu_r' = 2^r\Gamma(r + \tfrac{1}{2}\nu) \sum_{j=0}^{\infty} \binom{r}{j} \frac{(\lambda/2)^j}{\Gamma(j + \frac{1}{2}\nu)}.$$

The moment generating function of $\frac{1}{2}\log_e [\chi'^2(\lambda)/\nu]$ was used by Bennett [6] to evaluate the moments of this variable. It is evident that

$$E[(\tfrac{1}{2}\log_e [\chi_\nu'^2(\lambda)/\nu])^r] = 2^{-r} \sum_{j=0}^{\infty} \left[\frac{(\frac{1}{2}\lambda)^j}{j!} e^{-\frac{1}{2}\lambda}\right] E[(\log_e [\chi^2_{\nu+2j}/\nu])^r]$$

and the values of $E[(\log_e [\chi^2_{\nu+2j}/\nu])^r]$ can be obtained from

$$\kappa_r(\log_e \chi_\nu^2) = \psi^{(r-1)}(\tfrac{1}{2}\nu) + \epsilon_r \log 2$$

with $\epsilon_1 = \frac{1}{2}$, $\epsilon_r = 0$ for $r > 1$. (See Equations (10) and (14)′, Chapter 26.)

5. Properties of the Distribution

Reproductivity: From the definition it is clear that if $\chi_{\nu_1}'^2(\lambda_1)$ and $\chi_{\nu_2}'^2(\lambda_2)$ are independent then the sum $(\chi_{\nu_1}'^2(\lambda_1) + \chi_{\nu_2}'^2(\lambda_2))$ is distributed as $\chi_{\nu_1+\nu_2}'^2(\lambda_1 + \lambda_2)$. This may be described verbally by saying that the noncentral χ^2 distribution is reproductive under convolution, and that the degrees of freedom, and also the noncentralities, are additive under convolution.

Characterization: If Y has a $\chi_\nu'^2(\lambda)$ distribution and $Y = Y_1 + Y_2 + \cdots + Y_\nu$ where the Y_j's are independent and identically distributed then each Y_j has a $\chi_1'^2(\lambda/\nu)$ distribution. The special case $\nu = 2$ has been studied by McNolty [33].

$F(x;\nu,\lambda)$ is, of course, an increasing function of x for $x > 0$. It is a decreasing function of ν, and of λ. In fact for any fixed value of x

$$\lim_{\nu\to\infty} F(x;\nu,\lambda) = \lim_{\lambda\to\infty} F(x;\nu,\lambda) = 0.$$

The distribution of the standardized variable

$$\frac{\chi'^2 - (\nu + \lambda)}{[2(\nu + 2\lambda)]^{\frac{1}{2}}}$$

tends to normality as $\nu \to \infty$, λ remaining fixed or as $\lambda \to \infty$, ν remaining fixed.

Unimodality: The distribution of $\chi_\nu'^2(\lambda)$ is unimodal. The mode occurs at the intersection of the probability density functions of $\chi_\nu'^2(\lambda)$ and $\chi_{\nu-2}'^2(\lambda)$, i.e. at the value x satisfying the equation

$$p_{\chi_\nu'^2(\lambda)}(x) = p_{\chi_{\nu-2}'^2(\lambda)}(x).$$

6. Estimation

The noncentral χ^2 distribution depends on two parameters — ν, the degrees of freedom, and λ, the noncentrality. If ν is known then the maximum likelihood estimator, $\hat{\lambda}$, of λ, given values of n independent random variables X_1, X_2, $\ldots$, X_n each having density function (3), must satisfy the equation

$$\sum_{i=1}^n \left[\frac{\sum_{j=0}^\infty e^{-\frac{1}{2}\hat{\lambda}} \left\{ \frac{(\frac{1}{2}\hat{\lambda})^{j-1}}{(j-1)!} - \frac{(\frac{1}{2}\hat{\lambda})^j}{j!} \right\} p_{\chi_{\nu+2j}'^2}(X_i)}{\sum_{j=0}^\infty e^{-\frac{1}{2}\hat{\lambda}} \frac{(\frac{1}{2}\hat{\lambda})^j}{j!} p_{\chi_{\nu+2j}'^2}(X_i)} \right] = 0,$$

that is

$$\sum_{i=1}^n \left[\frac{\sum_{j=0}^\infty \frac{(\frac{1}{2}\hat{\lambda})^{j-1}}{(j-1)!} p_{\chi_{\nu+2j}'^2}(X_i)}{\sum_{j=0}^\infty \frac{(\frac{1}{2}\hat{\lambda})^j}{j!} p_{\chi_{\nu+2j}'^2}(X_i)} \right] = n$$

if this equation has a positive root. This equation is usually not easy to solve. For the case $\nu = 2$, Meyer [34] has shown that the equation has a positive solution if $\sum_{i=1}^n X_i > 2n$; otherwise the maximum likelihood estimator takes the value zero. He furthermore shows that

$$\lim_{n\to\infty} \Pr\left[\sum_{i=1}^n X_i > 2n\right] = 1.$$

When $\nu = 1$, results for estimating parameters of folded normal distributions (Chapter 13, Section 7.3) are applicable, because $\chi_1'(\lambda) = +\sqrt{\chi_1'^2(\lambda)}$ has the same distribution as $|U + \sqrt{\lambda}|$, where U has a unit normal distribution.

The general folded normal distribution (Leone *et al.* [30]) is the distribution of $|U\sigma + \xi|$ and has density function

$$p(x) = \sqrt{2/\pi}\,\sigma^{-1}[\cosh(\xi x\sigma^{-2})]\exp[-\tfrac{1}{2}(x^2+\xi^2)\sigma^{-2}] \qquad (0 < x). \tag{16}$$

(Evidently $|U\sigma + \xi|$ has the same distribution as $\sigma\chi_1'(\xi^2/\sigma^2)$.)

The first and second moments about zero of distribution (16) are

$$(\sqrt{2/\pi})\sigma\exp(-\tfrac{1}{2}\xi^2\sigma^{-2}) + \xi[1 - \Phi(-\xi/\sigma)] \tag{17.1}$$

and

$$\xi^2 + \sigma^2 \tag{17.2}$$

respectively. Leone *et al.* [30] give tables of expected value (μ) and standard deviation (σ') for $\mu/\sigma' = 1.33(0.01)1.50(0.02)1.70(0.05)2.40(0.1)3.0$. (Note that the least possible value of μ/σ' is $(\frac{1}{2}\pi - 1)^{-\frac{1}{2}} = 1.3237$.) Leone *et al.* [30] also give values of the cumulative distribution function to 4 decimal places for $\mu/\sigma' = 1.4(0.1)3.0$ and arguments at intervals of 0.01. Some values of the moment ratios (β_1 and β_2) are given by Elandt [12].

The parameters ξ and σ can be estimated by equating sample first and second moments to (17.1), (17.2) respectively. The tables of Leone *et al.* [30] facilitate solution. Simple explicit solutions are obtained by using second and fourth moments. Here $\theta = \xi/\sigma$ is estimated by $\tilde{\theta}$, the solution of the equation

$$\frac{\text{sample 4th moment}}{(\text{sample 2nd moment})^2} = \frac{3 + 6\tilde{\theta}^2 + \tilde{\theta}^4}{(1 + \tilde{\theta}^2)^2}. \tag{18}$$

Elandt [12] has obtained expansions to terms of order n^{-3} for the variances of the estimators of θ by the two methods. There appears to be little difference for θ less than 0.75; the method using first and second sample moments is about 40% more efficient when $\theta = 3$.

The maximum likelihood equations for estimators $\hat{\xi}$, $\hat{\sigma}$ of ξ and σ can be expressed in the form

$$\hat{\xi}^2 + \hat{\sigma}^2 = n^{-1} \sum_{j=1}^{n} X_j^2 \tag{19.1}$$

$$\hat{\xi} = n^{-1} \sum_{j=1}^{n} X_j \tanh(\hat{\xi} X_j/\hat{\sigma}^2). \tag{19.2}$$

Johnson [24] obtained asymptotic formulas for the variances of $\hat{\theta}$ ($= \hat{\xi}/\hat{\sigma}$) and $\hat{\sigma}$. They are rather complicated, but for large values of θ,

$$n\,\text{var}(\hat{\theta}) \doteqdot 1 + \tfrac{1}{2}\theta^2;$$
$$n\,\text{var}(\hat{\sigma}) \doteqdot \tfrac{1}{2};$$
$$\text{corr}(\hat{\theta},\hat{\sigma}) \doteqdot -\theta(2 + \theta^2)^{-\frac{1}{2}}.$$

Relative to the maximum likelihood estimators, the efficiency of estimation of θ from first and second sample moments is about 95% when $\theta = 1$, and increases with θ. For small θ, the efficiency is low.

If σ is known (e.g. if we have a $\chi_1'^2(\lambda)$ distribution and wish to estimate λ) the maximum likelihood equation for $\hat{\xi}$ is (19.2) with $\hat{\sigma}$ replaced by σ.

7. Tables

The most extensive tables of the noncentral χ^2 distribution are those of Haynam *et al.* [21]. These tables are especially intended to facilitate calculations involving the power of various χ^2 tests. Using (as in Chapter 17) $x(\nu,\alpha)$ to denote the upper $100\alpha\%$ point of the *central* χ^2 distribution with ν degrees of freedom and

$$\beta(\nu,\lambda,\alpha) = \Pr[\chi_\nu'^2(\lambda) > x(\nu,\alpha)] \tag{20}$$

to denote power with respect to noncentrality λ, the values tabulated are:

Table I: β to 4 decimal places for

$$\alpha = 0.001, 0.005, 0.01, 0.025, 0.05, 0.1$$
$$\lambda = 0(0.1)1.0(0.2)3.0(0.5)5(1)40(2)50(5)100$$
$$\nu = 1(1)30(2)50(5)100.$$

Table II: λ to 3 decimal places for the same values of α and ν as in Table I, and

$$1 - \beta = 0.1(0.02)0.7(0.01)0.99.$$

Table III: ν to 3 decimal places for the same values of α, λ and β as in Tables I and II.

The first tables (apart from special calculations) relating to the noncentral χ^2 distribution were complied by Fix [14]. These give λ to 3 decimal places for

$$\alpha = 0.01, 0.05$$
$$1 - \beta = 0.1(0.1)0.9$$
$$\nu = 6(1)20(2)40(5)60(10)100.$$

This table is also reproduced in the Bol'shev-Smirnov tables [7]. A similar table also is included in Owen's tables [36].

Bark *et al.* [4] have given tables of $\Pr[\chi_2'^2(v^2) \geq u^2] = Q(u,v)$ to six decimal places for $v = 0(0.02)3.00$ and $u = 0(0.02)$ until $Q(u,v)$ is less than 0.0000005. For cases when $v > 3$ and $u \leq 3$ they suggest using the formula

$$Q(u,v) = 1 - Q(v,u) + Q(v - u,0)e^{-uv}I_0(uv) \tag{A}$$

and give tables of the function $e^{-x}I_0(x)$. For $u \geq v > 3$, the formula suggested is

$$Q(u,v) = q - R(q,\epsilon) \tag{B}$$

where $q = 1 - \Phi(u - v - (2v)^{-1})$; $\epsilon = (1 + v^2)^{-1}$ and $R(q,\epsilon)$ is also given in these tables. Using formula (B), formula (A) can also be used for $v > u > 3$.

Johnson [25] has given tables of percentile points — values $x(\nu,\lambda,\alpha)$ such that $\Pr[\chi_\nu'^2(\lambda) > x(\nu,\lambda,\alpha)] = \alpha$ — to 4 significant figures for $\sqrt{\lambda} = 0.2(0.2)6.0$; $\nu = 1(1)12, 15, 20$; $\alpha = 0.001, 0.0025, 0.005, 0.01, 0.025, 0.1, 0.25, 0.5, 0.75, 0.9, 0.95, 0.975, 0.99, 0.995, 0.9975, 0.999$. Tables of $\sqrt{x(\nu,\lambda,\alpha)}$ to 4 significant figures for the same values of $\sqrt{\lambda}$ and ν, but only for $\alpha = 0.01, 0.025, 0.05, 0.95, 0.975$, and 0.99 are given by Johnson and Pearson [27].

Zolnowska [54] describes a method of generating random numbers from 'Rayleigh-Rice' distributions.

A computer program for calculating $p_{\chi_\nu'^2}(x)$ and $F(x,\nu,\lambda)$ has been published by Bargmann and Ghosh [3] (also Robertson [43]). They use formulas (1) and (3) and, provided parameters are in the range 10^{-8} to 10^{+8}, obtain accuracy to five significant figures. More detailed tables for the cases $\nu = 2, 3$, calculated in connection with "coverage" problems, are described in Section 9.

8. Approximations

Many approximations to the noncentral χ^2 distribution — in particular to the value of $\Pr[\chi_\nu'^2(\lambda) \leq x]$ have been suggested. In selecting an approximation both simplicity and accuracy should be considered, although these tend to be contrary requirements.

Both the form of (3), and the inequalities in (15) lead one to expect that a gamma distribution should give a useful approximation. The simplest approximation consists of replacing χ'^2 by a multiple of central χ^2, $c\chi_f^2$ say, with c and f so chosen that the first two moments of the two variables, $\chi_\nu'^2(\lambda)$ and $c\chi_f^2$, agree. The appropriate values of c and f are

$$c = \frac{\nu + 2\lambda}{\nu + \lambda}; \qquad f = \frac{(\nu + \lambda)^2}{\nu + 2\lambda} = \nu + \frac{\lambda^2}{\nu + 2\lambda}. \tag{21}$$

This approximation was suggested by Patnaik [38]. (Two additional corrective terms to Patnaik's approximation were derived by Roy and Mohamad [44].) Pearson [39] suggested an improvement of this approximation, introducing an additional constant b, and choosing b, c and f so that the first three moments are $\chi_\nu'^2(\lambda)$ and $(c\chi_f^2 + b)$ agree. The appropriate values of b, c and f are

$$\begin{aligned} b &= -\frac{\lambda^2}{\nu + 3\lambda}; \\ c &= \frac{\nu + 3\lambda}{\nu + 2\lambda}; \\ f &= \frac{(\nu + 2\lambda)^3}{(\nu + 3\lambda)^2} = \nu + \frac{\lambda^2(3\nu + 8\lambda)}{(\nu + 3\lambda)^2}. \end{aligned} \tag{22}$$

This gives a better approximation to $F(x;\nu,\lambda)$ than does Patnaik's approximation, for x large enough. But since the Pearson approximation ascribes a non-zero value to $\Pr[-\lambda^2(\nu + 3\lambda)^{-1} < \chi'^2 \leq 0]$, it is not as good an approximation when x is small.

It can be shown that, for x and ν fixed, the error of Patnaik's approximation to $F(x;\nu,\lambda)$ is $O(\lambda^2)$ as $\lambda \to 0$, $O(\lambda^{-\frac{1}{2}})$ as $\lambda \to \infty$; the error of Pearson's approximation is also $O(\lambda^2)$ as $\lambda \to 0$ but $O(\lambda^{-1})$ as $\lambda \to \infty$. In both cases the error bounds are uniform in x.

In both Patnaik's and Pearson's approximations f is usually fractional, so that interpolation is needed if standard χ^2 tables are used.

Approximations to the central χ^2 distribution (Chapter 17, Section 5) may be applied to the approximating central χ^2's in Patnaik's and Pearson's approximations. If the Wilson-Hilferty approximations be applied then the approximation

(23.1) "$\left(\frac{\chi'^2}{\nu + \lambda}\right)^{\frac{1}{3}}$ normal with expected value $1 - \frac{2(\nu + 2\lambda)}{9(\nu + \lambda)}$ and variance $\frac{2(\nu + 2\lambda)}{\nu + \lambda}$"

is obtained (Abdel-Aty [1]). Sankaran ([46], [47]) discussed a number of such further approximations including

(23.2) "$\{\chi'^2 - \frac{1}{2}(\nu - 1)\}^{\frac{1}{2}}$ approximately normal with expected value $\{1 + \frac{1}{2}(\nu - 1)\}^{\frac{1}{2}}$ and variance 1"

(23.3) "$\{[\chi'^2 - \frac{1}{3}(\nu - 1)]/(\nu + \lambda)\}^{\frac{1}{2}}$ approximately normal with expected value $\left\{1 - \frac{\nu - 1}{3(\nu + \lambda)}\right\}^{\frac{1}{2}}$ and variance $(\nu + \lambda)^{-1}$"

(23.4) "$\left(\frac{\chi'^2}{\nu + \lambda}\right)^h$ approximately normal with expected value

$$1 + h(h - 1)\frac{\nu + 2\lambda}{(\nu + \lambda)^2} - h(h - 1)(2 - h)(1 - 3h)\frac{(\nu + 2\lambda)^2}{2(\nu + \lambda)^4}$$

and variance $h^2\frac{2(\nu + 2\lambda)}{(\nu + \lambda)^2}\left[1 - (1 - h)(1 - 3h)\frac{\nu + 2\lambda}{(\nu + \lambda)^2}\right]$,

where $h = 1 - \frac{2}{3}(\nu + \lambda)(\nu + 3\lambda)(\nu + 2\lambda)^{-2}$."

Of these, (23.2) is not good for small values of λ; (23.4) is remarkably accurate for all λ but is rather complicated, and not much more accurate than Pearson's approximation. (See the comparisons in Table 1.)

Any of these approximations might be improved by using an Edgeworth expansion, but the need for calculating higher cumulants makes them unattractive.

Rice [42] has given an expansion (as a series in powers of ν^{-1}) for the cumulative distribution function which should give uniform accuracy over the whole range of values of the argument.

Other approximations, valid for small values of λ, may be obtained from the Laguerre series expansion (8). Better results are obtained by expanding the distribution of an appropriate *linear function* of χ'^2 in a Laguerre series (Tang [49], Tiku [50]).

Bol'shev and Kuznetzov [8] use a method in which the distribution of $\chi_\nu'^2(\lambda)$ is related to the distribution of a central χ^2 with the *same* number of degrees of freedom. They wish to determine a function $w(x;\nu,\lambda)$ such that

$$\Pr[\chi_\nu'^2(\lambda) \leq x] = \Pr[\chi_\nu'^2 \leq w(x;\nu,\lambda)].$$

This is equivalent to requiring that $w(\chi'^2;\nu,\lambda)$ shall be distributed as a central χ^2 with ν degrees of freedom.

For small λ,

$$w(x;\nu,\lambda) = w^*(x;\nu,\lambda) + O(\lambda^3) \tag{24}$$

where

$$w^*(x;\nu,\lambda) = x - x\lambda\nu^{-1} + \tfrac{1}{2}x\{1 + (\nu + 2)^{-1}x\}\lambda^2\nu^{-2}$$

and $O(\lambda^3)$ is uniform in any finite interval of x. Hence, as $\lambda \to 0$

$$F(x;\nu,\lambda) = \Pr[\chi^2 \leq w^*(x;\nu,\lambda)] + O(\lambda^3). \tag{25}$$

In order to estimate percentage points, i.e. solutions $x(\alpha,\nu,\lambda)$ of the equations

$$F(x;\nu,\alpha) = \alpha \tag{26}$$

the inverse function

$$x(w^*;\nu,\lambda) = w^* + w^*\nu^{-1} + \tfrac{1}{2}w^*\{1 - (\nu + 2)^{-1}w^*\}\lambda^2\nu^{-2} \tag{27}$$

is used. If $\chi^2_{\nu,\alpha}$ is the (tabulated) $100\alpha\%$ point of the (central) χ^2_ν distribution,

$$x^* = \chi^2_{\nu,\alpha} + \chi^2_{\nu,\alpha}\lambda\nu^{-1} + \tfrac{1}{2}\chi^2_{\nu,\alpha}\{1 - (\nu + 2)^{-1}\chi^2_{\nu,\alpha}\}\lambda^2\nu^{-2} \tag{28}$$

is used as an approximation to $x(\alpha,\nu,\lambda)$.

Finally, we mention two formulas obtained by direct normal approximation. If a normal distribution is fitted to the $\chi_\nu'^2(\lambda)$ distribution, we obtain

$$F(x;\nu,\lambda) \doteqdot \Phi\left[\frac{x - \nu - \lambda}{\{2(\nu + 2\lambda)\}^{\frac{1}{2}}}\right] \tag{29}$$

where

$$\Phi(y) = \frac{1}{\sqrt{2\pi}}\int_{-\infty}^{y} e^{-\frac{1}{2}u^2}\,du.$$

Applying a normal approximation to the right hand side of (4), Johnson [23] obtained

$$F(x;\nu,\lambda) \doteqdot \Phi\left[\frac{x - \nu - \lambda + 1}{\{2(\nu + 2\lambda)\}^{-\frac{1}{2}}}\right]. \tag{30}$$

In each case the error (as $\lambda \to \infty$) is $O(\lambda^{-\frac{1}{2}})$, uniformly in x.

These approximations are simple, but not very accurate. The relative accuracy of a number of approximations can be judged from Table 1. It can be seen that only Pearson (22) and Sankaran (23.4) are reliable over a wide range of values of λ. Patnaik's and Abdel-Aty's formulas deteriorate as λ increases while the other formulas improve.

Germond and Hastings [16] gave the approximation

$$\Pr[\chi_2'^2(r^2) \leq R^2] \doteqdot \frac{R^2}{2 + R^2/2}\exp\left[-\frac{r^2}{2 + R^2/2}\right] \tag{31}$$

which is correct to 4 decimal places for $R \leq 0.4$. They also gave a table of

TABLE 1

Errors of Approximation

(Upper 5% points) Errors

ν	λ	Exact Value	Johnson (30)	Patnaik (21)	Pearson (22)	Abdel-Aty (23.1)	Sankaran (23.2)	Sankaran (23.3)	Sankaran (23.4)
2	1	8.642	0.92	−0.01	−0.04	−0.08	0.09	0.23	−0.06
	4	14.641	0.55	0.08	−0.06	0.02	0.04	0.04	−0.01
	16	33.054	0.28	0.29	−0.03	0.27	0.02	0.01	0.02
	25	45.308	0.23	0.35	−0.03	0.33	0.01	0.00	0.00
4	1	11.707	0.88	0.01	−0.02	−0.04	0.20	0.26	−0.02
	4	17.309	0.57	0.07	−0.04	0.03	0.11	0.08	−0.03
	16	35.427	0.30	0.26	−0.03	0.23	0.04	0.01	0.00
	25	47.613	0.25	0.33	−0.02	0.30	0.03	0.01	0.01
7	1	16.003	0.83	0.01	−0.01	−0.02	0.28	0.24	−0.02
	4	21.228	0.59	0.05	−0.02	0.02	0.18	0.10	0.03
	16	38.970	0.33	0.19	−0.02	0.19	0.08	0.02	−0.01
	25	51.061	0.26	0.28	−0.02	0.27	0.05	0.01	0.00

(Lower 5% points) Errors

ν	λ	Exact Value	Johnson (30)	Patnaik (21)	Pearson (22)	Abdel-Aty (23.1)	Sankaran (23.2)	Sankaran (23.3)	Sankaran (23.4)
2	1	0.17	*	0.03	−0.09	0.00	*	*	−0.05
	4	0.65	−0.43	0.29	−0.12	0.24	0.08	−0.01	0.01
	16	6.32	−0.25	0.57	−0.02	0.55	0.02	0.00	0.02
	25	12.08	−0.21	0.60	−0.01	0.59	0.01	0.00	0.03
4	1	0.91	−0.07	0.02	−0.03	0.00	*	0.14	−0.04
	4	1.77	−0.24	0.18	−0.06	0.17	0.23	0.01	−0.03
	16	7.88	−0.20	0.48	−0.02	0.47	0.06	0.00	0.02
	25	13.73	−0.17	0.53	−0.01	0.53	0.04	0.00	0.05
7	1	2.49	0.10	0.02	0.00	0.00	0.64	0.11	−0.02
	4	3.66	−0.07	0.12	−0.02	0.10	0.34	0.03	−0.01
	16	10.26	−0.15	0.38	−0.01	0.37	0.11	0.00	0.01
	25	16.23	−0.14	0.45	−0.01	0.44	0.07	0.00	0.02

Notes: The tabled quantity is the approximate value less the exact value. Exact values of upper 5% points taken from Fisher [13]; lower 5% points from Garwood [15]. The stability of the errors in Pearson's approximation to the upper 5% points is noteworthy. A correction of $0.16(\nu + 2)^{-1}$ would produce remarkably accurate results — this would also apply to the lower 5 points for $\nu = 4$ and $\nu = 7$.

(small) corrections to this formula, giving 4 decimal place accuracy up to $R = 1.2$; and further tables for larger values of R. For $R > 5$ the formula

$$\Pr[\chi_2'^2(r^2) \leq R^2] \doteq \frac{1}{\sqrt{2\pi}} \int_{r-\sqrt{R^2-1}}^{\infty} e^{-\frac{1}{2}t^2}\, dt \tag{32}$$

gives useful results. We also note a simple empirical approximation to the upper 5% point of the χ'^2 (and χ') distribution due to Tukey [51] which appears to be quite accurate.

9. Applications

One use of the noncentral χ^2 distribution — that of representing the distribution of a sample variance from a normal population with unstable expected value — has been described in Section 1. A rather more generally useful application is in approximating to the power of χ^2-tests applied to contingency tables (tests of 'goodness of fit'). In one of simplest of such tests the data consist of N observations divided among k classes $\prod_1, \prod_2, \ldots, \prod_k$ with N_i observations in class $\prod_i$ $(i = 1, \ldots, k)$. If H_0 is the hypothesis that the probability of an observation falling into $\prod_i$ is π_{i0} $(i = 1, 2, \ldots, k)$ and the alternative hypotheses specify other values for these probabilities then an approximation to the likelihood ratio test is one with a critical region of form

$$T = \sum_{i=1}^{k} \frac{(N_i - N\pi_{i0})^2}{N\pi_{i0}} > K_\alpha$$

where K_α is a suitably chosen constant. If the true values of the probabilities are π_i $(i = 1, 2, \ldots, k)$ with, of course, $\sum_{i=1}^{k} \pi_i = 1$, then T is approximately distributed as $\chi_{k-1}'^2(\lambda)$ with

$$\lambda = N \sum_{i=1}^{k} \frac{(\pi_i - \pi_{i0})^2}{\pi_{i0}}.$$

If $\pi_i = \pi_{i0}$ for all i, that is, if H_0 is valid, then $\lambda = 0$ and the approximate distribution is that of a central χ_{k-1}^2. So, to obtain a significance level approximately equal to α, one may take

$$K_\alpha = \chi_{k-1,\alpha}^2.$$

The power, when the true values of the probabilities are $\pi_1, \pi_2, \ldots, \pi_k$ is then approximately

$$\Pr[\chi_{k-1}'^2(\lambda) > \chi_{k-1,\alpha}^2].$$

There is a good discussion of more complex forms of χ^2 tests in Patnaik [38].

Noncentral χ^2 also appears in the calculation of approximate powers of certain nonparametric tests (Andrews [2], Lehmann [29] (pp. 302–306)).

Extending an argument of Wilks [53] (p. 419) it can be shown that, when the data can be represented by n independent random variables with identical distributions depending on parameters $(\theta_1, \theta_2, \ldots, \theta_k)$ then the limiting distribution (as $n \to \infty$) of

$$-2 \log_e \text{(likelihood ratio)}$$

is, under certain sequences of alternative hypotheses converging to the null hypothesis, a noncentral χ^2 distribution. The 'likelihood ratio' here is the ratio of two maximized values of the likelihood function, the numerator being restricted by certain of the θ's being assigned fixed values, while the denominator is not restricted.

Sugiura [48] has recently obtained an asymptotic expansion (up to order n^{-1}) of the non-null distribution of the logarithm of the likelihood ratio statistic for tests of multivariate linear hypotheses (Chapter 35) in the form of a linear combination of non-central χ^2 probabilities with increasing numbers of degrees of freedom and the same non-centrality parameter.

Noncentral χ^2 also appears in a slightly disguised form in calculating the probability that a random point $(X_1, X_2, \ldots, X_\nu)$, with the X's mutually independent normal variables, each having expected value 0 and standard deviation σ (the same for all i), falls within an 'offset' hypersphere $\sum_{i=1}^{\nu} (X_i - \xi_i)^2 \leq R^2$. This probability is evidently

$$\Pr\left[\chi_\nu'^2 \left(\sigma^{-2} \sum_{i=1}^{\nu} \xi_i^2\right) \leq (R/\sigma)^2\right].$$

A number of tables of this quantity have been produced, especially for the physically interesting cases $\nu = 2$ and $\nu = 3$. An extensive summary is given by Guenther and Terragno [20], who also give a useful bibliography. For the case $\nu = 2$ very detailed tables are available (Bell Aircraft Co. [5], DiDonata and Jarnagin [10], Marcum [32]). The most easily available short summary tables are in Burington and May [9] (pp. 102–5), DiDonato and Jarnagin [11], and Owen [36] (pp. 178–80). References [10] and [11] contain values of R/σ to give specified values of probabilities; the other references give values of probabilities for specified values of R/σ and $(\xi_1^2 + \xi_2^2)/\sigma^2$.

For the case $\nu = 3$ there is the simple relation

$$\Pr[\chi_3'^2(\lambda) \leq (R/\sigma)^2] = \frac{1}{\sqrt{2\pi}} \int_{\lambda^{\frac{1}{2}} - R/\sigma}^{\lambda^{\frac{1}{2}} + R/\sigma} \exp\left(-\tfrac{1}{2}u^2\right) du - \frac{1}{\sqrt{2\pi\lambda}} \left[\exp\left\{-\tfrac{1}{2}(\lambda^{\frac{1}{2}} - R/\sigma)^2\right\} - \exp\left\{-\tfrac{1}{2}(\lambda^{\frac{1}{2}} + R/\sigma)^2\right\}\right]. \tag{33}$$

There is thus less need for extensive tables in this case, but there is a short table available (Guenther [18]).

For general conditions under which a quadratic form in normal variables is distributed as non-central χ^2, see Chapter 29.

10. Relationship to Other Distributions

We have noted in Section 6, that $\chi_1'(\lambda)$ is a *folded* normal variable.

Equation (4) represents a connection between the noncentral χ^2 and Poisson distributions. Other relationships, already mentioned in this chapter, are:

(i) If $\lambda = 0$ the noncentral χ^2 becomes a central χ^2.

(ii) The limiting distribution of a standardized $\chi_\nu'^2(\lambda)$ variable is the unit normal distribution if either (a) $\nu \to \infty$, λ remaining constant or (b) $\lambda \to \infty$, ν remaining constant.

Further relationships are:

(iii) The limiting distribution of a standardized (singly or doubly) noncentral F variable, as the denominator degrees of freedom tends to infinity (noncentralities remaining constant) is the distribution of multiple of a noncentral χ^2 variable (Chapter 30, Section 5).

(iv) The noncentral χ^2 distributions may be regarded as members of the class of distributions which are mixtures of central χ^2 distributions. Moreover, Press [40] has shown that the distribution of linear functions of independent noncentral χ^2 variates with positive coefficients can be expressed as mixtures of distributions of central χ^2's. This is part of the theory of quadratic forms in normal variables, which is the subject of Chapter 29.

REFERENCES

[1] Abdel-Aty, S. H. (1954). Approximate formulae for the percentage points and the probability integral of the non-central χ^2-distribution, *Biometrika*, **41,** 538–540.

[2] Andrews, F. C. (1954). Asymptotic behavior of some rank tests for analysis of variance, *Annals of Mathematical Statistics*, **25,** 724–736.

[3] Bargmann, R. E. and Ghosh, S. P. (1964). *Noncentral statistical distribution programs for a computer language*, I.B.M. Research Report R.C.-1231.

[4] Bark, L. S., Bol'shev, L. N., Kuznetzov, P. I. and Cherenkov, A. P. (1964). *Tables of the Rayleigh-Rice Distribution*, Computation Center, Academy of Sciences, USSR, Moscow.

[5] Bell Aircraft Corporation (1956). *Tables of Circular Normal Probabilities*, Report No. 02-949-106, Operations Analysis Group, Dynamics Section, Bell Aircraft Corporation, Buffalo, New York.

[6] Bennett, B. M. (1955). Note on the moments of the logarithmic noncentral χ^2 and z distributions, *Annals of Institute of Statistical Mathematics, Tokyo*, **7,** 57–61.

[7] Bol'shev, L. N. and Smirnov, N. V. (1965). *Tables of Mathematical Statistics*, Moscow: Akademia Nauk SSSR.

[8] Bol'shev, L. N. and Kuznetzov, P. I. (1963). On computing the integral $p(x,y) = \dots$, *Zhurnal Vychislitelnoj Matematiki i Matematicheskoi Fiziki*, **3,** 419–430. (In Russian)

[9] Burington, R. S. and May, D. C. (1953). *Handbook of Probability and Statistics with Tables*, Sandusky, Ohio: Handbook Publishers.

[10] DiDonato, A. R. and Jarnagin, M. P. (1962). *A method for computing the generalized circular error function and circular coverage functions*, N. W. L. Report No. 1768, U. S. Naval Weapons Laboratory, Dahlgren, Virginia.

[11] DiDonato, A. R., and Jarnagin, M. P. (1962). A method for computing the circular coverage function, *Mathematics of Computation*, **16,** 347–355.

[12] Elandt, Regina C. (1961). The folded normal distribution: Two methods of estimating parameters from moments, *Technometrics*, **3,** 551–562.

[13] Fisher, R. A. (1928). The general sampling distribution of the multiple correlation coefficient, *Proceedings of the Royal Society of London*, **121A,** 654–673.

[14] Fix, Evelyn (1949). Tables of non-central χ^2, *University of California Publications in Statistics*, **1,** No. 2, 15–19.

[15] Garwood, F., Referred to in [1]. Unpublished Ph.D. thesis, University of London, 1934.

[16] Germond, H. H. and Hastings, Cecil (1944). *Scatter Bombing of a Circular Target*, A report submitted by the Bombing Research Group, Columbia University and the Applied Mathematics Group, Columbia University to the Applied Mathematics Panel, National Defense Research Committee, May 1944.

[17] Graybill, F. A. (1961). *An Introduction to Linear Models*, Vol. **1,** New York: McGraw-Hill, Inc.

[18] Guenther, W. C. (1961). On the probability of capturing a randomly selected point in three dimensions, *SIAM Review*, **3,** 247–250.

[19] Guenther, W. C. (1964). Another derivation of the non-central chi-square distribution, *Journal of the American Statistical Association*, **59,** 957–960.

[20] Guenther, W. C. and Terragno, P. J. (1964). A review of the literature on a class of coverage problems, *Annals of Mathematical Statistics*, **35,** 232–260.

*[21] Haynam, G. E., Govindarajulu, Z. and Leone, F. C. (1962). *Tables of the Cumulative Non-Central Chi-square Distribution*, Case Statistical Laboratory, Publication No. 104.

[22] Helstrom, C. W. (1960). *Statistical Theory of Signal Detection*, Oxford: Pergamon Press.

[23] Johnson, N. L. (1959). On an extension of the connexion between Poisson and χ^2-distributions, *Biometrika*, **46,** 352–363.

[24] Johnson, N. L. (1962). The folded normal distribution: Accuracy of estimation by maximum likelihood, *Technometrics*, **4,** 249–256.

[25] Johnson, N. L. (1968). *Tables of Percentile Points of Noncentral Chi-square Distributions*, Mimeo Series No. 568, Institute of Statistics, University of North Carolina.

[26] Johnson, N. L. and Leone, F. C. (1964). *Statistics and Experimental Design in Engineering and the Physical Sciences*, Volume **I,** New York: John Wiley & Sons, Inc.

[27] Johnson, N. L. and Pearson, E. S. (1969). Tables of percentage points of non-central χ, *Biometrika*, **56,** 315–333.

[28] Kerridge, D. (1965). A probabilistic derivation of the non-central χ^2 distribution, *Australian Journal of Statistics*, **7,** 37–39, (Correction **7,** 114).

[29] Lehmann, E. L. (1959). *Testing Statistical Hypotheses*, New York: John Wiley & Sons, Inc.

[30] Leone, F. C., Nelson, L. S. and Nottingham, R. B. (1961). The folded normal distribution, *Technometrics*, **3,** 543–550.

[31] Lowe, J. R. (1960). A table of the integral of the bivariate normal distribution over an offset circle, *Journal of the Royal Statistical Society, Series B*, **22,** 177–187.

[32] Marcum, J. I. (1950). *Tables of Q-functions*, RAND Report No. RM-339, RAND Corporation, Santa Monica, California.

[33] McNolty, F. (1962). A contour-integral derivation of the non-central chi-square distribution, *Annals of Mathematical Statistics*, **33,** 796–800.

[34] Meyer, P. L. (1967). The maximum likelihood estimate of the non-centrality parameter of a non-central χ^2 variate, *Journal of the American Statistical Association*, **61,** 1258–1264.

[35] Miller, K. S., Bernstein, R. I., and Blumenson, L. E. (1958). Generalized Rayleigh processes, *Quarterly of Applied Mathematics*, **16,** 137–145 (and *Note*, *Ibid*, **20,** Jan. 1963).

[36] Owen, D. B. (1962). *Handbook of Statistical Tables*, Reading, Mass.: Addison-Wesley Publishing Company, Inc.

[37] Park, J. H. (1961). Moments of the generalized Rayleigh distribution, *Quarterly of Applied Mathematics*, **19,** 45–49.

[38] Patnaik, P. B. (1949). The non-central χ^2- and F-distributions and their applications, *Biometrika*, **36,** 202–232.

[39] Pearson, E. S. (1959). Note on an approximation to the distribution of non-central χ^2, *Biometrika*, **46,** 364.

[40] Press, S. J. (1966). Linear combinations of non-central chi-square variates, *Annals of Mathematical Statistics*, **37,** 480–487.

[41] Rainville, E. D. (1960). *Special Functions*, New York: The Macmillan Company.

[42] Rice, S. O. (1968). Uniform asymptotic expressions for saddle point integrals — application to a probability distribution occurring in noise theory, *Bell System Technical Journal*, **47,** 1971–2013.

[43] Robertson, G. H. (1969). Computation of the noncentral chi-square distribution, *Bell System Technical Journal*, **48,** 201–207.

[44] Roy, J., and Mohamad, J. (1964). An approximation to the non-central chi-square distribution, *Sankhyā, Series A*, **26,** 81–84.

[45] Ruben, H. (1960). Probability content of regions under spherical normal distributions, I., *Annals of Mathematical Statistics*, **31,** 598–618.

[46] Sankaran, M. (1959). On the non-central chi-square distribution, *Biometrika*, **46,** 235–237.

[47] Sankaran, M. (1963). Approximations to the non-central chi-square distribution, *Biometrika*, **50,** 199–204.

[48] Sugiura, N. (1968). *Asymptotic expansions of the power functions of the likelihood ratio tests for multivariate linear hypotheses and independence*, Mimeo Series No. 563, Institute of Statistics, University of North Carolina.

[49] Tang, P. C. (1938). The power function of the analysis of variance tests with tables and illustrations of their use, *Statistical Research Memoirs*, **2,** 126–149.

[50] Tiku, M. L. (1965). Laguerre series forms of non-central χ^2 and F distributions, *Biometrika*, **52,** 415–426.

[51] Tukey, J. W. (1957). Approximations to the upper 5% points of Fisher's B distribution and non-central χ^2, *Biometrika*, **44,** 528–530.

[52] Vaart, H. R. van der (1967). A note on the derivation of the non-central chi-square density function, *Statistica Neerlandica*, **21,** 99–100.

[53] Wilks, S. S. (1962). *Mathematical Statistics*, New York: John Wiley & Sons, Inc.

[54] Zolnowska, Halina (1965). Generators of random numbers of Rayleigh and Rice's distributions, *Algorytmy*, **3,** 73–94.

*Part of the tables in [21] have been published in *Selected Tables in Mathematical Statistics* (Ed. H. L. Harter and D. B. Owen), Chicago: Markham, 1970.

29

Quadratic Forms in Normal Variables

1. Introduction and Definition

Suppose $\mathbf{X}' = (X_1,X_2,\ldots,X_n)$ is a random vector following a multivariate normal distribution with expected value vector $(\xi_1,\xi_2,\ldots,\xi_n)$ and variance-covariance matrix $\mathbf{V}$ (Chapter 35). The quadratic form $Q(X_1,\ldots,X_n)$ associated with the symmetric matrix A defined as

$$Q(\mathbf{X}) = Q(X_1,\ldots,X_n) = \sum_{i=1}^{n} \sum_{j=1}^{n} a_{ij}X_iX_j.$$

Quadratic forms enter into many statistics associated with normally distributed random variables. Formal analysis of variance, for example, is entirely concerned with statistics constructed from quadratic forms in random variables representing original observations (or transformations thereof). Here we try to provide an exhaustive discussion of representations of the exact distributions of quadratic forms.

Various methods of deriving the distribution of Q will be discussed. Applications of the distribution, and relationship with other distributions will be mentioned. There is a separate section on conditions under which two quadratic forms are independent. These are relevant to problems in the analysis of covariance.

Some particularly simple cases (central and noncentral χ^2) have been discussed in earlier chapters. In the present chapter there will be less emphasis on approximations to the distributions, though there will be a number of references to this topic.

Use of characteristic functions unifies the various forms of representation of the distributions, so particular attention is devoted to this aspect.

2. Notation and Interpretations of the Distribution

We first consider a *central* quadratic form in which $\xi_1 = \xi_2 = \cdots = \xi_n = 0$. For distinctiveness, we will use the symbols $Z_1, Z_2, \ldots, Z_n$ to denote elements of a random column vector $\mathbf{Z}$ with probability density given by

$$p_{\mathbf{Z}}(\mathbf{z}) = (2\pi)^{-\frac{1}{2}n}|\mathbf{V}|^{-\frac{1}{2}} \exp(-\tfrac{1}{2}\mathbf{z}'\mathbf{V}^{-1}\mathbf{z}) \tag{1}$$

($|\mathbf{V}|$ is the determinant of $\mathbf{V}$, and $\mathbf{Z}'$, $\mathbf{z}'$ are the transposes of $\mathbf{Z}$, $\mathbf{z}$, respectively.) Any homogeneous quadratic form in $Z_1, Z_2, \ldots, Z_n$ may be expressed as

$$Q(\mathbf{Z}) = \mathbf{Z}'\mathbf{A}\mathbf{Z} = \sum_{i=1}^{n} \sum_{j=1}^{n} a_{ij}Z_iZ_j \tag{2}$$

where $\mathbf{A}$ is a $n \times n$ symmetric matrix with elements a_{ij}. We wish to find the distribution of $Q(\mathbf{Z})$, that is

$$\Pr[Q(\mathbf{Z}) \leq y] \qquad (-\infty < y < \infty). \tag{3}$$

This is formally equivalent to evaluating the n-dimensional integral

$$\int_{\mathbf{Z}'\mathbf{A}\mathbf{Z} \leq y} p_{\mathbf{Z}}(\mathbf{z})\, d\mathbf{z} \tag{4}$$

where $d\mathbf{z}$ denotes $dz_1\, dz_2 \cdots dz_n$.

We may reduce this to a convenient standard form. Since $\mathbf{V}$ is positive definite and symmetrical, it can be factored as

$$\mathbf{V} = \mathbf{L}\mathbf{L}' \tag{5}$$

where $\mathbf{L}$ is a non-singular lower triangular matrix (that is, a matrix with zero entries everywhere above the main diagonal). Since $\mathbf{L}$ is non-singular, the change of variables

$$\mathbf{z}_{(1)} = \mathbf{L}^{-1}\mathbf{z}, \tag{6}$$

is permissible. Since $\mathbf{z}'_{(1)}\mathbf{z}_{(1)} = \mathbf{z}'\mathbf{V}^{-1}\mathbf{z}$, and $|\mathbf{V}| = |\mathbf{L}|^2$ formula (4) becomes

$$\Pr[Q(\mathbf{Z}) \leq y] = (2\pi)^{\frac{1}{2}n} \int_{\mathbf{z}'_{(1)}(\mathbf{L}'\mathbf{A}\mathbf{L})\mathbf{z}_{(1)} \leq y} \exp(-\tfrac{1}{2}\mathbf{z}'_{(1)}\mathbf{z}_{(1)})\, d\mathbf{z}_{(1)}. \tag{7}$$

The matrix $\mathbf{L}'\mathbf{A}\mathbf{L}$ is symmetric, so if $\mathbf{\Lambda}$ is the diagonal matrix of eigenvalues of $\mathbf{L}'\mathbf{A}\mathbf{L}$ and $\mathbf{P}$ the associated orthogonal matrix of eigenvectors, then

$$\mathbf{P}'\mathbf{L}'\mathbf{A}\mathbf{L}\mathbf{P} = \mathbf{\Lambda}.$$

Hence if the further linear transformation $\mathbf{W} = \mathbf{P}'\mathbf{z}_{(1)}$ is applied, (7) becomes

$$\Pr[Q(\mathbf{Z}) \leq y] = (2\pi)^{-\frac{1}{2}n} \int_{\mathbf{w}'\mathbf{\Lambda}\mathbf{w} \leq y} \exp(-\tfrac{1}{2}\mathbf{w}'\mathbf{w})\, d\mathbf{w}. \tag{8}$$

Thus, the distribution of Q is the same as that of $\mathbf{W}'\mathbf{\Lambda}\mathbf{W} = \sum_{i=1}^{n} \lambda_i W_i^2$ where the variables W_i are independent unit normal ($N(0,1)$) variables, and the numbers $\lambda_1 \geq \lambda_2 \geq \cdots \geq \lambda_n$ are the eigenvalues of $\mathbf{L}'\mathbf{A}\mathbf{L}$, or equivalently of $\mathbf{VA}$.

It follows that attention can be focussed on the distribution of

$$Q(\mathbf{W}) = \sum_{i=1}^{n} \lambda_i W_i^2 \tag{9}$$

where the W_i's are mutually independent unit normal variables. We now introduce the notation

$$F_n(y;\lambda_1,\lambda_2,\ldots,\lambda_n) = \Pr\left[\sum_{i=1}^{n} \lambda_i W_i^2 \leq y\right] \tag{10}$$

to denote the cumulative distribution function of Q.*

In the general (noncentral) case, when the ξ's are not all zero, an exactly similar reduction is possible, leading to

$$Q(\mathbf{W}) = \sum_{i=1}^{n} \lambda_i (W_i - \omega_i)^2$$

where ω_i is the same function of the ξ's as W_i is of the Z's. We introduce the notation

$$F_n(y;\lambda_1,\lambda_2,\ldots,\lambda_n;\omega_1,\omega_2,\ldots,\omega_n) = \Pr\left[\sum_{i=1}^{n} \lambda_i (W_i - \omega_i)^2 \leq y\right] \tag{11}$$

to denote the (noncentral) cumulative distribution of Q in this case.*

For brevity, we will often write

$$F_n(y;\lambda_1,\lambda_2,\ldots,\lambda_n) = F_n(y;\boldsymbol{\lambda}) \tag{12.1}$$

$$F_n(y;\lambda_1,\lambda_2,\ldots,\lambda_n;\omega_1,\omega_2,\ldots,\omega_n) = F_n(y;\boldsymbol{\lambda};\boldsymbol{\omega}). \tag{12.2}$$

For the corresponding probability density functions, the notations

$$p_n(y;\boldsymbol{\lambda}) = \frac{dF_n(y;\boldsymbol{\lambda})}{dy} \tag{13.1}$$

$$p_n(y;\boldsymbol{\lambda};\boldsymbol{\omega}) = \frac{dF_n(y;\boldsymbol{\lambda};\boldsymbol{\omega})}{dy} \tag{13.2}$$

will be used.

Note that, by definition

$$F_n(y;\boldsymbol{\lambda}) = F_n(y;\boldsymbol{\lambda};\mathbf{0})$$

and

$$p_n(y;\boldsymbol{\lambda}) = p_n(y;\boldsymbol{\lambda};\mathbf{0}).$$

*Note that we omit the subscript Q after F, for convenience.

Geometrical interpretation of $F_n(y;\boldsymbol{\lambda};\boldsymbol{\omega})$ is sometimes helpful. As a simple example, suppose $\lambda_n \geq 0$ (so that no λ's are negative). Then the region

$$\sum_{i=1}^{n} \lambda_i(W_i - \omega_i)^2 < y$$

is an ellipsoid with center at $(\omega_1,\omega_2,\ldots,\omega_n)$ and semi-axes $(y/\lambda_1)^{\frac{1}{2}}, (y/\lambda_2)^{\frac{1}{2}}, \ldots, (y/\lambda_n)^{\frac{1}{2}}$, and $F_n(y;\boldsymbol{\lambda};\boldsymbol{\omega})$ is the probability content of a spherical normal distribution over this ellipsoid. Alternatively, the ellipsoid may be transformed into a hypersphere, at the expense of the variables $\mathbf{W}'_{(1)} = \mathbf{W}'\boldsymbol{\Lambda}^{\frac{1}{2}}$ having unequal variances ($\operatorname{var}(W_{(1)i}) = \lambda_i, i = 1,2,\ldots,n$). In fact

(14) $F_n(y;\boldsymbol{\lambda};\boldsymbol{\omega})$

$$= (2\pi)^{-\frac{1}{2}n} \int_{(\mathbf{w}-\boldsymbol{\omega})'\boldsymbol{\Lambda}(\mathbf{w}-\boldsymbol{\omega})\leq y} \exp\left(-\tfrac{1}{2}\mathbf{w}'\mathbf{w}\right) d\mathbf{w}$$

$$= (2\pi)^{-\frac{1}{2}n}(\lambda_1\lambda_2\cdots\lambda_n)^{-\frac{1}{2}} \int_{(\mathbf{w}_{(1)}-\boldsymbol{\omega}_{(1)})'(\mathbf{w}_{(1)}-\boldsymbol{\omega}_{(1)})} \exp\left(-\tfrac{1}{2}\mathbf{w}'_{(1)}\boldsymbol{\Lambda}^{-1}\mathbf{w}_{(1)}\right) d\mathbf{w}_{(1)}$$

where $\boldsymbol{\omega}'_{(1)} = \boldsymbol{\omega}'\boldsymbol{\Lambda}^{\frac{1}{2}}$.

There is a good discussion of the geometrical approach in Ruben [63] and [64].

3. Characteristic Function and Moments

Since the W_j's are mutually independent unit normal random variables, the characteristic function of $Q(\mathbf{W}) = \sum_{j=1}^{n} \lambda_j(W_j - \omega_j)^2$ is

(15)

$$E(e^{itQ(\mathbf{W})}) = \exp\left(-\frac{1}{2}\sum_{j=1}^{n}\omega_j^2\right)\exp\left\{\frac{1}{2}\sum_{j=1}^{n}\frac{\omega_j^2}{1-2it\lambda_j}\right\}\prod_{j=1}^{n}(1-2it\lambda_j)^{-\frac{1}{2}}.$$

In the central case, each ω_j is equal to zero, and the characteristic function is

$$\prod_{j=1}^{n}(1-2it\lambda_j)^{-\frac{1}{2}}. \tag{16}$$

We will use the symbols $\phi_n(t;\boldsymbol{\lambda})$ and $\phi_n(t;\boldsymbol{\lambda};\boldsymbol{\omega})$ to denote the characteristic function in the central and noncentral cases respectively.

The cumulants are easily obtained from the cumulant generating function

(17)

$$K_n(\theta;\boldsymbol{\lambda};\boldsymbol{\omega}) = \log E(e^{\theta Q(\mathbf{W})})$$

$$= -\frac{1}{2}\sum_{j=1}^{n}\omega_j^2 + \frac{1}{2}\sum_{j=1}^{n}\frac{\omega_j^2}{1-2\theta\lambda_j} - \frac{1}{2}\sum_{j=1}^{n}\log(1-2\theta\lambda_j).$$

Denoting the sth cumulant of $Q(\mathbf{W})$ by $\kappa_s(Q;\boldsymbol{\lambda};\boldsymbol{\omega})$, we have

(18) $$K_n(\theta;\boldsymbol{\lambda};\boldsymbol{\omega}) = \sum_{s=1}^{\infty} \kappa_s(Q;\boldsymbol{\lambda};\boldsymbol{\omega}) \frac{\theta^s}{s!}\cdot$$

Comparing coefficients in (17) and (18)

(19) $$\kappa_s(Q;\boldsymbol{\lambda};\boldsymbol{\omega}) = 2^{s-1}(s-1)! \sum_{j=1}^{n} (s\omega_j^2 + 1)\lambda_j^s.$$

In particular in the central case

(20) $$\kappa_s(Q;\boldsymbol{\lambda}) = \kappa_s(Q;\boldsymbol{\lambda};\mathbf{0}) = 2^{s-1}(s-1)! \sum_{j=1}^{n} \lambda_j^s.$$

Returning to the notation of Section 3, we observe that the distribution of $\mathbf{Z'AZ}$ is the same as that of $\mathbf{W'\Lambda W}$ where $\mathbf{\Lambda}$ is the diagonal matrix of eigenvalues of $\mathbf{VA}$. Explicit determination of $\lambda_1, \lambda_2, \ldots, \lambda_n$ is usually troublesome, and it is worth while remembering that

(21) $$\sum_{j=1}^{n} \lambda_j^s = tr(\mathbf{VA})^s.$$

Hence (20) may be written

(22) $$\kappa_s(Q;\boldsymbol{\lambda}) = 2^{s-1}(s-1)!\,tr(\mathbf{VA})^s.$$

If necessary, central moments, and moments about zero can be determined from the cumulants in the usual way (Chapter 1, Section 5).

Harvey [31] gives formulas for fractional moments of Q in terms of integrals of functions containing a finite number of elementary terms. Series expansions for fractional moments can also be obtained from the expansions developed in this chapter.

4. Some Important Special Cases

If $\lambda_j = 1$ for $j = 1, 2, \ldots, n$, then $Q(\mathbf{W}) = \sum_{j=1}^{n} (W_j - \omega_j)^2$ and has the noncentral χ^2 distribution with n degrees of freedom and noncentrality parameter $\sum_{j=1}^{n} \omega_j^2$ (Chapter 28). If the eigenvalues are not all distinct but there are m distinct values $\lambda_1', \lambda_2', \ldots, \lambda_m'$ with multiplicities $n_1, n_2, \ldots, n_m$ respectively then

(23) $$Q(\mathbf{W}) = \sum_{r=1}^{m} \lambda_r' \chi_{n_r}'^2(\eta_r)$$

where the $\chi_{n_r}'^2$ are mutually independent and $\eta_r = \sum_{(r)} \omega_j^2$, the summation being over all j for which $\lambda_j = \lambda_r'$.

5. Positive Definite Forms — Central Case

5.1 *General Formulas*

We will first describe the derivation of the distribution for the central case. It might be thought more desirable to study the general case first, obtaining the central case finally by putting $\omega_j = 0$. However, in the central case, the formulas are not so complicated and so the essential features of the various methods will appear more clearly.

For similar reasons, we first consider the special case when the eigenvalues are all of even multiplicity. In this case n must be even, and in the most general situation there are $\frac{1}{2}n$ distinct eigenvalues $\beta_1, \beta_2, \ldots, \beta_m$ (putting $m = \frac{1}{2}n$) each of multiplicity two. In this case the characteristic function of $Q(\mathbf{W})$ is (from (15))

$$\phi_n(t;\boldsymbol{\lambda}) = \prod_{j=1}^{m} (1 - 2it\beta_j)^{-1}. \tag{24}$$

Assuming for the moment that $\beta_1 > \beta_2 \cdots > \beta_n > 0$ (i.e. that $\mathbf{A}$ is positive definite, and all multiplicities are exactly two), the partial fraction expansion

$$\phi_n(t;\boldsymbol{\lambda}) = \sum_{j=1}^{m} \frac{b_j}{1 - 2it\beta_j} \tag{25}$$

where

$$b_j = \beta_j^{m-1} \sum_{\substack{l=1 \\ l \neq j}}^{m} (\beta_j - \beta_l)^{-1}$$

leads to the formula

$$p_n(y;\boldsymbol{\lambda}) = \frac{1}{2} \sum_{j=1}^{m} (b_j/\beta_j) \exp\left(-\tfrac{1}{2}y/\beta_j\right) \qquad (y > 0). \tag{26}$$

This formula was given by Robbins [60].

The restriction that the β_i's be distinct is not essential since we used this only to obtain (25). If, for example $\beta_1 = \beta_2 = \cdots = \beta_r$, then the first r terms of the right-hand side of (25) will be of the form

$$\sum_{j=1}^{r} \frac{B_j}{(1 - 2it\beta)^j} \tag{27}$$

where β is the common value of $\beta_1, \beta_2, \ldots, \beta_r$ and the B_j's are constants (not depending on t). Inverting this part of the characteristic function gives

$$\sum_{j=1}^{r} B_j p_{\beta\chi^2_{2j}}(y). \tag{28}$$

We now restrict ourselves to the case when $\mathbf{A}$ (and so $Q(\mathbf{W})$) is positive definite, in addition to all ω's being zero, but the eigenvalues of $\mathbf{VA}$ are of unspecified multiplicity. As above, without loss of generality we assume that the oots $\lambda_1, \lambda_2, \ldots, \lambda_n$ of **VA** are numbered so that $\lambda_1 \geq \lambda_2 \geq \cdots \geq \lambda_n > 0$. In this case, the region $Q(\mathbf{W}) \leq y$ is an ellipsoid.)

A number of different expansions for $p_n(y;\boldsymbol{\lambda})$ and $F_n(y;\boldsymbol{\lambda})$ will be presented. n order to demonstrate the essential identity of the various expansions we will erive each of them directly from the characteristic function $\phi_n(t;\boldsymbol{\lambda})$. This nethod of approach also leads to simple recurrence formulas for calculating oefficients in the expansion.

The characteristic function of $Q(\mathbf{W})$, under the present conditions is (from 16)),

$$\phi_n(t;\boldsymbol{\lambda}) = \prod_{j=1}^{n} (1 - 2it\lambda_j)^{-\frac{1}{2}}.$$

For $n > 2$, $\phi_n(t;\boldsymbol{\lambda})$ is integrable, so that for $n > 2$,

29) $$p_n(y;\boldsymbol{\lambda}) = (2\pi)^{-1} \int_{-\infty}^{\infty} e^{-ity} \phi_n(t;\boldsymbol{\lambda})\, dt.$$

For $n = 2$, this may be interpreted as a Cauchy principal value, that is

30) $$p_2(y;\boldsymbol{\lambda}) = \lim_{R\to\infty} (2\pi)^{-1} \int_{-\infty}^{\infty} e^{-ity} \phi_2(t;\boldsymbol{\lambda})\, dt.$$

$\phi_n(t;\boldsymbol{\lambda})e^{-ity}$ is an analytic function (of t) except at the branch points $-i/2\lambda_j$, $j = 1,2,\ldots,n$). Grad and Solomon [21] show that

31) $$p_n(y;\boldsymbol{\lambda}) = \frac{1}{2\pi} \int_{-\infty}^{\infty} e^{-ity} \phi_n(y;\boldsymbol{\lambda})\, dt = \frac{1}{2\pi} \int_C e^{-ity} \phi_n(t;\boldsymbol{\lambda})\, dt$$

vhere C is the contour shown (for $n = 5$) in Figure 1 (with no loop to $-i\infty$ f n is even). Evaluating this last integral gives

32.1) $$p_{2k}(y;\boldsymbol{\lambda}) = \frac{1}{\pi} \sum_{j=1}^{k} (-1)^{j-1} \int_{-(2\lambda_{2j})^{-1}}^{-(2\lambda_{2j-1})^{-1}} e^{y\tau} |\phi_{2k}(i\tau;\boldsymbol{\lambda})|\, d\tau;$$

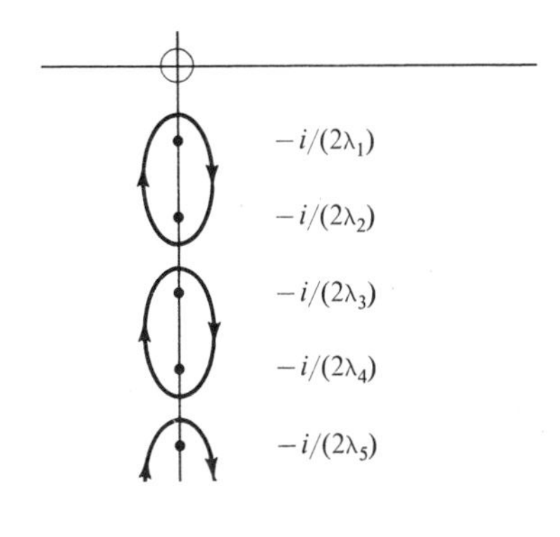

FIGURE 1

$$(32.2)\qquad p_{2k+1}(y;\boldsymbol{\lambda}) = \frac{1}{\pi}\sum_{j=1}^{k}(-1)^{j-1}\int_{-(2\lambda_{2j})^{-1}}^{-(2\lambda_{2j-1})^{-1}} e^{y\tau}|\phi_{2k+1}(i\tau;\boldsymbol{\lambda})|\,d\tau$$
$$+\frac{(-1)^k}{\pi}\int_{-\infty}^{-(2\lambda_{2k+1})^{-1}} e^{y\tau}|\phi_{2k+1}(i\tau;\boldsymbol{\lambda})|\,d\tau.^*$$

Putting $\theta = -\tau$, we note that

$$(33)\qquad \phi_n(-i\theta;\boldsymbol{\lambda}) = E(e^{\theta Q(\mathbf{W})}) = M_n(\theta;\boldsymbol{\lambda})$$

which is the moment generating function of $Q(\mathbf{W})$ (Section 4). Making this substitution in (32.1) and (32.2) gives

$$(34.1)\qquad p_{2k}(y;\boldsymbol{\lambda}) = \frac{1}{\pi}\sum_{j=1}^{k}(-1)^{j-1}\int_{(2\lambda_{2j-1})^{-1}}^{(2\lambda_{2j})^{-1}} e^{-y\theta}|M_{2k}(\theta;\boldsymbol{\lambda})|\,d\theta$$

$$(34.2)\qquad p_{2k+1}(y;\boldsymbol{\lambda}) = \frac{1}{\pi}\sum_{j=1}^{k}(-1)^{j-1}\int_{(2\lambda_{2j-1})}^{(2\lambda_{2j})^{-1}} e^{-y\theta}|M_{2k+1}(\theta;\boldsymbol{\lambda})|\,d\theta$$
$$+\frac{(-1)^k}{\pi}\int_{(2\lambda_{2k+1})^{-1}}^{\infty} e^{-y\theta}|M_{2k+1}(\theta;\boldsymbol{\lambda})|\,d\theta.$$

Grad and Solomon [21] made a special study of these formulas for the cases $n = 2$ and $n = 3$.

5.2 *Expansion as Mixtures of Central χ^2 Distributions*

We now discuss a direct method of obtaining expansion as a linear function (or 'mixture') of central χ^2 probability functions. The expansion sought is of the form

$$(35)\qquad F_n(y;\boldsymbol{\lambda}) = \sum_{j=0}^{\infty} e_j \Pr[\chi^2_{n+2j} \le y/\beta]$$

where β is an arbitrary positive constant, and the e_j's depend on β as well as on $\lambda_1, \lambda_2, \ldots, \lambda_n$. We start from formula (16)

$$\phi_n(t;\lambda) = \prod_{j=1}^{n}(1 - 2it\lambda_j)^{-\frac{1}{2}}$$

and rewrite this in the form

$$(36)\qquad \phi_n(t;\lambda) = u^{\frac{1}{2}n}\prod_{j=1}^{n}(\beta/\lambda_j)^{\frac{1}{2}}\{1-(1-\beta/\lambda_j)u\}^{-\frac{1}{2}}$$

where $u = (1 - 2it\beta)^{-1}$. For $|u| < \min_j |1-\beta/\lambda_j|^{-1}$ the product in (36) can

*The moduli signs in the integrand are needed because ϕ_n is imaginary over the ranges of integration.

be expanded as a power series in u:

$$\prod_{j=1}^{n} (\beta/\lambda_j)^{\frac{1}{2}}[1 - (1 - \beta/\lambda_j)u]^{-\frac{1}{2}} = \sum_{j=0}^{\infty} e_j u^j. \tag{37}$$

Hence

$$\phi_n(t;\lambda) = \sum_{j=0}^{\infty} e_j(1 - 2it\beta)^{-\frac{1}{2}(n+2j)}. \tag{38}$$

Thus, if the required representation exists, the values of e_j are determined by (37). Ruben [65] showed that the series of (35) converges uniformly over every finite interval of y.

To obtain useful expressions for the e_j's we note that $\prod_{j=1}^{n} [1 - (1 - \beta/\lambda_j)u]^{-\frac{1}{2}}$ is the moment generating function (with argument u) of

$$Q(\mathbf{W}) = \frac{1}{2}\sum_{j=1}^{n} (1 - \beta/\lambda_j)W_j^2.$$

Hence

$$e_r = \left(\frac{\beta^{\frac{1}{2}n}}{\prod_{j=1}^{n} \lambda_j^{\frac{1}{2}}}\right) \frac{\mu_r'(Q;1 - \beta/\lambda_1, 1 - \beta/\lambda_2, \ldots, 1 - \beta/\lambda_n)}{2^r r!} \tag{39}$$

where $\mu_r'(Q;1 - \beta/\lambda_1, \ldots, 1 - \beta/\lambda_n) = E[\{Q(\mathbf{W})\}^r]$.

The quantities μ_r' can be calculated from the corresponding cumulants, from (20)

$$\kappa_r(Q;1 - \beta/\lambda_1, \ldots, 1 - \beta/\lambda_n) = 2^{r-1}(r - 1)! \sum_{j=1}^{n} (1 - \beta/\lambda_j)^r. \tag{40}$$

Putting $H_r = \sum_{j=1}^{n} (1 - \beta/\lambda_j)^r$, (40) becomes

$$\kappa_r(Q;1 - \beta/\lambda_1, \ldots, 1 - \beta/\lambda_n) = 2^{r-1}(r - 1)!\, H_r. \tag{40$'$}$$

Using the formula $\mu_r' = \sum_{j=0}^{r-1} \binom{r-1}{j} \kappa_{r-j}\mu_j'$, we obtain the formulas

$$e_0 = \prod_{j=1}^{n} (\beta/\lambda_j)^{\frac{1}{2}} \tag{41}$$

$$e_r = (2r)^{-1} \sum_{j=0}^{r-1} H_{r-j}e_j \qquad (r \geq 1)$$

from which the e_j's can conveniently be calculated.

Robbins [60] gives a special case of (35) obtained by putting β equal to the geometric mean of $\lambda_1, \lambda_2, \ldots, \lambda_n$. Robbins and Pitman [61] give another special case, in which $\beta = \min_j \lambda_j$. If we wish to make $e_1 = 0$, then H_1 must be zero, and so β must be equal to the harmonic mean of $\lambda_1, \lambda_2, \ldots, \lambda_n$:

$$\beta = n\left[\sum_{j=1}^{n} \lambda_j^{-1}\right]^{-1}.$$

For certain values of β, (35) gives a mixture distribution (i.e. all $e_j > 0$). Ruben [65] has studied the problem of determining for what values of β the expansion (35) is a mixture. For $\beta > n\left[\sum_{j=1}^{n} \lambda_j^{-1}\right]^{-1}$, (35) is not a mixture distribution; for $0 < \beta \leq \min_j \lambda_j$, (35) is a mixture distribution. For other values of β, the answer depends on the relative magnitudes of the λ's.

Even when (35) is not a mixture, it is sometimes possible to find an upper bound for the error committed in using a finite number (N) of terms of (35) to approximate $F_n(y;\boldsymbol{\lambda})$. If

(42)
$$M = \max_j |1 - \beta/\lambda_j| < 1$$

then

(43)
$$\left|\sum_{j=N}^{\infty} e_j \Pr[\chi^2_{n+2j} \leq y/\beta]\right| \leq e_0 \cdot \frac{\Gamma(\frac{1}{2}n + N)}{\Gamma(\frac{1}{2}n)} \cdot \frac{M^N}{N!} (1 - M)^{-(\frac{1}{2}n+N)} \Pr[\chi^2_{n+2N} \leq (1 - M)y/\beta]$$

(see Ruben [63]).

For $\beta \leq \lambda_n$ (so that it is certain that there is a mixture), the bound is least when $\beta = \lambda_n$. For (non-mixture) distributions for which $\beta \geq \lambda_1$, the bound is least when $\beta = \lambda_1$. However $\beta = 2\lambda_1\lambda_n/(\lambda_1 + \lambda_n)$ gives much smaller values than $\beta = \lambda_1$ or $\beta = \lambda_n$. Ruben [63] gives a full discussion of these and allied points.

Formulas (41) make it possible to compute the values of the e_j's from the quantities

(44)
$$H_r = tr(\mathbf{I} - \beta\mathbf{A}^{-1}\mathbf{V}^{-1})^r$$

(cf. (21)) without explicit evaluation of the eigenvalues $\lambda_1, \lambda_2, \ldots, \lambda_n$. It is true that using (44) requires inversion of the matrix **VA**, but this is usually less trouble than evaluation of $\lambda_1, \lambda_2, \ldots, \lambda_n$.

5.3 *Laguerre Series Expansions*

If $\lambda_1 = \lambda_2 = \cdots = \lambda_n = \lambda$ then $Q(\mathbf{W}) = \lambda \sum_{j=1}^{n} W_j^2$ is distributed as $\lambda \cdot$ (central χ^2 with n degrees of freedom). If the ratios $\lambda_1 : \lambda_2 : \ldots : \lambda_n$ are not far from 1, it would seem natural to use expansions for $p_n(y;\boldsymbol{\lambda})$ and $F_n(y;\boldsymbol{\lambda})$ with the first term corresponding to the distribution of a multiple of a single central χ^2. We now describe how such expansions may be constructed by expressing the ratio

$$p_n(y;\lambda_1,\lambda_2,\ldots,\lambda_n)/p_n(y;\beta,\beta,\ldots,\beta)$$

as a series of Laguerre polynomials associated with the denominator, where β is a constant which can be chosen arbitrarily. (Note that $p_n(y;\beta,\beta,\ldots,\beta) = \beta p_n(y/\beta;\mathbf{1})$ where $\mathbf{1}$ denotes the vector $(1,1,\ldots,1)$ and $p_n(y;\mathbf{1})$ is the probability density function of χ_n^2.)

Hotelling [33] indicated, in an abstract, that he had used this method, and Grad and Solomon [21] gave a few more details of Hotelling's method. However, neither publication gives explicit formulas for the coefficients, nor are conditions for convergence of the expansions noted.

Conditions for convergence can be obtained from formulas given by Gurland [27], whose approach we will follow, though we express the coefficients in a different form which seems to be rather more convenient for our purposes.

The expansion will be in terms of *generalized Laguerre polynomials* associated with the weight function $y^\gamma e^{-y}$ as defined (Szegö [81], p. 370) by the equation

$$L_r^{(\gamma)}(y) = \frac{1}{r!}y^{-\gamma}e^y\frac{d^r}{dy^r}(y^{r+\gamma}e^{-y}) = \sum_{j=0}^{r}\frac{(-y)^j}{j!(r-j)!}\cdot\frac{\Gamma(r+\gamma+1)}{\Gamma(j+\gamma+1)} \quad \text{for } \gamma > -1. \tag{45}$$

The polynomials possess the orthogonality properties

$$\int_0^\infty y^\gamma e^{-y}L_r^{(\gamma)}(y)L_{r'}^{(\gamma)}(y)\,dy = \begin{cases} 0 & (r \neq r') \\ \dfrac{\Gamma(r+\gamma+1)}{r!} & (r = r'). \end{cases} \tag{46}$$

We seek an expansion of $p_n(y;\lambda)$ in the form

$$p_n(y;\boldsymbol{\lambda}) = \beta^{-1}p_n(y/\beta;\mathbf{1})\sum_{j=0}^{\infty}b_jL_j^{(\frac{1}{2}n-1)}(\tfrac{1}{2}y/\beta). \tag{47}$$

We first rearrange $\phi_n(t;\boldsymbol{\lambda})$ in the following way:

$$\begin{aligned}\phi_n(t;\boldsymbol{\lambda}) &= \prod_{j=1}^{n}(1-2it\lambda_j)^{-\frac{1}{2}} \\ &= (1-2it\beta)^{-\frac{1}{2}n}\prod_{j=1}^{n}\left[1-\frac{2it(\beta-\lambda_j)}{1-2it\beta}\right]^{-\frac{1}{2}} \\ &= \sum_{j=0}^{\infty}c_j(1-2it\beta)^{-\frac{1}{2}(n+2j)}(-2it)^j.\end{aligned} \tag{48}$$

The coefficients c_j are obtained from the identity (in s)

$$\prod_{j=1}^{n}[1-(\beta-\gamma_j)s]^{-\frac{1}{2}} \equiv \sum_{j=0}^{n}c_js^j. \tag{49}$$

The expansion (48) is valid, and the series is uniformly convergent for all t for which

(50) $$\left|\frac{2it(\beta - \lambda_j)}{1 - 2it\beta}\right| < 1$$

for all λ_j. This is true for *all* real t if $\beta > \frac{1}{2}\max_j \lambda_j = \frac{1}{2}\lambda_1$. If this be so, then Fourier's inversion formula can be applied to (48), leading to

(51) $$p_n(y;\boldsymbol{\lambda}) = \frac{1}{2\pi}\int_{-\infty}^{\infty} e^{-ity}\left[\sum_{j=0}^{\infty} c_j(1 - 2it\beta)^{-\frac{1}{2}(n+2j)}(-2it)^j\right] dt$$
$$= \frac{1}{\beta} p_n(y/\beta;\mathbf{1}) \sum_{j=0}^{\infty} c_j\beta^{-j} \frac{j!\Gamma(\frac{1}{2}n)}{\Gamma(\frac{1}{2}n + j)} L_j^{(\frac{1}{2}n-1)}(\tfrac{1}{2}y/\beta).$$

(The interchange is justified by the absolute uniform convergence of (48).) This is of the required form (47) with

(52) $$b_j = c_j\beta^{-j}j!\Gamma(\tfrac{1}{2}n)/\Gamma(\tfrac{1}{2}n + j).$$

To obtain convenient formulas from which the c_j's (and hence the b_j's) may be calculated we use a method similar to that employed in Section 5.2. Analogously to (39) we have

(53) $$\prod_{j=1}^{n} [1 - (\beta - \lambda_j)s]^{\frac{1}{2}} = \sum_{j=0}^{\infty} \frac{\mu_j'(Q;\beta - \lambda_1,\ldots,\beta - \lambda_n)}{2^j j!} s^j,$$

where $\mu_j'(Q;\beta - \lambda_1,\ldots,\beta - \lambda_n) = E\left[\left\{\sum_{j=1}^{n} (\beta - \lambda_j)W_j^2\right\}^j\right]$. Equating coefficients in (49) and (53),

(54) $$c_r = \frac{\mu_r'(Q;\beta - \lambda_1,\ldots,\beta - \lambda_n)}{2^r r!}$$

where we must have $\beta > \frac{1}{2}\lambda_1$. Using the relation between moments about zero, and cumulants as in Section 5.2 we obtain the formulas:

(55) $$\begin{cases} c_0 = 1 \\ c_r = (2r)^{-1}\sum_{j=0}^{r-1} S_{r-j}c_j \qquad (r \geq 1) \end{cases}$$

where

$$S_r = \sum_{j=1}^{n} (\beta - \lambda_j)^r.$$

Computation is considerably simplified if

$$\kappa_1(Q;\beta - \lambda_1,\ldots,\beta - \lambda_n) = \mu_1'(Q;\beta - \lambda_1,\ldots,\beta - \lambda_n) = 0$$

so that the moments (μ_r') about zero are also central moments (μ_r). In order to arrange that this is so, it is necessary to make

$$\sum_{j=1}^{n} (\beta - \lambda_j) = 0$$

that is

$$\beta = n^{-1} \sum_{j=1}^{n} \lambda_j. \tag{56}$$

This simplification can be used only if

$$n^{-1} \sum_{j=1}^{n} \lambda_j > \tfrac{1}{2}\lambda_1. \tag{57}$$

Formula (56) is a natural choice for the value of β. If all λ's are equal then, with this choice,

$$\beta - \lambda_j = 0 \text{ for all } j$$

and the series expansion reduces to a single term (of central χ^2 form) which in this case corresponds to the exact distribution. However, sometimes (for example, when the λ values are widely scattered) choice of β according to (56) may lead to slower convergence than would some other value of β.

If β does satisfy (56), then putting $d_r = r!c_r$, we have the simple formulas

$$\begin{aligned} &d_0 = 1; \quad d_1 = 0; \quad d_2 = \tfrac{1}{2}S_2; \quad d_3 = S_3; \\ &d_4 = 3S_4 + \tfrac{3}{4}S_2^2; \quad d_5 = 12S_5 + 5S_3S_2. \end{aligned} \tag{58}$$

Grad and Solomon [21] used these six values to approximate to $F_n(y;\lambda)$ for $n = 2$ and $n = 3$. They rearranged the terms so that available χ^2 tables could be used efficiently. They found that with $n = 3$ and $(\lambda_1,\lambda_2,\lambda_3) = (1.2,0.9,0.9)$ the approximation is accurate to 4 decimal places. With $(\lambda_1,\lambda_2,\lambda_3) = (2.1,0.6,0.3)$ condition (57) is not satisfied, and, as might be expected, the approximation was found to be quite poor.

The quantities S_r can be evaluated without calculating the individual eigenvalues, since

$$S_r = tr(\beta\mathbf{I} - \mathbf{VA})^r. \tag{59}$$

An expression for $F_n(y;\boldsymbol{\lambda})$ can be obtained by using the relationship

$$\begin{aligned} (2\beta)^{-1} \int_0^y (\tfrac{1}{2}x/\beta)^{\frac{1}{2}n-1} \exp(-\tfrac{1}{2}x/\beta) L_j^{(\frac{1}{2}n-1)}(\tfrac{1}{2}x/\beta)\, dx \\ = j^{-1}(\tfrac{1}{2}y/\beta)^{\frac{1}{2}n} \exp(-\tfrac{1}{2}y/\beta) L_{j-1}^{(\frac{1}{2}n)}(\tfrac{1}{2}y/\beta) \qquad (j \geq 1). \end{aligned} \tag{60}$$

The expansion is

$$\begin{aligned} F_n(y;\boldsymbol{\lambda}) = \frac{1}{2^{\frac{1}{2}n}\Gamma(\frac{1}{2}n)} \int_0^{y/\beta} x^{\frac{1}{2}n-1} e^{-\frac{1}{2}x}\, dx \\ + \sum_{j=1}^{\infty} \frac{(j-1)!c_j}{\beta^j \Gamma(\frac{1}{2}n + j)} \left(\frac{y}{2\beta}\right)^{\frac{1}{2}n} e^{-\frac{1}{2}y/\beta} L_{j-1}^{(\frac{1}{2}n)}(\tfrac{1}{2}y/\beta) \qquad (y > 0). \end{aligned} \tag{61}$$

This formula was given by Gurland [27], who showed that (51) and (61) are uniformly convergent for all $y \geq 0$ if $\lambda_1 \leq 3\lambda_n$. Kotz *et al.* [41] removed this last condition, and also gave bounds for function error.

5.4 *Expansion in Maclaurin Series*

$F_n(y;\boldsymbol{\lambda})$ and $p_n(y;\boldsymbol{\lambda})$ can each be expanded in a series of powers of y. These are obtained from the Laplace transform of $p_n(y;\lambda)$. Provided the real part of τ exceeds $\max_j |(2\lambda_j)^{-1}|$,

$$\text{(62)} \qquad \int_0^\infty e^{-\tau y} p_n(y;\boldsymbol{\lambda})\,dy = \prod_{j=1}^n (1 + 2\tau\lambda_j)^{-\frac{1}{2}}$$
$$= \prod_{j=1}^n (2\tau\lambda_j)^{-\frac{1}{2}} \prod_{j=1}^n (1 + (2\tau\lambda_j)^{-1})^{-\frac{1}{2}}.$$

Again, using the method of Sections 5.2 and 5.3, identifying the last product with a moment generating function, we have:

$$\text{(63)} \qquad \int_0^\infty e^{-\tau y} p_n(y;\boldsymbol{\lambda})\,dy$$
$$= \left(\prod_{j=1}^n \lambda_j\right)^{-\frac{1}{2}} \sum_{j=0}^\infty \frac{(-1)^j \mu_j'(Q;\lambda_1^{-1},\ldots,\lambda_n^{-1})}{j!\,2^{\frac{1}{2}(n+4j)}\tau^{\frac{1}{2}(n+2j)}}$$
$$= \left(\prod_{j=1}^n \lambda_j\right)^{-\frac{1}{2}} \sum_{j=0}^\infty \frac{(-1)^j \mu_j'(Q;\lambda_1^{-1},\ldots,\lambda_n^{-1})}{j!\,2^{\frac{1}{2}(n+4j)}} \int_0^\infty \frac{e^{-\tau y} y^{\frac{1}{2}(n+2j)-1}}{\Gamma(\frac{1}{2}n + j)}\,dy$$
$$= \int_0^\infty e^{-\tau y}\left[\left(\prod_{j=1}^n \lambda_j\right)^{-\frac{1}{2}} \sum_{j=0}^\infty \frac{(-1)^j \mu_j'(Q;\lambda_1^{-1},\ldots,\lambda_n^{-1})}{j!\,2^{\frac{1}{2}n+2j}\Gamma(\frac{1}{2}n + j)}\, y^{\frac{n}{2}+j-1}\right] dy$$

where the interchange of order of summation and integration is justified by Fubini's theorem (remembering $\operatorname{Re}\tau > \max_j |(2\lambda_j)^{-1}|$; see also (71)). By the uniqueness theorem for Laplace transforms

$$\text{(64)} \qquad p_n(y;\boldsymbol{\lambda}) = \left(\prod_{j=1}^n \lambda_j\right)^{-\frac{1}{2}} \sum_{j=0}^\infty \frac{(-1)^j \mu_j'(Q;\lambda_1^{-1},\ldots,\lambda_n^{-1})}{j!\,2^{j+1}\Gamma(\frac{1}{2}n + j)} \left(\frac{y}{2}\right)^{\frac{1}{2}n+j-1}.$$

The series in (64) converges uniformly in any bounded interval of y, and so

$$\text{(65)} \qquad F_n(y;\boldsymbol{\lambda}) = \left(\prod_{j=1}^n \lambda_j\right)^{-\frac{1}{2}} (\tfrac{1}{2}y)^{\frac{1}{2}n} \sum_{j=0}^\infty \frac{A_j}{\Gamma(\frac{1}{2}n + j + 1)} \left(-\frac{y}{2}\right)^j$$

where

$$\text{(66)} \qquad A_j = \mu_j'(Q;\lambda_1^{-1},\ldots,\lambda_n^{-1})/(2^j j!).$$

Pachares [56] derived (65) by direct calculation. Both (64) and (65) were given by Robbins [60], though the coefficients are in a more complicated form.

As in Sections 5.2 and 5.3, recurrence relationships may be obtained for the coefficients in (64) and (65). We have

$$A_0 = 1;\ A_r = (2r)^{-1} \sum_{j=0}^{r-1} V_{r-j} A_j \qquad (r \geq 1) \tag{67}$$

where

$$V_r = \sum_{j=1}^{n} \lambda_j^{-r} = tr(\mathbf{VA})^{-r}.$$

Bounds on A_r may be obtained as follows: Since $\lambda_n = \min_j \lambda_j$,

$$\sum_{j=1}^{n} \lambda_j^{-1} W_j^2 \leq \lambda_n^{-1} \sum_{j=1}^{n} W_j^2 \tag{68}$$

so that

$$\mu_r'(Q;\lambda_1^{-1},\ldots,\lambda_n^{-1}) \leq \lambda_n^{-r}\lambda_r'(Q;1,\ldots,1) = \left(\frac{2}{\lambda_n}\right)^r \cdot \frac{\Gamma(\frac{1}{2}n + r)}{\Gamma(\frac{1}{2}n)}. \tag{69}$$

Hence

$$A_r \leq \frac{\Gamma(\frac{1}{2}n + r)}{r!\,\Gamma(\frac{1}{2}n)\lambda_n^r}. \tag{70}$$

From this inequality it follows that

$$p_n(y;\boldsymbol{\lambda}) \leq \frac{1}{2\Gamma(\frac{1}{2}n)} \left(\prod_{j=1}^{n} \lambda_j\right)^{-\frac{1}{2}} \left(\frac{y}{2}\right)^{\frac{1}{2}n-1} \exp\left(\frac{y}{2\lambda_n}\right) \tag{71}$$

(justifying the application of Fubini's theorem, and hence taking the Laplace transform term-by-term in (63)).

The series expansion (65) appears to be useful for computation only for small values of y.

5.5 *Integral Equations for $p_n(y;\lambda)$ and $F_n(y;\lambda)$*

The following analysis is due to Grenander *et al.* [22]. Differentiating (16) logarithmically,

$$-i\phi'(y;\boldsymbol{\lambda}) = \phi(y;\boldsymbol{\lambda}) \sum_{j=1}^{n} \frac{\lambda_j}{1 - 2it\lambda_j}. \tag{72}$$

Applying the inverse Fourier transform,

$$y p_n(y;\boldsymbol{\lambda}) = \int_0^y p_n(y - x) h(x)\, dx \tag{73}$$

where

$$h(x) = \frac{1}{2} \sum_{j=1}^{n} e^{-x/(2\lambda_j)},$$

or equivalently,

$$(74) \qquad yF_n(y;\boldsymbol{\lambda}) = \int_0^y F_n(y - x;\boldsymbol{\lambda})H(x)\,dx$$

where

$$H(x) = 1 + h(x),$$

we obtain integral equations which may be used as a basis for numerical methods of calculating $p_n(y;\boldsymbol{\lambda})$ and $F_n(y;\boldsymbol{\lambda})$. Details are given by Grenander *et al.* [22].

5.6 *A Recurrence Relation*

It follows from (14) that

$$F_n(y;\boldsymbol{\lambda}) = (2\pi)^{-\frac{1}{2}n} \int_{\Sigma\lambda_j w_j^2 \le y} \exp\left(-\frac{1}{2}\sum_{j=1}^n w_j^2\right) d\mathbf{w}.$$

The intersection of the hyperplane $w_n = w$ with the n-dimensional ellipsoid $\sum_{j=1}^n \lambda_j w_j^2 \le y$ is an $(n-1)$-dimensional ellipsoid defined by

$$(75) \qquad \sum_{j=1}^{n-1} \lambda_j w_j^2 \le y - \lambda_n w^2.$$

Hence

$$(76) \qquad F_n(y;\boldsymbol{\lambda}) = (2\pi)^{-\frac{1}{2}} \int_{-\sqrt{y/\lambda_n}}^{\sqrt{y/\lambda_n}} e^{-\frac{1}{2}w^2} F_{n-1}(y - \lambda_n w^2;\boldsymbol{\lambda})\,dw.$$

Applying the transformation $u = w\sqrt{\lambda_n/y}$, we obtain the relation:

$$(77) \qquad F_n(y;\boldsymbol{\lambda}) = 2\sqrt{y/\lambda_n} \int_0^1 (2\pi)^{-\frac{1}{2}} e^{-\frac{1}{2}(yu^2/\lambda_n)} F_{n-1}(y(1-u^2);\boldsymbol{\lambda})\,du.$$

5.7 *Probability Content of an Ellipse*

The distribution function $F_2(y;\lambda_1,\lambda_2)$ has important applications, as it is the probability that the point (W_1,W_2) falls within the ellipse $\lambda_1 W_1^2 + \lambda_2 W_2^2 = y$, when (W_1,W_2) are independent unit normal variables. There is an interesting and useful relationship between $F_2(y;\lambda_1,\lambda_2)$ and the certain offset-circle probabilities (discussed in Section 8), and so with the noncentral χ^2 distribution (Chapter 28). The relation is

$$(78) \qquad F_2(y;\lambda_1,\lambda_2) = \Pr[\chi_2'^2(r^2) \le R^2] - \Pr[\chi_2'^2(R^2) \le r^2]$$

where

$$R = \tfrac{1}{2}[(y/\lambda_1)^{\frac{1}{2}} + (y/\lambda_2)^{\frac{1}{2}}];\ r = \tfrac{1}{2}[(y/\lambda_2)^{\frac{1}{2}} - (y/\lambda_1)^{\frac{1}{2}}].$$

(We take $\lambda_1 \geq \lambda_2$.)

To establish this result, first note that the ellipse $\lambda_1 W_1^2 + \lambda_2 W_2^2 = y$ has semi-axes $(y/\lambda_2)^{\frac{1}{2}} = R + r$ and $(y/\lambda_1)^{\frac{1}{2}} = R - r$ respectively. Hence (78) states that the probabilities of falling within E (an ellipse) and A (a circle of radius R *minus* a circle of radius r) in Figures 2 and 3 are equal. We now note the length of arc of the circle with center at the origin and radius ρ which falls in E is equal to the length which falls in A. (For example, for $\rho < R - r$ the length is $2\pi\rho$ and for $R + r < \rho$ the length is zero.) Hence

$$(79)\qquad (2\pi)^{-1}\iint_E \exp\left(-\tfrac{1}{2}(W_1^2 + W_2^2)\right) dW_1\, dW_2 = (2\pi)^{-1}\iint_A \exp\left(-\tfrac{1}{2}(W_1^2 + W_2^2)\right) dW_1\, dW_2.$$

The result (78) is stated without proof by Ruben [63] and appears in Grubbs [24]. Ruben states that there is a proof due to D. C. Kleinecke [63] (p. 612).

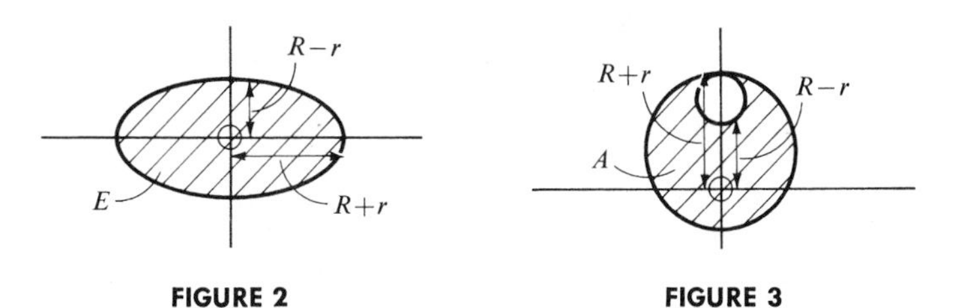

FIGURE 2 FIGURE 3

5.8 *Approximations*

If the λ_j's are bounded, then by the central limit theorem $F_n(y;\boldsymbol{\lambda})$ (standardized) is approximated by a normal distribution as n tends to infinity. Such approximation is usually not very good. A better approximation is $Q(\mathbf{W})$ approximately distributed as $\beta\chi_\nu^2$ with β, ν chosen to make the first two moments agree with those of $F_n(y;\boldsymbol{\lambda})$. The degrees of freedom, ν, are usually fractional. Satterthwaite [66], [67] has checked the accuracy of this approximation numerically. If $\alpha + \beta\chi_\nu^2$ be used instead of $\beta\chi_\nu^2$, definite further improvement might be expected, analogous to the results of Pearson for the noncentral χ^2 distribution (Chapter 28, Section 8). Harvey [31] extended Patnaik's and Pearson's forms of approximation by replacing the central χ^2 by a noncentral χ^2 (introducing one extra parameter). Harvey suggests using the extension of Patnaik's approximation when it is possible to find values β, ν and λ such that a $\chi_\nu'^2(\lambda)$ has the same first three moments as Q. The condition for this is

$$2[\mu_2(Q)]^2 - \mu_1'(Q)\mu_3(Q) \geq 0.$$

The appropriate values of the parameters are

$$\beta = \tfrac{1}{2}\mu_3[2\mu_2 + \sqrt{4\mu_2^2 - 2\mu_1'\mu_3}]^{-1}$$

$$\nu = \frac{1}{2\beta}\left[4\mu_1' - \frac{\mu_2}{\beta}\right]$$

$$\lambda = \frac{1}{2\beta}\left[\frac{\mu_2}{\beta} - 2\mu_1'\right].$$

The condition for applying the *original* form of Pearson's approximation $(a\chi_\nu^2 + b)$ with $b > 0$ is the complement of this, and Harvey suggests using the extended Patnaik, or original Pearson, approximation according as

$$2[\mu_2(Q)]^2 - \mu_1'(Q)\mu_3(Q) \gtrless 0.$$

Other approximations are based on the fact that when the λ_j's are equal in pairs (i.e. n even and $\lambda_1 = \lambda_2 \geq \lambda_3 = \lambda_4 \geq \cdots \geq \lambda_{n-1} = \lambda_n$) then $F(y;\boldsymbol{\lambda})$ can be expressed as a finite sum of elementary functions. Now, remembering that (in the general case) $\lambda_1 \geq \lambda_2 \geq \cdots \geq \lambda_n$, we have (for n even)

$$\sum_{j=1}^{n/2} \lambda_{2j-1}(W_{2j-1}^2 + W_{2j}^2) \geq \sum_{j=1}^{n} \lambda_j W_j^2 \geq \sum_{j=1}^{n/2} \lambda_{2j}(W_{2j-1}^2 + W_{2j}^2). \tag{80}$$

It follows that

$$F_n(y;\lambda_1,\lambda_1,\lambda_3,\lambda_3,\ldots,\lambda_{n-1},\lambda_{n-1}) \leq F_n(y;\lambda) \leq F_n(y;\lambda_2,\lambda_2,\lambda_4,\lambda_4,\ldots,\lambda_n,\lambda_n) \tag{81}$$

The two extreme terms in (81) can be evaluated as finite sums of elementary functions (e.g. Soest [75]). A similar analysis is possible if n is odd. The accuracy of approximation decreases with the size of the differences $\lambda_1 - \lambda_2$, $\lambda_3 - \lambda_4, \ldots$.

An improvement of these approximations has been suggested by Siddiqui [72]. He proposed (for n even) the approximation

$$\tilde{F}_n(y;\boldsymbol{\lambda}) = \theta F_n(y;\lambda_1,\lambda_1,\ldots,\lambda_{n-1},\lambda_{n-1}) + (1-\theta)F_n(y;\lambda_2,\lambda_2,\ldots,\lambda_n,\lambda_n) \tag{82}$$

where $\theta(0 \leq \theta \leq 1)$ is chosen in a suitable way. Siddiqui suggests choosing θ to minimize

$$\int_0^\infty |\tilde{F}_n(y;\boldsymbol{\lambda}) - F_n(y;\boldsymbol{\lambda})|^2\, dy.$$

The necessary calculations are straightforward, though tedious.

If the matrix **VA** is a Toeplitz matrix, or approximates such a matrix, Szegö's theorem on the asymptotic distribution of the eigenvalues of Toeplitz matrices can be used to some advantage [22].

Okamoto [53] has shown that

$$F_n(y;\boldsymbol{\lambda}) \leq F_n(y/G(\boldsymbol{\lambda});\mathbf{1})$$

where $G(\boldsymbol{\lambda})$ is the geometric mean $\left[\prod_{j=1}^{n} \lambda_j\right]^{1/n}$. Numerical investigation (for the cases $n = 2$ and $n = 3$) by Siotani [73] showed that this upper bound is, in fact, quite a good approximation, particularly when the value of $F_n(y;\boldsymbol{\lambda})$ is not too large (≤ 0.5) and when the λ_i's do not differ too much among themselves. On the basis of his numerical investigations, Siotani suggested the approximation

$$F_n(y;\boldsymbol{\lambda}) \doteqdot F_n(yh(\boldsymbol{\lambda});\mathbf{1})$$

where $h(\boldsymbol{\lambda}) = [G(\boldsymbol{\lambda})]^{-1}[H(\boldsymbol{\lambda})/G(\boldsymbol{\lambda})]^2$, $H(\boldsymbol{\lambda})$ being the harmonic mean

$$n\left[\sum_{j=1}^{n} \lambda_j^{-1}\right]^{-1}.$$

This approximation is especially useful for larger values of $F_n(y;\boldsymbol{\lambda})(\geq 0.8$ say). For smaller values, the approximation

$$F_n(y;\boldsymbol{\lambda}) \doteqdot F_n(yh_1(\boldsymbol{\lambda});\mathbf{1})$$

with $h_1(\lambda) = [G(\boldsymbol{\lambda})]^{-1}[H(\boldsymbol{\lambda})/G(\boldsymbol{\lambda})]$, appears preferable. These approximations are based on empirical rather than analytical, arguments (though Okamoto's inequality is rigorously established).

5.9 *Limiting Distribution*

We now consider the limiting distribution, as n tends to infinity, of

$$Q_{(n)}(\mathbf{Z}) = \sum_{i=1}^{n} \sum_{j=1}^{n} a_{ij}Z_iZ_j$$

where the Z's are independent unit normal variables. Varberg [84] has pointed out that if the limit of $\sum_{i=1}^{n} \sum_{j=1}^{n} a_{ij}^2$ is finite, then the $Q_{(n)}(\mathbf{Z})$'s do have a limiting distribution.

He further obtains the formula

$$\Pr\left[\lim_{n\to\infty} Q_{(n)}(\mathbf{Z}) \geq z\right]$$
$$= \frac{\prod_{r=2}^{\infty} (1 - \sigma_r/\sigma_1)^{-\frac{1}{2}n_r}}{\Gamma(\frac{1}{2}n_1)} \left(\frac{z}{2\sigma_1}\right)^{\frac{1}{2}n_1-1} \left[\exp\left(-\frac{z}{2\sigma_1}\right)\right][1 + \epsilon(z)]$$

where $\lim_{z\to\infty} \epsilon(z) = 0$; $\sigma_1, \sigma_2, \ldots$ are distinct eigenvalues of a certain infinite matrix, (in effect a matrix with a_{ij} in the i-th row and j-th column) and $n_1, n_2, \ldots$ are their respective multiplicities.

Whittle [87] has given the following sufficient set of conditions for the limiting distribution of $Q_{(n)}(\mathbf{Z})$ to be normal when the Z_i's are not necessarily unit normal variables (though they are independent and each has zero mean).

(i) $E[|Z_j|^{4+2\delta}]$ is finite and $E[|Z_j|^{4+2\delta}]/\{\mathrm{var}\,(Z_j)\}^{2+\delta}$ is uniformly bounded for some δ, $0 < \delta < 1$.

(ii) $\beta_2(Z_j) \geq 1 + d$ for some $d > 0$ and all j.

(iii) The integers $j = 1, 2, \ldots, n$ can be grouped into a finite number of sets $g_1, g_2, \ldots, g_r$ such that

$$\text{(a)} \lim_{n\to\infty} \frac{\|A\|^2 - \sum_{j=1}^{r} \|A_j\|^2}{\|A\|^2} = 0 \qquad \left(\lim_{n\to\infty} \frac{\sum_{j=1}^{r} \|A_j\|^2}{\|A\|^2} = 1 \right)$$

and

$$\text{(b)} \lim_{r\to\infty} \sum_{j=1}^{r} \left[\frac{\|A_j\|}{\|A\|} \right]^{2+\delta} = 0$$

where $\|A_j\|$, $\|A\| = (\sum \sum a_{il}^2 \,\mathrm{var}(Z_i)\,\mathrm{var}(Z_l))^{\frac{1}{2}}$, the summations being over i and l in g_j, and over all i and l, respectively.

5.10 *Another Inversion of the Characteristic Function*

By inversion of the characteristic function it is possible to express the cumulative distribution as a definite integral in a single variable:

$$\Pr[Y \leq y] = 2\pi^{-1} \int_0^\infty t^{-1} \sin\left(\tfrac{1}{2}ty\right) \cos\left(-ty + \tfrac{1}{2} \sum_{j=1}^{n} \tan^{-1}(2\lambda_j t)\right) \prod_{j=1}^{n} (1 + 4\lambda_j^2 t^2)^{-\frac{1}{4}}\, dt.$$

Feiveson and Delaney [15], who give this formula, used it to calculate exact values by quadrature for comparison with the approximation by a gamma distribution starting at zero and having the same first two moments.

6. Positive Definite Forms — Noncentral Case

We now consider the distribution of $Q(\mathbf{W}) = \sum_{j=1}^{n} \lambda_j (W_j - \omega_j)^2$ with some, at least, of the ω_j's not equal to zero. As in Section 5, we require $\lambda_1 \geq \lambda_2 \geq \cdots \geq \lambda_n > 0$, that is, that $Q(\mathbf{W})$ be positive definite. We seek convenient expressions for the cumulative distribution function $F_n(y;\boldsymbol{\lambda};\boldsymbol{\omega})$ and the probability density function $p_n(y;\boldsymbol{\lambda};\boldsymbol{\omega})$.

The types of expansion (Sections 5.2–5.4) which have been discussed for the central case can also be used in the noncentral case. Expansions of $F_n(y;\boldsymbol{\lambda};\boldsymbol{\omega})$ in a series of central χ^2 probabilities, and also in a series of noncentral χ^2 probabilities, have been given by Ruben [64], [65], who also provides convenient recurrence relations for the coefficients. The power series expansion (65) of Pachares was extended to the noncentral case by Shah and Khatri [70], but the expressions they give for the coefficients do not seem very convenient for computation. Gurland's Laguerre series expansion (52) was extended to the noncentral case by Shah [68] but here again it appears troublesome to calculate the coefficients. However, employing the characteristic function in the same way as in Section 5, it is possible to obtain convenient formulas for calculating coefficients in these two series. The discussion which follows is much briefer than in the corresponding parts of Section 5, which should be consulted for more detail.

6.1 *Expansion as Mixtures of Central χ^2 Distributions*

We first seek to obtain expansions of the form

$$F_n(y;\boldsymbol{\lambda};\boldsymbol{\omega}) = \sum_{j=0}^{\infty} e'_j \Pr[\chi^2_{n+2j} < y/\beta] \tag{83}$$

where $\beta(> 0)$ is a suitably chosen constant. From (15) the characteristic function of $F_n(y;\boldsymbol{\lambda};\boldsymbol{\omega})$ is

(84.1)

$$\phi_n(t;\boldsymbol{\lambda};\boldsymbol{\omega}) = \exp\left(-\frac{1}{2}\sum_{j=1}^{n}\omega_j^2\right)\exp\left(\frac{1}{2}\sum_{j=1}^{n}\frac{\omega_j^2}{1-2it\lambda_j}\right)\prod_{j=1}^{n}(1-2it\lambda_j)^{-\frac{1}{2}}$$

or equivalently

$$\phi_n(t;\boldsymbol{\lambda};\boldsymbol{\omega}) = \exp\left(\sum_{j=1}^{n}\frac{it\lambda_j\omega_j^2}{1-2it\lambda_j}\right)\prod_{j=1}^{n}(1-2it\lambda_j)^{-\frac{1}{2}}. \tag{84.2}$$

To obtain an expansion of form (83), $\phi_n(t;\boldsymbol{\lambda};\boldsymbol{\omega})$ must be expanded in a series of powers of $(1 - 2it\beta)^{-1}$. We put $\theta = (1 - 2it\beta)^{-1}$ in (84.2) giving

$$\begin{aligned}\phi_n\left(\frac{\theta-1}{2i\beta\theta};\boldsymbol{\lambda};\boldsymbol{\omega}\right) &= \exp\left[-\frac{1-\theta}{2}\sum_{j=1}^{n}\frac{\omega_j^2}{1-\theta(1-\beta/\lambda_j)}\right] \\ &\quad\times\prod_{j=1}^{n}\left(\frac{\beta\theta}{\lambda_j}\right)^{\frac{1}{2}}\left\{1-\left(1-\frac{\beta}{\lambda_j}\right)\theta\right\}^{-\frac{1}{2}} \\ &= \theta^{\frac{1}{2}n}\sum_{j=0}^{\infty}e'_j\theta^j\end{aligned} \tag{85}$$

provided the right-hand side can be expanded as a power series in θ, that is

$$|\theta| < \min_j |1 - \beta/\lambda_j|^{-1}. \tag{86}$$

This condition is satisfied for all real t. Inverting (85) leads to the required expansion (83).

In order to obtain convenient formulas for the e'_j, we write (from (85))

(87)

$$\begin{aligned}\log\phi_n\left(\frac{\theta-1}{2i\beta\theta};\boldsymbol{\lambda};\boldsymbol{\omega}\right) &= -\frac{1-\theta}{2}\sum_{j=1}^{n}\frac{\omega_j^2}{1-(1-\beta/\lambda_j)\theta}+\tfrac{1}{2}\sum_{j=1}^{n}\log(\beta/\lambda_j) \\ &\quad-\tfrac{1}{2}\sum_{j=1}^{n}\log\{1-(1-\beta/\lambda_j)\theta\} = \sum_{j=0}^{\infty}G_j\theta^j\end{aligned}$$

with

$$G_r = \sum_{j=1}^{n} (1 - \beta/\lambda_j)^r + r\beta \sum_{j=1}^{n} (\omega_j^2/\lambda_j)(1 - \beta/\lambda_j)^{r-1}. \tag{88}$$

The e_j' may be calculated from the formulas

$$\begin{cases} e_0' = \left[\exp\left(-\tfrac{1}{2}\sum_{j=1}^{n}\omega_j^2\right)\right]\prod_{j=1}^{n}(\beta/\lambda_j)^{\frac{1}{2}} \\ e_r' = \dfrac{1}{2r}\sum_{j=0}^{r-1} G_{r-j}e_j' \qquad (r \geq 1). \end{cases} \tag{89}$$

These formulas were given by Ruben [64], who also showed that the series (83) is uniformly convergent over any finite interval of y.

Ruben also gave the following bound for error in truncating the series (83) at the Nth term:

$$\begin{aligned} &\left|\sum_{j=N}^{\infty} e_j' \Pr[\chi_{n+2j}^2 \leq y/\beta]\right| \\ &\qquad \leq e_0' \frac{\Gamma(\frac{1}{2}n + N)}{\Gamma(\frac{1}{2}n)} \frac{M^N}{N!} (1 - M)^{-\frac{1}{2}(n+2N)} \Pr[\chi_{n+2N}^2 \leq (1 - M)y/\beta] \end{aligned} \tag{90}$$

where $M = \frac{1}{2}\beta \sum_{j=1}^{n} (\omega_j^2/\lambda_j) + \max_j |1 - \beta/\lambda_j|$, provided $M < 1$.

6.2 *Expansion as Mixtures of Noncentral* χ^2 *Distributions*

We now write $\phi_n(t;\boldsymbol{\lambda};\boldsymbol{\omega})$ in the form

(91)

$$\begin{aligned} \phi_n(t;\boldsymbol{\lambda};\boldsymbol{\omega}) &= \exp\left(\frac{it\beta\sum_{j=1}^{n}\omega_j^2}{1 - 2it\beta}\right)\exp\left[\frac{it}{1 - 2it\beta}\sum_{j=1}^{n}\frac{\omega_j^2(\lambda_j - \beta)}{1 - 2it\lambda_j}\right]\prod_{j=1}^{n}(1 - 2it\lambda_j)^{-\frac{1}{2}} \\ &= \exp\left(\frac{it\beta\sum_{j=1}^{n}\omega_j^2}{1 - 2it\beta}\right)\sum_{j=0}^{\infty} e_j''(1 - 2it\beta)^{-\frac{1}{2}(n+2j)} \end{aligned}$$

where the e_j'' are defined by the identity

$$\begin{aligned} \theta^{\frac{1}{2}n}\sum_{j=0}^{\infty} e_j''\theta^j \equiv \exp&\left[-\frac{\theta(1 - \theta)}{2}\sum_{j=1}^{n}\frac{\omega_j^2(1 - \beta/\lambda_j)}{1 - (1 - \beta/\lambda_j)\theta}\right] \\ &\times \prod_{j=1}^{n}\left(\frac{\beta\theta}{\lambda_j}\right)^{\frac{1}{2}}[1 - \theta(1 - \beta/\lambda_j)]^{-\frac{1}{2}}. \end{aligned} \tag{92}$$

Putting

(93.1) $$h_1 = \sum_{j=1}^{n} (1 - \omega_j^2)(1 - \beta/\lambda_j)$$

(93.2) $$h_r = G_r \qquad (r \geq 2)$$

(G_r being defined in (88)), and using an argument similar to that used in 6.1, we obtain

(94) $$\begin{cases} e_0'' = \prod_{j=1}^{n} (\beta/\lambda_j)^{\frac{1}{2}} \\ e_r'' = (2r)^{-1} \sum_{j=0}^{r-1} h_{r-j} e_j'' \qquad (r \geq 1). \end{cases}$$

Inverting (91) we obtain

(95) $$F_n(y;\boldsymbol{\lambda};\boldsymbol{\omega}) = \sum_{j=0}^{\infty} e_j'' \Pr\left[\chi_{n+2j}'^2\left(\sum_{j=1}^{n} \omega_j^2\right) \leq y/\beta\right].$$

Ruben [64] gave formulas (94) and (95), and showed that the series in (95) is uniformly convergent over any finite interval of y. He also gave the following bound for error in truncating the series (95) at the Nth term:

(96) $$\left|\sum_{j=N}^{\infty} e_j'' \Pr\left[\chi_{n+2j}'^2\left(\sum_{j=1}^{n} \omega_j^2\right) \leq y/\beta\right]\right|$$
$$\leq e_0'' \frac{\Gamma(\frac{1}{2}n + N)}{\Gamma(\frac{1}{2}n)} \frac{M'^N}{N!} (1 - M')^{-\frac{1}{2}(n+2N)}$$
$$\times \Pr\left[\chi_{n+2N}'^2\left(\sum_{j=1}^{n} \omega_j^2\right) \leq (1 - M')y/\beta\right]$$

where $M' = M + \frac{1}{2}\sum_{j=1}^{n} \omega_j^2$, provided $M' < 1$. (M is as defined in (90).)

6.3 *Laguerre Series Expansions*

We now seek an expansion (analogous to (47)) of the form

(97) $$p_n(y;\boldsymbol{\lambda};\boldsymbol{\omega}) = \beta^{-1} p_n(y/\beta;\mathbf{1}) \sum_{j=0}^{\infty} b_j' L_j^{(\frac{1}{2}n-1)}(\tfrac{1}{2}y/\beta)$$

where β is a positive constant.

Putting $\theta' = -2it\beta(1 - 2it\beta)^{-1}$ in (84.2) we obtain:

(98) $$\phi_n(t;\boldsymbol{\lambda};\boldsymbol{\omega}) = (1 - 2it\beta)^{-\frac{1}{2}n} \sum_{j=0}^{\infty} c_j' \beta^{-j} \theta'^j$$

(provided $|\theta'| < \min_j |1 - \lambda_j/\beta|^{-1}$), where

$$\sum_{j=0}^{\infty} c'_j \beta^{-j} \theta'^j \equiv \exp\left[-\frac{\theta'}{2\beta} \sum_{j=1}^{n} \frac{\lambda_j \omega_j^2}{1 - (1 - \lambda_j/\beta)\theta'}\right] \prod_{j=1}^{n} \{1 - (1 - \lambda_j/\beta)\theta'\}^{-\frac{1}{2}}.$$

Following the same type of argument as before, we find that the c'_j can be calculated from the formulas

$$\text{(99)} \qquad \begin{cases} c'_0 = 1 \\ c'_r = (2r)^{-1} \sum_{j=0}^{r-1} S'_{r-j} c'_j \qquad (r \geq 1) \end{cases}$$

where

$$\text{(100)} \qquad S'_r = -(r/\beta) \sum_{j=1}^{n} \lambda_j \omega_j^2 (1 - \lambda_j/\beta)^{r-1} + \sum_{j=1}^{n} (1 - \lambda_j/\beta)^r.$$

Inverting (98), we obtain

(101)

$$p_n(y;\boldsymbol{\lambda},\boldsymbol{\omega}) = \beta^{-1} p_n(y/\beta;\mathbf{1}) \sum_{j=0}^{\infty} c'_j \frac{j!\Gamma(\frac{1}{2}n)}{\Gamma(\frac{1}{2}n + j)} L_j^{(\frac{1}{2}n-1)}(\tfrac{1}{2}y/\beta) \qquad (y \geq 0).$$

Integration of (101) yields

(102)

$$F_n(y;\boldsymbol{\lambda};\boldsymbol{\omega}) = \Pr[\chi_n^2 \leq y/\beta] + \left(\frac{y}{2\beta}\right)^{\frac{1}{2}n} e^{-\frac{1}{2}y/\beta} \sum_{j=1}^{\infty} c'_j \frac{(j-1)!}{\Gamma(\frac{1}{2}n + j)} L_{j-1}^{(\frac{1}{2}n)}(\tfrac{1}{2}y/\beta).$$

These series are uniformly convergent for all $y \geq 0$, provided $\beta > \frac{1}{2}\lambda_1$. The following bound for error holds:

$$\text{(103)} \qquad \left| \sum_{j=N+1}^{\infty} c'_j \frac{(j-1)!}{\Gamma(\frac{1}{2}n + j)} L_{j-1}^{(\frac{1}{2}n)}(\tfrac{1}{2}y/\beta) \right| \leq R^{N+2}(1 - R)^{-(n+1)} \exp\left[\tfrac{1}{2} \sum_{j=1}^{n} \omega_j^2 / R^3\right] e^{\frac{1}{2}y/\beta}$$

for any R, $0 < R < 1$. (Kotz *et al.* [42].)

An expression for $F_n(y;\boldsymbol{\lambda};\boldsymbol{\omega})$ in terms of Laguerre polynomials was obtained by Shah [68], using Gurland's method, though the coefficients are expressed in a rather more complicated form.

6.4 *Expansion in Maclaurin Series*

We proceed in an exactly similar fashion as for the central case (Section 5.4) and find

$$\text{(104)} \qquad \int_0^{\infty} e^{-\tau y} p_n(y;\boldsymbol{\lambda};\boldsymbol{\omega})\, dy = \tau^{-\frac{1}{2}n} \sum_{j=0}^{\infty} A'_j (-\tau)^{-j} \quad \left(\operatorname{Re} \tau > \frac{1}{2\lambda_n}\right).$$

The A'_j are determined by the equations

$$(105)\qquad \begin{cases} A'_0 = \prod_{j=1}^{n} (2\lambda_j)^{-\frac{1}{2}} \exp\left(-\frac{1}{2}\sum_{j=1}^{n} \omega_j^2\right) \\ A'_r = \sum_{j=0}^{r-1} V'_{r-j}A'_j \qquad (r \geq 1) \end{cases}$$

where

$$V'_r = \sum_{j=1}^{n} (2\lambda_j)^{-r}(1 - r\omega_j^2).$$

Omitting details of the justification for interchanging summation and integration, we have

$$(106)\qquad p_n(y;\lambda;\omega) = y^{\frac{1}{2}n-1} \sum_{j=0}^{\infty} A'_j \frac{(-y)^j}{\Gamma(\frac{1}{2}n + j)}$$

$\left(\text{taking Re}\,\tau \geq (2\lambda_n)^{-1}\left\{n + \sum_{j=1}^{n} \omega_j^2\right\}\right)$. The series is uniformly absolutely convergent over any finite interval of y. Integrating term by term gives

$$(107)\qquad F_n(y;\lambda;\omega) = y^{\frac{1}{2}n} \sum_{j=0}^{\infty} A'_j \frac{(-y)^j}{\Gamma(\frac{1}{2}n + j + 1)}\cdot$$

Shah and Khatri [70] give a double series in powers of y, as a generalization of Pachares' [56] result for the central case. Formulas (105) make the calculation of the coefficients A'_j much simpler than it would be using the formulas in [70].

It is possible to obtain upper bounds for the error in truncating the series (107) to the first N terms (see Kotz *et al.* [42]). However, except for very small values of y, Ruben's series (83) is preferable to (107).

7. Indefinite Forms

Relatively little attention has been devoted to the problem of obtaining the distribution of indefinite quadratic forms. Gurland [28] obtained a complicated expansion in terms of Laguerre polynomials for the central case; which was extended to the noncentral case by Shah [68]. They are neither theoretically nor practically useful, because as Gurland remarks in [28], the weight function for the Laguerre polynomials vanishes for negative values of its argument, while, in general,

$$p_n(y;\boldsymbol{\lambda};\omega) > 0, \qquad F_n(y;\boldsymbol{\lambda};\omega) > 0$$

for all values of y, positive or negative.

Suppose that the first m of the λ's are positive and the remainder negative so that

$$(108)\qquad Q(\mathbf{W}) = Q_1(\mathbf{W}) - Q_2(\mathbf{W})$$

where

$$Q_1(\mathbf{W}) = \sum_{j=1}^{m} \lambda_j(W_j - \omega_j)^2 \tag{109.1}$$

$$Q_2(\mathbf{W}) = \sum_{j=m+1}^{n} (-\lambda_j)(W_j - \omega_j)^2 \tag{109.2}$$

Q_1 and Q_2 are independent positive definite quadratic forms, and each may be handled separately by the methods of Sections 5 and 6. It is clear for example that if each of Q_1 and Q_2 be represented as a mixture of central χ^2's (as in Section 6.1), then $Q_1 - Q_2$ can be represented as a mixture of differences between pairs of independent central χ^2's.

This type of analysis has been worked out in some detail by Imhof [35], and later by Press [57], who used confluent hypergeometric functions to give a convenient form for his results.

The methods of Grad and Solomon ([21], see also Section 5.1) can be extended to indefinite central quadratic forms. The characteristic function of $Q(\mathbf{W})$ (with $\boldsymbol{\omega} = 0$) is

$$\phi_n(t;\boldsymbol{\lambda}) = \prod_{j=1}^{m} (1 - i\lambda_j t)^{-\frac{1}{2}} \prod_{j=m+1}^{n} [1 + i(-\lambda_j)t]^{-\frac{1}{2}}. \tag{110}$$

By Jordan's theorem

$$p_n(y;\boldsymbol{\lambda}) = \lim_{R\to\infty} \frac{1}{2\pi} \int_{-R}^{R} e^{-iyt}\phi_n(t;\boldsymbol{\lambda})\,dt. \tag{111}$$

To apply contour integration we introduce cuts from $-i/(2\lambda_{2j-1})$ to $-i(2\lambda_{2j})$ for $2 \le 2j \le m$, with a cut from $-i/(2\lambda_m)$ to $-i\infty$ if m is odd; and similarly, on the positive imaginary axis, cuts from $-i(2\lambda_{n-2j})$ to $-i/(2\lambda_{n-2j-1})$ for $0 \le 2j \le n - m - 2$ with a cut from $-i/(2\lambda_{m+1})$ to $i\infty$ if $(n - m)$ is odd. We choose the branch of $\phi_n(t;\lambda)$ which is positive on the imaginary axis between $-i/(2\lambda_1)$ and $-i/(2\lambda_n)$, and consider the integral of $e^{-iyt}\phi_n(t;\lambda)$ round the contours in Figures 4 and 5 according as y is positive or negative.

FIGURE 4 FIGURE 5

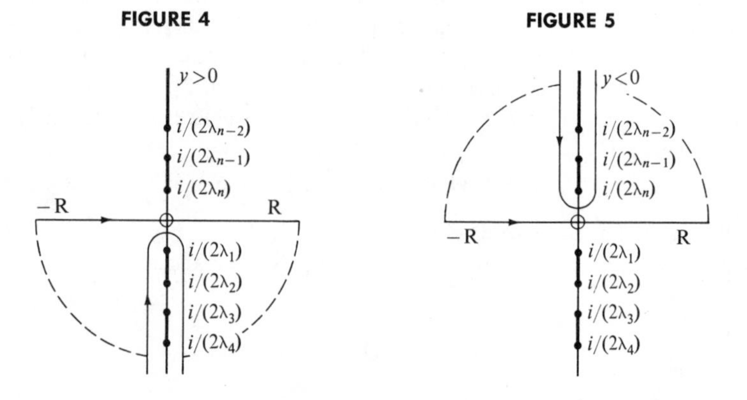

We find

$$(112)\qquad p_n(y;\boldsymbol{\lambda}) = \begin{cases} \dfrac{1}{\pi}\sum(-1)^{j-1}\displaystyle\int_{(2\lambda_{2j-1})^{-1}}^{(2\lambda_{2j})^{-1}} e^{-y\theta}|\phi_n(-i\theta;\boldsymbol{\lambda})|\,d\theta \text{ for } y > 0 \\ \dfrac{1}{\pi}\sum(-1)^{j-1}\displaystyle\int_{(-2\lambda_{n-2j})^{-1}}^{(-2\lambda_{n-2j-1})^{-1}} e^{y\theta}|\phi_n(i\theta;\boldsymbol{\lambda})|\,d\theta \text{ for } y < 0. \end{cases}$$

For $y > 0$ the summation is over $2 \leq 2j \leq m$; for $y < 0$ the summation is over $0 \leq 2j \leq (n - m - 2)$. Note that

$$(113)\qquad \phi_n(-i\theta;\boldsymbol{\lambda}) = \prod_{j=1}^{m}|1 - 2\lambda_j\theta|^{-\frac{1}{2}} \prod_{j=m+1}^{n}|1 + 2(-\lambda_j)\theta|^{-\frac{1}{2}}.$$

It should be borne in mind that it is necessary that no two λ's should be equal for formulas (112) to be valid.

Formulas (112) may be used for numerical computation. Grad and Solomon did so for the positive definite case ($m = n$).

The cumulative distribution function is obtained by integration of (112) which yields

(114)

$$F_n(y;\boldsymbol{\lambda}) = \begin{cases} 1 - \dfrac{1}{\pi}\sum(-1)^{j-1}\displaystyle\int_{(2\lambda_{2j-1})^{-1}}^{(2\lambda_{2j})^{-1}} \theta^{-1}e^{-y\theta}|\phi_n(-i\theta;\boldsymbol{\lambda})|\,d\theta & \text{for } y > 0 \\ \dfrac{1}{\pi}\sum(-1)^{j-1}\displaystyle\int_{(-2\lambda_{n-2j})^{-1}}^{(-2\lambda_{n-2j-1})^{-1}} \theta^{-1}e^{y\theta}|\phi_n(i\theta;\boldsymbol{\lambda})|\,d\theta & \text{for } y < 0. \end{cases}$$

Formulas (112) and (114) seem well suited to obtaining asymptotic estimates as $y \to \infty$ by the use of Watson's Lemma (see, e.g. Carrier *et al.* [8], p. 253).

Robinson [62] obtained the distribution of the difference of two independent χ^2 variables and expanded the distribution of a general quadratic form as a mixture of distributions of this type.

If χ_p^2 and χ_q^2 are independent χ^2 variables then the probability density of

$$X = \chi_p^2 - \chi_q^2$$

can be expressed in terms of the 'second solution of Kummer's equation' (Slater [74]), which is itself expressible in terms of two confluent hypergeometric functions.

If both p $(= 2k)$ and q $(= 2l)$ are even integers, the probability density can be expressed as a mixture of a finite number of central χ^2 distributions:

$$p_X(x \mid p,q) = \begin{cases} \displaystyle\sum_{s=0}^{k-1} 2^{-(s+l)}\frac{l^{[s]}}{s!}\, p_{\chi^2_{2k-2s}}(x) & x \geq 0 \\ \displaystyle\sum_{s=0}^{l-1} 2^{-(s+k)}\frac{k^{[s]}}{s!}\, p_{\chi^2_{2l-2s}}(x) & x \leq 0 \end{cases}$$

where $l^{[s]}$, $k^{[s]}$ are ascending factorials as defined in Chapter 1 (equation (22)).

(See also Wang [85] where the distribution of $\alpha\chi_p^2 - \beta\chi_q^2$ is given, for p, q even.)

Now let

$$X = a(\chi_n^2 + a_1\chi_{n_1}^2 + \cdots + a_r\chi_{n_r}^2 - b_1\chi_{m_1}^2 - \cdots - b_s\chi_{m_s}^2)$$

where the χ^2's are independent and the coefficients are positive constants such that $a_i \geq 1$ $(i = 1,\ldots,r)$, $b_i \geq 1$ $(i = 1,\ldots,s)$. Define $\{d_j\}$ and $\{d_k^*\}$ by the identities

$$\prod_{i=1}^{r} \{a_i^{-\frac{1}{2}n_i}[1 - (1 - a_i^{-1})z]^{-\frac{1}{2}n_i}\} = \sum_{j=0}^{\infty} d_j z_j$$

$$\prod_{i=1}^{s} \{b_i^{-\frac{1}{2}m_i}[1 - (1 - b_i^{-1})z]^{-\frac{1}{2}m_i}\} = \sum_{k=0}^{\infty} d_k^* z^k;$$

then we will have

$$d_j \geq 0, \quad \sum_{j=0}^{\infty} d_j = 1; \quad d_k^* \geqq 0, \quad \sum_{k=0}^{\infty} d_k = 1.$$

Hence

$$\Pr[X \leq x] = \sum_{j=0}^{\infty} \sum_{k=0}^{\infty} d_j d_k^* F_X(x/a \mid N + 2j, M + 2k)$$

where

$$F_X(x \mid p,q) = \int_0^x p_X(t \mid p,q)\, dt; \quad N = n + \sum_{j=1}^{r} n_r; \quad M = \sum_{j=1}^{s} m_j.$$

As a special case, let

$$X = a\chi_n^2 - b\chi_m^2, \qquad 0 < a \leq b;$$

then

$$\Pr(X \leq x) = \sum_{j=0}^{\infty} \left(\frac{a}{b}\right)^{\frac{1}{2}m} \frac{(m/2)!}{j!} \left(1 - \frac{a}{b}\right)^j F(x/a \mid n, m + 2j),$$

which includes the more recent result of Wang [85].

8. Some Theoretical Results

8.1 *Independence of Two Quadratic Forms*

For discussions of conditions for independence of two quadratic forms only *central* forms need be considered, for if $\mathbf{X}'\mathbf{A}\mathbf{X}$ and $\mathbf{X}'\mathbf{B}\mathbf{X}$ are mutually independent, so will be $(\mathbf{X} + \boldsymbol{\xi})'\mathbf{A}(\mathbf{X} + \boldsymbol{\xi})$ and $(\mathbf{X} + \boldsymbol{\xi})'\mathbf{B}(\mathbf{X} + \boldsymbol{\xi})$ and conversely.

Cochran [10] showed that $\mathbf{X}'\mathbf{A}\mathbf{X}$ and $\mathbf{X}'\mathbf{B}\mathbf{X}$ are mutually independent if and only if

(115) $$|\mathbf{I} - \theta\mathbf{A} - \phi\mathbf{B}| = |\mathbf{I} - \theta\mathbf{A}|\,|\mathbf{I} - \phi\mathbf{B}|$$

for all (real) values of θ and ϕ. This result is often difficult to apply. The following (necessary and sufficient) condition

(116) $$\mathbf{AB} = 0$$

stated by Craig [11] is easier to use. (For various derivations of this result see Aitken [1], Hotelling [32], and Lancaster [43].)

An alternative form of (115) is

(117) $$tr(\theta\mathbf{A})^s + tr(\theta\mathbf{B})^s = tr(\theta\mathbf{A} + \phi\mathbf{B})^s$$

for $s = 1, 2, \ldots$, and all (real) θ and ϕ (Lancaster [43]). This form of the criterion is useful in establishing certain results of Matérn and Kawada. Matérn [50] assumes that **A** and **B** are positive semi-definite. Then $\mathbf{X}'\mathbf{AX}$ and $\mathbf{X}'\mathbf{BX}$ are independent if they are uncorrelated. Kawada [39] shows that, for general **A** and **B**, if all four cross-correlations among first and second powers of $\mathbf{X}'\mathbf{AX}$ and $\mathbf{X}'\mathbf{BX}$ are zero, then $\mathbf{X}'\mathbf{AX}$ and $\mathbf{X}'\mathbf{BX}$ are independent. An alternative proof is given by Lancaster [43].

A number of results of this kind can be summarized in the diagram below, suggested by Hultquist [34].

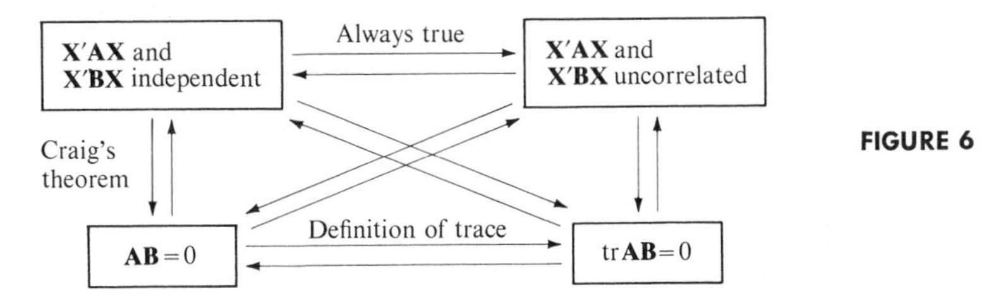

FIGURE 6

Although the general case, in which the dispersion matrix of **X** is **V** (non-singular) can be reduced to the $\mathbf{V} = \mathbf{I}$ case, there are theorems, checking the independence directly, which are of some intrinsic interest. Good [19], [20] and Shanbhag [71] proved that for **V** not necessarily non-singular, $\mathbf{X}'\mathbf{AX}$ and $\mathbf{X}'\mathbf{BX}$ are independent if and only if $\mathbf{VAVBV} = 0$. This condition is equivalent to $\mathbf{AVBV} = 0$ when **A** is non-negative or non-positive, and to $\mathbf{AVB} = 0$ when both **A** and **B** are non-negative or non-positive. (Some additional results were reported in the recent communication by Styan [78].)

8.2 *Conditions for a Quadratic Form to be Distributed as* χ^2

The central quadratic form $\mathbf{X}'\mathbf{AX}$ follows a χ^2 distribution if and only if the matrix **A** is idempotent ($\mathbf{A}^2 = \mathbf{A}$). This can be established by showing that all eigenvalues of **A** must be equal to 1 or 0.

An essential feature of (univariate) analysis of variance consists of partitioning the sum of squares of values $X_1, X_2, \ldots, X_n$ into a sum of (positive definite) quadratic forms. Thus

(118) $$\sum_{i=1}^{n} X_i^2 = \mathbf{X}'\mathbf{X} = Q_1(\mathbf{X}) + \cdots + Q_m(\mathbf{X})$$

where

(119) $$Q_k(\mathbf{X}) = \mathbf{X}'\mathbf{A}_k\mathbf{X} \qquad (k = 1,2,\ldots,m).$$

An important theorem due to Cochran [10] states that a necessary and sufficient condition for $Q_1(\mathbf{X}), \ldots, Q_m(\mathbf{X})$ to be independent χ^2 variables is that

$$\sum_{k=1}^{m} \text{rank } (\mathbf{A}_k) = n.$$

James [36] showed that the three properties:

(i) $\sum_{k=1}^{m} \text{rank } (\mathbf{A}_k) = n$;
(ii) Each $\mathbf{A}_k$ is idempotent;
(iii) $\mathbf{A}_j\mathbf{A}_k = 0$ for all $j \neq k$;

are equivalent.

Another result of interest in connection with analysis of variance is (Lancaster [43]): Suppose that $\mathbf{X}'\mathbf{A}\mathbf{X}$ and $\mathbf{X}'\mathbf{B}\mathbf{X}$ are each distributed as χ^2, with ν_A, ν_B degrees of freedom ($\nu_A > \nu_B$) respectively. Then $\mathbf{X}'\mathbf{B}\mathbf{X}$ is a *component* of $\mathbf{X}'\mathbf{A}\mathbf{X}$ (i.e., $\mathbf{X}'\mathbf{A}\mathbf{X} - \mathbf{X}'\mathbf{B}\mathbf{X}$ is a χ^2 variable with $(\nu_A - \nu_B)$ degrees of freedom, independent of $\mathbf{X}'\mathbf{B}\mathbf{X}$) if and only if the correlation between $\mathbf{X}'\mathbf{A}\mathbf{X}$ and $\mathbf{X}'\mathbf{B}\mathbf{X}$ is $m/\sqrt{mn}$.

In a recent paper, Good [20] proved that a necessary and sufficient condition for the form $\mathbf{X}'\mathbf{A}\mathbf{X}$, with the dispersion matrix $\mathbf{V}$, to follow a χ^2 distribution with r degrees of freedom is that $\mathbf{AV}$ has r unit characteristic roots, with the rest being zero. This implies (Styan [78]) that the condition

$$(\mathbf{AV})^2 = \mathbf{AC}$$

is also a necessary and sufficient condition for $X'AX$ to follow a χ^2 distribution with r degrees of freedom if and only if

$$\text{rank } (\mathbf{AV}) = \text{tr}(\mathbf{AV}) = r,$$

or

$$\text{rank } (\mathbf{AV}) = \text{rank } (\mathbf{VAV}) = r.$$

8.3 *Conditions for a Quadratic Form to be Distributed as Non-central* χ^2

We now generalize the problem of the last subsection.

Let $\mathbf{X}' = (X_1,\ldots,X_n)$ be a random normal vector with mean vector $\boldsymbol{\xi}$ and dispersion matrix $\mathbf{V}$ (not necessarily non-singular). Let $\mathbf{A}$ be an $n \times n$ real symmetric matrix and assume that $\text{tr}(\mathbf{AV}) = r \neq 0$.

A theorem by Mäkeläinen [47] states that $Q(\mathbf{X}) = \mathbf{X}'\mathbf{A}\mathbf{X}$ has a noncentral χ^2 distribution if and only if

$$\mathbf{VAVAV} = \mathbf{VAV};$$

$$\boldsymbol{\xi}'\mathbf{AVA}\boldsymbol{\xi} = \boldsymbol{\xi}'\mathbf{A}\boldsymbol{\xi} \qquad \text{and} \qquad \mathbf{VAVA}\boldsymbol{\xi} = \mathbf{VA}\boldsymbol{\xi}.$$

The degrees of freedom and the noncentrality parameter are r and $2^{-1}\boldsymbol{\xi}'\mathbf{A}\boldsymbol{\xi}$, respectively.

This theorem includes important cases as corollaries:

(i) $Q(\mathbf{X})$ is distributed as a central χ^2 variable if and only if

$$\mathbf{VAVAV} = \mathbf{VAV}; \ \boldsymbol{\xi}'\mathbf{A}\boldsymbol{\xi} = 0; \qquad \text{and} \qquad \mathbf{VA}\boldsymbol{\xi} = \mathbf{0}.$$

(The degrees of freedom are $tr(\mathbf{AV})$).

(ii) If $\mathbf{A}$ is positive semidefinite, then $Q(\mathbf{X})$ is distributed as a noncentral χ^2 variable if and only if

$$\mathbf{AV} \text{ is idempotent and } \mathbf{A}\boldsymbol{\xi} = \mathbf{AVA}\boldsymbol{\xi}.$$

The parameters of the distribution are $tr(\mathbf{AV}) = \text{rank}\ (\mathbf{AV})$ and $2^{-1}\boldsymbol{\xi}'\mathbf{A}\boldsymbol{\xi}$. $Q(\mathbf{X})$ is a central χ^2 variable if and only if $\mathbf{AV}$ is idempotent and $\mathbf{A}\boldsymbol{\xi} = \mathbf{0}$.

(iii) If $\mathbf{A}$ is positive definite, $Q(\mathbf{X})$ is a noncentral χ^2 variable if and only if $\mathbf{VAV} = \mathbf{V}$ and $\boldsymbol{\xi} = \mathbf{VA}\boldsymbol{\xi}$, and the parameters of the distribution are rank $(\mathbf{V})$ and $2^{-1}\boldsymbol{\xi}'\mathbf{A}\boldsymbol{\xi}$. The condition $\mathbf{VAV} = \mathbf{V}$ can be replaced by $\mathbf{AVA} = \mathbf{A}$.

The results in (iii) were originally obtained by Carpenter [7].

9. Tables

Tabulation of $F_n(y;\boldsymbol{\lambda};\boldsymbol{\omega})$ presents difficulties on account of the number of variables involved. There are $(2n + 1)$ variables in general; even in the central case $(\boldsymbol{\omega} = \mathbf{0})$ there are $(n + 1)$ variables. One variable can be removed by introducing the standardization (or normalization) condition $\sum_{j=1}^{n} \lambda_j = 1$, but even so, it is very difficult to construct adequate tables for $n \geq 4$. It is not surprising that until recently the available tables were mostly for $n = 2$ and $n = 3$.

From (78) it can be seen that values of $F_2(y;\lambda_1,\lambda_2;0,0)$ can be obtained from a suitable table of the noncentral χ^2 distribution with two degrees of freedom (as in tables of 'offset circle' probabilities referred to in Section 5.7).

Solomon [77] gives extensive tables of $F_2(y;\lambda_1,\lambda_2;0,0)$.

Owen [54] gives extracts from Solomon's tables. He tabulates $F_2(y;\lambda_1,\lambda_2;0,0)$ for

$$\begin{aligned} y &= 0.1(0.1)1.0(0.5)2.0(1)5; \\ \lambda_1 &= 0.5,\ 0.6,\ 2/3,\ 0.7,\ 0.75,\ 0.8,\ 0.875,\ 0.9,\ 0.95,\ 0.99 \\ \lambda_2 &= 1 - \lambda_1; \end{aligned}$$

and also values of y satisfying the equation $F_2(y;\lambda_1,\lambda_2;0,0) = p$ for the same values of λ_1 and λ_2 and

$$p = 0.05(0.05)0.30(0.10)0.70(0.05)0.95.$$

Owen also gives $F_3(y;\lambda_1,\lambda_2,\lambda_3;0,0,0)$ for

$$y = 0.1(0.1)1.0(0.5)2.0(1)5$$

$$(\lambda_1,\lambda_2,\lambda_3) \equiv (0.9,0.05,0.05), (0.8,0.1,0.1), (0.7,0.2,0.1), (0.6,0.3,0.1), (0,6,0.2,0.2), (0.5,0.4,0.1), (0.4,0.4,0.2), (0.4,0.3,0.3), (\tfrac{1}{3},\tfrac{1}{3},\tfrac{1}{3}).$$

and also values of y satisfying the equation $F_3(y;\lambda_1,\lambda_2,\lambda_3;0,0,0) = p$ for these sets of values of $(\lambda_1,\lambda_2,\lambda_3)$ and

$$p = 0.05(0.05)0.30(0.10)0.70(0.05)0.95.$$

Marsaglia's tables [49] of $F_3(y;\lambda_1,\lambda_2,\lambda_3;0,0,0)$ are more extensive.

Johnson and Kotz [37] give values of $F_n(y;\boldsymbol{\lambda},\mathbf{0})$ for $n = 4$ and $n = 5$ to 5 significant figures for $y = 0.1(0.1)10.2(0.2)$ and $F_n \leq 0.999$, for selected sets of values of the λ's (36 for $n = 4$ and 78 for $n = 5$). A paper in Sankhyā ([37]) by the same authors contains a table of *percentile points* for $n = 4$ in each of the thirty-six cases mentioned above and a computer program (based on formulas (18) and (71) of [41] for calculation of $F_n(y;\boldsymbol{\lambda},\mathbf{0})$) written in Fortran IV.

For the noncentral case, Lowe [44] and DiDonato and Jarnagin [13], [14] have provided tables of $F_2(y;\boldsymbol{\lambda},\boldsymbol{\omega})$. These latter authors have also given tables of solutions (for y) of the equation $F_2(y;\boldsymbol{\lambda},\boldsymbol{\omega}) = \alpha$ for $\alpha = 0.05$, 0.2, 0.5, 0.7, 0.9, 0.95. Lowe's tables give three decimal accuracy for $\lambda_1/\lambda_2 = 4$, 16, 64 and various values of y and $\boldsymbol{\omega}$.

Much more extensive tables, of the same kind, are contained in Groenewoud *et al.* [23].

Shah [69] has announced tables of $F_n(y;\boldsymbol{\lambda},\boldsymbol{\omega})$ for unspecified values of n (presumably greater than 2).

10. Applications

10.1 *Weapons and Optimal Control Problems*

The use of the noncentral χ^2 distribution in evaluating probabilities of hitting circular targets has been discussed in Chapter 28, Section 8. In general, calculating probabilities for plane circular targets with a bivariate normal distribution representing the terminal errors, is equivalent to evaluating the cumulative distribution function of a quadratic form in two normal variables. Similarly, if the target be a sphere, and the error distribution represented by a trivariate normal distribution, then a quadratic form in three normal variables needs to be studied (e.g. Grubbs [24]).

A formally similar situation arises in certain problems of optimal control. Suppose that the 'state' variables are represented by an n-dimensional vector $\mathbf{X}$, and the desired values of these variables by the vector $\mathbf{X}^*$, which is estimated on the basis of sample values. Suppose, further, that this estimate $\hat{\mathbf{X}}^*$ is multinormally distributed with expected value vector $\mathbf{X}^*$ and variance-covariance matrix $\mathbf{V}_1$ (of rank n). By changing control variables it is hoped to make $\mathbf{X}$ approach $\hat{\mathbf{X}}^*$, but because of incomplete control, $\mathbf{X} - \hat{\mathbf{X}}^*$ is not, in fact, exactly zero. We suppose that $\mathbf{X} - \mathbf{X}^*$ is independent of $\hat{\mathbf{X}}^* - \mathbf{X}^*$, and is multinormally distributed with expected value $\mathbf{0}$ and variance-covariance matrix $\mathbf{V}_2$ (of rank n). The deviations from optimal state

$$\mathbf{d} = \mathbf{X} - \mathbf{X}^* = (\mathbf{X} - \hat{\mathbf{X}}^*) + (\hat{\mathbf{X}}^* - \mathbf{X}^*)$$

have a joint multinormal distribution with expected value $\mathbf{0}$ and variance-covariance matrix $\mathbf{V} = \mathbf{V}_1 + \mathbf{V}_2$. Finally, we suppose that the loss function is the quadratic form $\mathbf{d}'\mathbf{A}\mathbf{d}$, where $\mathbf{A}$ is positive definite. If all these conditions are satisfied, then the probability of loss not exceeding y is equal to $F_n(y;\boldsymbol{\lambda};\mathbf{0})$ where $\boldsymbol{\lambda}$ is the vector of eigenvalues of $\mathbf{VA}$ (see Ruben [63]).

10.2 *Estimation of Power Spectra*

Suppose $\ldots X_{-1}, X_0, X_1, \ldots$ is a stationary time series originated by a process with zero mean. Since the process is stationary the covariance

$$v_{|i-j|} = E[X_i X_j] = E[X_{|i-j|} X_0] \tag{120}$$

depends only on $|i - j|$. The covariances v_k can be represented in the form

$$v_k = \frac{1}{2\pi} \int_{-\pi}^{\pi} e^{ik\theta}\, dF(\theta) \tag{121}$$

where $F(\theta)$ is a cumulative distribution function, (with $F(-\pi) = 0$, $F(\pi) = 1$) and is called the *spectral distribution function*. If $F(\theta)$ is absolutely continuous, it can be represented in the form

$$F(\theta) = \int_{-\pi}^{\theta} f(y)\, dy. \tag{122}$$

$f(\theta)$ is called the *spectral density function*. In the estimation of $f(\theta)$ from observed values of the process, use is made of the function

$$Q_w = \sum_{i=1}^{n} \sum_{j=1}^{n} w_{|i-j|} X_i X_j \tag{123}$$

where

$$w_k = \frac{1}{2\pi} \int_{-\pi}^{\pi} e^{ik\theta} w(\theta)\, d\theta. \tag{124}$$

The function $w(\theta)$, called the *spectral window*, is chosen arbitrarily, with a view to suitability for the particular estimation required. If the value of $f(\theta)$ for a particular value, θ_0, of θ is to be estimated, $w(\theta)$ is chosen to be relatively large in a small neighborhood including θ_0, and relatively small everywhere else. This is called a "narrow window" — it should not, in fact, be too "narrow" as this would increase the variance of Q_w.

If the process be normal, then Q_w has the cumulative distribution function $F_n(y;\boldsymbol{\lambda};\mathbf{0})$ where $\boldsymbol{\lambda}$ is the vector of eigenvalues of the matrix $\mathbf{VW}$ (where $\mathbf{V}(n \times n) \equiv (v_{|i-j|})$; $\mathbf{W}(n \times n) = (w_{|i-j|})$). If the ratio $Q_w/f(\theta_0)$ has a distribution which is insensitive to the shape of $f(\theta)$, then comparison of the observed value of Q_w with the (approximate) distribution of $Q_w/f(\theta_0)$ provides a basis for estimating $f(\theta_0)$. In practice $\lambda_1, \lambda_2, \ldots, \lambda_n$ are replaced by the eigenvalues of $\mathbf{W}$ (since $\mathbf{V}$ is usually unknown). Details of this application are given in Grenander *et al.* [22].

10.3 *Distribution of Quadratic Forms in Variate Differences*

When it is suspected that a trend underlies a sequence of observed values $X_1, X_2, \ldots X_n$, the usual mean square estimator of variance,

$$(n-1)^{-1} \sum_{i=1}^{n} (X_i - \bar{X})^2$$

is often modified. One such modified estimator is the *mean square successive difference*

$$\delta^2 = \frac{1}{2(n-1)} \sum_{i=1}^{n-1} (X_{i+1} - X_i)^2. \tag{125}$$

While this modification does not eliminate the effects of trends (even for a simple linear trend, there is a positive bias), it does reduce their effects considerably, as compared with the usual mean square estimator.

If, in fact, there is no trend, and the X_i's can be represented by independent normal variables with common expected value and with common variance σ^2, then δ^2 is an unbiased estimator of σ^2. In this case the distribution of δ^2 can be derived from the fact that δ^2 is a quadratic form in the X_i's with matrix

$$\begin{pmatrix} 1 & -1 & 0 & \ldots & 0 & 0 \\ -1 & 2 & -1 & \ldots & 0 & 0 \\ 0 & -1 & 2 & \ldots & 0 & 0 \\ \vdots & \vdots & \vdots & & \vdots & \vdots \\ 0 & 0 & 0 & \ldots & 2 & -1 \\ 0 & 0 & 0 & \ldots & -1 & 1 \end{pmatrix}.$$

This matrix has rank $(n-1)$. The $(n-1)$ nonzero eigenvalues are

$$\lambda_j = 4 \sin^2 \left(\frac{(n-j)\pi}{2n} \right) \qquad (j = 1, \ldots (n-1)). \tag{126}$$

For further details see von Neumann [52].

A generalized form of (125) is used in application of the variate difference method (Tintner [82]). If it be supposed that the trend can be represented by a polynomial (in time) of order $k - 1$, and that $X_1, X_2, \ldots, X_n$ are observed at equal intervals of time then $\Delta^k X_i$ will be trend-free. If it further be supposed that the *deviations* of the X's from the trend are independent normal variables each with zero expected value, and variance σ^2 then

$$\delta^2_{[k]} = \frac{1}{(n-k)\binom{2k}{k}} \sum_{i=1}^{n-k} (\Delta^k X_i)^2 \tag{127}$$

is an unbiased estimator of σ^2. This is a quadratic form in the X's, with matrix of rank $(n - k)$. The distribution of $\delta^2_{[2]}$ has been considered by Kamat [38].

10.4 χ^2-*tests*

Some aspects of the theory of χ^2-tests have already been discussed in Chapter 17. Here we merely note that the general theory of positive definite quadratic forms enables us to approximate the distributions of χ^2 more closely, and under more varied conditions, than does the theory of central and non-central χ^2. For example, Chernoff and Lehmann [9] have shown that if a χ^2-statistic is calculated for data arranged in k groups, with s distinct parameters fitted by the maximum likelihood method, then its asymptotic distribution (as the number of observations tend to infinity) is not χ^2_{k-1}, but is the distribution of

$$\chi^2_{k-s-1} + \sum_{j=k-s}^{k-1} \lambda_j U_j^2$$

where the U's are independent unit normal random variables, each independent of χ^2_{k-s-1}

10.5 *Analysis of Variance*

Distributions of sums of squares occurring in the analysis of variance are discussed in Chapter 26, Section 5 and in Chapter 30, Section 1. Here, again, if it is desired to consider the distributions of sums of squares under general conditions of heteroscedasticity, the general theory of positive definite quadratic forms (as distinct from χ^2 and non-central χ^2) is needed (Box [5], [6].)

If the methods described by David and Johnson [12] are to be used in evaluating the properties of mean-square-ratio criteria under general conditions, some theory of *indefinite* quadratic forms is needed.

We conclude this chapter by noting investigations into the distributions of quadratic forms in certain non-normal distributions by Subrahmanian — in particular mixtures of normal distributions [79] and distributions represented by Edgeworth or Gram-Charlier series [80].

REFERENCES

[1] Aitken, A. C. (1950). On the statistical independence of quadratic forms in normal variates, *Biometrika*, **37,** 93–96.

[2] Anderson, T. W. and Darling, D. A. (1952). Asymptotic theory of certain "goodness of fit" criteria based on stochastic processes, *Annals of Mathematical Statistics*, **23,** 193–211.

[3] Bhat, B. R. (1962). On the distribution of certain quadratic forms in normal variates, *Journal of the Royal Statistical Society, Series B*, **24,** 148–151.

[4] Bhattacharyya, A. (1945). A note on the distribution of the sum of chi-squares, *Sankhyā*, **7,** 27–28.

[5] Box, G. E. P. (1954). Some theorems on quadratic forms applied in the study of analysis of variance problems, I. Effect of inequality of variance in the one-way classification, *Annals of Mathematical Statistics*, **25,** 290–302.

[6] Box, G. E. P. (1954). *Ibid.* II. Effects of inequality of variance and of correlation between errors in the two-way classification, *Annals of Mathematical Statistics*, **25,** 484–498.

[7] Carpenter, O. (1950). Note on the extension of Craig's theorem to non-central variates, *Annals of Mathematical Statistics*, **21,** 455–457.

[8] Carrier, G. E., Krook, M. and Pearson, C. E. (1966). *Functions of a Complex Variable*, New York: McGraw-Hill, Inc.

[9] Chernoff, H. and Lehmann, E. L. (1954). The use of maximum likelihood estimates in χ^2 tests for goodness of fit, *Annals of Mathematical Statistics*, **25,** 579–586.

[10] Cochran, W. G. (1934). The distribution of quadratic forms in a normal system, with applications to the analysis of covariance, *Proceedings of the Cambridge Philosophical Society*, **30,** 178–191.

[11] Craig, A. T. (1943). Note on the independence of certain quadratic forms, *Annals of Mathematical Statistics*, **14,** 195–197.

[12] David, F. N. and Johnson, N. L. (1951). A method of investigating the effect of non-normality and heterogeneity of variance on tests of the general linear hypothesis, *Annals of Mathematical Statistics*, **22,** 382–392.

[13] DiDonato, A. R. and Jarnagin, M. P. (1961). Integration of the general bivariate Gaussian distribution over an offset circle, *Mathematics of Computation*, **15,** 375–382.

[14] DiDonato, A. R. and Jarnagin, M. P. (1962). A method for computing the circular coverage function, *Mathematics of Computation*, **16,** 347–355.

[15] Feiveson, A. H. and Delaney, F. C. (1968). *The distribution and properties of a weighted sum of chi squares*, Report TN D-4575, National Aeronautics and Space Administration.

[16] Freiberger, W. F. and Jones, R. H. (1960). Computation of the frequency function of a quadratic form in random normal variables, *Journal of the Association of Computation Machinery*, **7,** 245–250.

[17] Geary, R. C. (1936). Distribution of "Student's" ratio for non-normal samples, *Journal of the Royal Statistical Society, Series B*, **3,** 178–184.

[18] Gil-Pelaez, J. (1951). Note on the inversion theorem, *Biometrika*, **38,** 481–482.

[19] Good, I. J. (1963). On the independence of quadratic expressions, *Journal of the Royal Statistical Society, Series B*, **25,** 377–382, (Correction: **28,** 584).

[20] Good, I. J. (1969). Conditions for a quadratic form to have a chi-squared distribution, *Biometrika*, **56,** 215–216.

[21] Grad, A. and Solomon, H. (1955). Distribution of quadratic forms and some applications, *Annals of Mathematical Statistics*, **26,** 464–477.

[22] Grenander, U., Pollak, H. O. and Slepian, D. (1959). The distribution of quadratic forms in normal variates: A small sample theory with applications to spectral analysis, *SIAM Journal*, **7,** 374–401.

[23] Groenewoud, C., Hoaglin, D. C. and Vitalis, J. A. (1967). *Bivariate Normal Offset Circle Probability Tables With Offset Ellipse Transformations, I, II.*, Cornell Aeronautical Laboratory, Inc., Buffalo, N. Y.

[24] Grubbs, F. E. (1964). Approximate circular and noncircular offset probabilities of hitting, *Operations Research*, **12,** 51–62.

[25] Guenther, W. C. and Terragno, P. J. (1964). A review of the literature on a class of coverage problems, *Annals of Mathematical Statistics*, **35,** 232–260.

[26] Gurland, J. (1948). Inversion formulae for the distribution of ratios, *Annals of Mathematical Statistics*, **19,** 228–237.

[27] Gurland, J. (1953). Distribution of quadratic forms and ratios of quadratic forms, *Annals of Mathematical Statistics*, **24,** 416–427.

[28] Gurland, J. (1955). Distribution of definite and of indefinite quadratic forms, *Annals of Mathematical Statistics*, **26,** 122–127.

[29] Gurland, J. (1956). Quadratic forms in normally distributed random variables, *Sankhyā*, **17,** 37–50.

[30] Harter, H. L. (1960). Circular probabilities, *Journal of the American Statistical Association*, **55,** 723–731.

[31] Harvey, J. R. (1965). *Fractional moments of a quadratic form in noncentral normal random variables*, Unpublished thesis, North Carolina State University.

[32] Hotelling, H. (1944). A note on a matric theorem of A. T. Craig, *Annals of Mathematical Statistics*, **15,** 427–429.

[33] Hotelling, H. (1948). Some new methods for distributions of quadratic forms (Abstract), *Annals of Mathematical Statistics*, **19,** 119.

[34] Hultquist, R. (1966). Diagramming theorems relative to quadratic forms, *American Statistician*, **20,** (5), 31.

[35] Imhof, J. P. (1961). Computing the distribution of quadratic forms in normal variables, *Biometrika*, **48,** 419–426.

[36] James, G. S. (1952). Notes on a theorem of Cochran, *Proceedings of the Cambridge Philosophical Society*, **48,** 443–446.

[37] Johnson, N. L. and Kotz, S. (1969). *Tables of distributions of quadratic forms in central normal variables*, I and II. Institute of Statistics, Mimeo Series No. 543 and No. 557, University of North Carolina at Chapel Hill, (also *Sankhyā, Series B*, **30,** 303–314).

[38] Kamat, A. R. (1954). Distribution theory of two estimates for standard deviation based on second variate differences, *Biometrika*, **41,** 1–11.

[39] Kawada, Y. (1950). Independence of quadratic forms in normally correlated variables, *Annals of Mathematical Statistics*, **21**, 614–615.

[40] Khatri, C. G. (1966). On certain distribution problems based on positive definite quadratic functions in normal vectors, *Annals of Mathematical Statistics*, **37**, 468–479.

[41] Kotz, S., Johnson, N. L. and Boyd, D. W. (1967). Series representations of distributions of quadratic forms in normal variables I. Central case, *Annals of Mathematical Statistics*, **38**, 823–837.

[42] Kotz, S., Johnson, N. L. and Boyd, D. W. (1967). Series representations of distributions of quadratic forms in normal variables II. Non-central case, *Annals of Mathematical Statistics*, **38**, 838–848.

[43] Lancaster, H. O. (1954). Traces and cumulants of quadratic forms in normal variables, *Journal of the Royal Statistical Society, Series B*, **16**, 247–254.

[44] Lowe, J. R. (1960). A table of the integral of the bivariate normal distribution over an offset circle, *Journal of the Royal Statistical Society, Series B*, **22**, 177–187.

[45] Lukacs, E. (1952). The stochastic independence of symmetric and homogeneous linear and quadratic statistics, *Annals of Mathematical Statistics*, **23**, 442–449.

[46] Luther, N. Y. (1965). Decomposition of symmetric matrices and distributions of quadratic forms, *Annals of Mathematical Statistics*, **36**, 683–690.

[47] Mäkeläinen, T. (1966). On quadratic forms in normal variables, *Commentationes Physico-Mathematicae, Societas Scientarium Fennica*; **31**, (12), 1–6.

[48] Marcum, J. I. (1950). *Table of Q functions*, RAND Report No. RM-339, The RAND Corporation, Santa Monica, California.

[49] Marsaglia, G. (1960). *Tables of the distribution of quadratic forms of ranks two and three*, Boeing Scientific Research Laboratories Report No. D1-82-0015-1.

[50] Matérn, B. (1949). Independence of non-negative quadratic forms in normally correlated variables, *Annals of Mathematical Statistics*, **20**, 119–120.

[51] McNolty, F. (1965). Kill probability when the lethal effect is variable, *Operations Research*, **13**, 478–482.

[52] Neumann, J. von., (1941). Distribution of the ratio of the mean square successive difference to the variance, *Annals of Mathematical Statistics*, **12**, 367–395.

[53] Okamoto, M. (1960). An inequality for the weighted sum of χ^2 variates, *Bulletin of Mathematical Statistics*, **9**, 69–70.

[54] Owen, D. B. (1962). *Handbook of Statistical Tables*, Reading, Massachusetts: Addison-Wesley Publishing Company, Inc.

[55] Pachares, J. (1952). The distribution of the difference of two independent chi-squares (Abstract), *Annals of Mathematical Statistics*, **23**, 639.

[56] Pachares, J. (1955). Note on the distribution of a definite quadratic form, *Annals of Mathematical Statistics*, **26**, 128–131.

[57] Press, S. J. (1966). Linear combinations of non-central chi-square variates, *Annals of Mathematical Statistics*, **37**, 480–487.

[58] RAND Report R-234 (1952). *Offset Circle Probabilities*. The RAND Corporation, Santa Monica, California.

[59] Rao, C. R. (1952). *Advanced Statistical Methods in Biometric Research*, New York: John Wiley & Sons, Inc.

[60] Robbins, H. E. (1948). The distribution of a definite quadratic form, *Annals of Mathematical Statistics*, **19,** 266–270.

[61] Robbins, H. E. and Pitman, E. J. G. (1949). Application of the method of mixtures to quadratic forms in normal variates, *Annals of Mathematical Statistics*, **20,** 552–560.

[62] Robinson, J. (1965). The distribution of a general quadratic form in normal variates, *Australian Journal of Statistics*, **7,** 110–114.

[63] Ruben, H. (1960). Probability contant of regions under spherical normal distributions, I., *Annals of Mathematical Statistics*, **31,** 598–619.

[64] Ruben, H. (1962). *Ibid.*, IV., *Annals of Mathematical Statistics*, **33,** 542–570.

[65] Ruben, H. (1963). A new result on the distribution of quadratic forms, *Annals of Mathematical Statistics*, **34,** 1582–1584.

[66] Satterthwaite, F. E. (1941). Synthesis of variance, *Psychometrika*, **6,** 309–316.

[67] Satterthwaite, F. E. (1946). An approximate distribution of estimates of variance components, *Biometrika*, **2,** 110–114.

[68] Shah, B. K. (1963). Distribution of definite and of indefinite quadratic forms from a non-central normal distribution, *Annals of Mathematical Statistics*, **34,** 186–190.

[69] Shah, B. K. (1968). Tables of positive definite quadratic forms in non-central normal variables (Abstract), *Annals of Mathematical Statistics*, **39,** 705–706.

[70] Shah, B. K. and Khatri, C. G. (1961). Distribution of a definite quadratic form for non-central normal variates, *Annals of Mathematical Statistics*, **32,** 883–887. (Correction **34,** 673).

[71] Shanbhag, D. N. (1966). On the independence of quadratic forms, *Journal of the Royal Statistical Society, Series B*, **28,** 582–583.

[72] Siddiqui, M. M. (1965). Approximations to the distribution of quadratic forms, *Annals of Mathematical Statistics*, **36,** 677–682.

[73] Siotani, M. (1964). Tolerance regions for a multivariate normal population, *Annals of the Institute of Mathematical Statistics, Tokyo*, **16,** 135–153.

[74] Slater, L. J. (1960). *Confluent Hypergeometric Functions*, Cambridge University Press.

[75] Soest, J. van (1968). Afschatting van verdelingen van kwadratische vormen met toepassingen op aanpassingstoetsen voor normaliteit en exponentialiteit, *Statistica Neerlandica*, **22,** 235–240.

[76] Solomon, H. (1960). On the distribution of quadratic forms in normal variates, *Proceedings of the 4th Berkeley Symposium on Mathematical Statistics and Probability*, **1,** 645–653.

[77] Solomon, H. (1960). *Distribution of Quadratic Forms: Tables and Applications*, Technical Report No. 45, Applied Mathematics and Statistics Laboratories, Stanford University.

[78] Styan, G. P. H. (1969). *Notes on the distribution of quadratic forms in singular normal variables*, Technical Report No. 122, University of Minnesota.

[79] Subrahmanian, K. (1968). A note on the distribution of quadratic forms for mixtures of normal distributions (Abstract), *Annals of Mathematical Statistics*, **39,** 706.

[80] Subrahmanian, K. (1966), (1968). Some contributions to the theory of non normality, I (univariate case), II, *Sankhyā*, *Series A*, **28**, 389–406, *Ibid*, **30** 411–432.

[81] Szegö, G. (1939). *Orthogonal Polynomials*, American Mathematical Society Colloquium Publication, **23**, American Mathematical Society, N. Y.

[82] Tintner, G. (1940). *The Variate Difference Method*, Cowles Commission Mono graph, No. **5**, Bloomington, Indiana: Principia Press.

[83] Titchmarsh, E. C. (1937). *Introduction to the Theory of Fourier Integrals*, London Oxford University Press.

[84] Varberg, D. E. (1966). Convergence of quadratic forms in independent random variables, *Annals of Mathematical Statistics*, **37**, 567–576.

[85] Wang, Y. Y. (1967). A comparison of several variance component estimators *Biometrika*, **54**, 301–305.

[86] Whittle, P. (1960). Bounds for the moments of linear and quadratic forms in independent variables, *Teoriya Veroyatnostei i ee Primeneniya*, **5**, 331–335 (English translation, same volume: 302–305.)

[87] Whittle, P. (1964). On the convergence to normality of quadratic forms in inde pendent variables, *Teoriya Veroyatnostei i ee Primeneniya*, **9**, 113–118.

30

Noncentral F Distribution

1. Definition and Genesis

In Chapter 26 the F distribution with ν_1, ν_2 degrees of freedom was defined as the distribution of the ratio $(\chi^2_{\nu_1}/\nu_1)(\chi^2_{\nu_2}/\nu_2)^{-1}$, where the two χ^2's are mutually independent. If both χ^2's are replaced by noncentral χ^2's, we obtain the *doubly noncentral F distribution with ν_1, ν_2 degrees of freedom and noncentrality parameters* λ_1, λ_2, defined as the distribution of the ratio

$$\{\chi'^2_{\nu_1}(\lambda_1)/\nu_1\}\{\chi'^2_{\nu_2}(\lambda_2)/\nu_2\}^{-1}. \tag{1}$$

In many applications, λ_2 is equal to zero, so that there is a central χ^2 in the denominator. This might be called a 'singly noncentral F', but the word 'singly' is usually omitted, and it is called a *noncentral F with ν_1, ν_2 degrees of freedom and noncentrality parameter* λ_1. The case $\lambda_1 = 0$, $\lambda_2 \neq 0$ is not usually considered separately, since this corresponds simply to the *reciprocal* of a noncentral F, as just defined.

We will use the notations $F''_{\nu_1,\nu_2}(\lambda_1,\lambda_2)$ for the doubly noncentral F variable defined by (1), and $F'_{\nu_1,\nu_2}(\lambda_1)$ for the noncentral F variable

$$\{\chi'^2_{\nu_1}(\lambda_1)/\nu_1\}(\chi^2_{\nu_2}/\nu_2)^{-1} \tag{2}$$

(and also for the corresponding distributions). Note that, using these symbols

$$F''_{\nu_1,\nu_2}(0,\lambda_2) = [F'_{\nu_2,\nu_1}(\lambda_2)]^{-1}.$$

In this chapter we will be mainly concerned with (singly) noncentral F distributions. Doubly noncentral F distributions will appear again in Section 6.

The noncentral F distribution is used in the calculation of the power functions of tests of general linear hypotheses. As pointed out in Chapter 26, these include standard tests used in the analysis of variance (e.g. Scheffé [22], Tang

[27], Madow [14], Lehmann [12] to mention just a few of the earlier papers and numerous basic texts).

More recent fields of application include communication theory (Price [21]), where it is used for determination of probabilities of errors in transmission over certain types of noise-perturbed channels.

2. Historical Remarks

The noncentral beta distribution, which is related to the noncentral F distribution (Section 6), was derived by Fisher [6] in 1928, in connection with research on the distribution of the multiple correlation coefficient. Its properties were discussed by Wishart [34] in 1932. The noncentral F distribution itself was derived by Tang [27] in 1938, though Patnaik [18] seems to have been the first (in 1949) to call the distribution by this name.

Tang [27] also used the doubly noncentral F distribution (though without actually using this name) in studies of the properties of analysis of variance tests under nonstandard conditions.

3. Properties

Since the numerator and denominator in (2) are independent, it follows that

$$\begin{aligned}\mu_r'(F'_{\nu_1,\nu_2}(\lambda_1)) &= (\nu_2/\nu_1)^r \mu_r'(\chi'^2_{\nu_1}(\lambda_1))\mu'_{-r}(\chi^2_{\nu_2}) \\ &= \left(\frac{\nu_2}{\nu_1}\right)^r \frac{\Gamma(\frac{1}{2}\nu_1 + r)\Gamma(\frac{1}{2}\nu_2 - r)}{\Gamma(\frac{1}{2}\nu_1)\Gamma(\frac{1}{2}\nu_2)} \sum_{j=0}^{r} \binom{r}{j} \frac{\Gamma(\frac{1}{2}r_1)}{\Gamma(\frac{1}{2}r_1 + j)} (\tfrac{1}{2}\lambda_1\nu_1)^j,\end{aligned}$$

whence

$$E[F'_{\nu_1,\nu_2}(\lambda_1)] = \frac{\nu_2(\nu_1 + \lambda_1)}{\nu_1(\nu_2 - 2)} \qquad (\nu_2 > 2); \tag{3.1}$$

(3.2)

$$\operatorname{var}(F'_{\nu_1,\nu_2}(\lambda_1)) = 2\left(\frac{\nu_2}{\nu_1}\right)^2 \frac{(\nu_1 + \lambda_1)^2 + (\nu_1 + 2\lambda_1)(\nu_2 - 2)}{(\nu_2 - 2)^2(\nu_2 - 4)} \qquad (\nu_2 > 4);$$

and the third central moment is

(3.3)

$$\begin{aligned}\mu_3(F'_{\nu_1,\nu_2}(\lambda_1)) &= \frac{4}{(\nu_2 - 4)(\nu_2 - 6)} \\ &\times \left[4\left(\frac{\nu_1 + \lambda_1}{\nu_2 - 2}\right)^3 + \frac{6(\nu_1 + \lambda_1)(\nu_1 + 2\lambda_1)}{(\nu_2 - 2)^2} + \frac{2(\nu_1 + 3\lambda_1)}{\nu_2 - 2}\right] \\ &\qquad (\nu_2 > 6).\end{aligned}$$

We also note the characteristic function:

$$\sum_{j=0}^{\infty} \left[\left(\frac{\lambda_1}{2}\right)^j \bigg/ j!\right] e^{-\frac{1}{2}\lambda_1} M\left(\frac{\nu_1}{2} + j;\ -\frac{\nu_2}{2};\ -(\nu_2/\nu_1)it\right)$$

(cf. Equation (7) of Chapter 26).

Remembering (see Chapter 28) that the distribution of $\chi'^2_{\nu_1}(\lambda_1)$ can be represented as a mixture of central $\chi^2_{\nu_1+2j}$ distributions in proportions $e^{-\frac{1}{2}\lambda_1}(\frac{1}{2}\lambda_1)^j/j!$ $(j = 0,1,2,\ldots)$ we see that

$$G'_{\nu_1,\nu_2}(\lambda_1) = \chi^2_{\nu_1}(\lambda_1)/\chi^2_{\nu_2} \tag{4}$$

is distributed as a mixture of central G_{ν_1+2j,ν_2} distributions (as defined in Chapter 26, equation (3)) in proportions $e^{-\frac{1}{2}\lambda_1}(\frac{1}{2}\lambda_1)^j/j!$ $(j = 0,1,2,\ldots)$. Hence the probability density function of $G'_{\nu_1,\nu_2}(\lambda_1)$ is (writing only G' for convenience)

$$\begin{aligned} p_{G'}(g) &= \sum_{j=0}^{\infty} \left(\frac{[\frac{1}{2}\lambda_1]^j}{j!} e^{-\frac{1}{2}\lambda_1} \right) \frac{g^{\frac{1}{2}\nu_1+j-1}}{B(\frac{1}{2}\nu_1 + j, \frac{1}{2}\nu_1)(1+g)^{\frac{1}{2}(\nu_1+\nu_2)+j}} \\ &= \frac{e^{-\frac{1}{2}\lambda_1}}{B(\frac{1}{2}\nu_1,\frac{1}{2}\nu_2)} \cdot \frac{g^{\frac{1}{2}\nu_1-1}}{(1+g)^{\frac{1}{2}(\nu_1+\nu_2)}} \sum \left[\frac{\frac{1}{2}\lambda_1 g}{1+g} \right]^j \\ &\quad \times \frac{(\nu_1+\nu_2)(\nu_1+\nu_2+2)\ldots(\nu_1+\nu_2+2\cdot\overline{j-1})}{j!\nu_1(\nu_1+2)\ldots(\nu_1+2\cdot\overline{j-1})} \\ &\qquad\qquad (0 < g). \end{aligned} \tag{5}$$

The probability density function of $F'_{\nu_1,\nu_2}(\lambda_1) = (\nu_2/\nu_1)G'_{\nu_1,\nu_2}(\lambda_1)$ is (now using the contraction F', for convenience):

$$\begin{aligned} p_{F'}(f) &= \frac{e^{-\frac{1}{2}\lambda_1}\nu_1^{\frac{1}{2}\nu_1}\nu_2^{\frac{1}{2}\nu_2}}{B(\frac{1}{2}\nu_1,\frac{1}{2}\nu_2)} \cdot \frac{f^{\frac{1}{2}\nu_1-1}}{(\nu_2+\nu_1 f)^{\frac{1}{2}(\nu_1+\nu_2)}} \sum_{j=0}^{\infty} \left(\left[\frac{\frac{1}{2}\lambda_1\nu_1 f}{\nu_2+\nu_1 f} \right]^j \right. \\ &\quad \left. \times \frac{(\nu_1+\nu_2)(\nu_1+\nu_2+2)\ldots(\nu_1+\nu_2+2\cdot\overline{j-1})}{j!\nu_1(\nu_1+2)\ldots(\nu_1+2\cdot\overline{j-1})} \right) \\ &= p_{F_{\nu_1,\nu_2}}(f) \cdot e^{-\frac{1}{2}\lambda_1} \sum_{j=0}^{\infty} \left(\left[\frac{\frac{1}{2}\lambda_1\nu_1 f}{\nu_2+\nu_1 f} \right]^j \right. \\ &\quad \left. \times \frac{(\nu_1+\nu_2)(\nu_1+\nu_2+2)\ldots(\nu_1+\nu_2+2\cdot\overline{j-1})}{j!\nu_1(\nu_1+2)\ldots(\nu_1+2\cdot\overline{j-1})} \right) \end{aligned} \tag{6}$$

where, of course, $p_{F_{\nu_1,\nu_2}}(f)$ is the density function of the central F distribution with ν_1, ν_2 degrees of freedom.

(Note that while

$$p_{G'}(g) = \sum_{j=0}^{\infty} [e^{-\frac{1}{2}\lambda_1}(\tfrac{1}{2}\lambda_1)^j/j!] p_{G_{\nu_1+2j,\nu_2}}(g)$$

it is *not* true that

$$p_{F'}(f) = \sum_{j=0}^{\infty} [e^{-\frac{1}{2}\lambda_1}(\tfrac{1}{2}\lambda_1)^j/j!] p_{F_{\nu_1+2j,\nu_2}}(f).)$$

The cumulative distribution can be expressed in terms of an infinite series of multiples of incomplete beta function ratios

(7) $$\Pr[F'_{\nu_1,\nu_2}(\lambda_1) \leq f_0] = \Pr\left[G'_{\nu_1,\nu_2}(\lambda_1) \leq \frac{\nu_1}{\nu_2} f_0\right]$$
$$= \sum_{j=0}^{\infty} \left(\frac{[\frac{1}{2}\lambda_1]^j}{j!} e^{-\frac{1}{2}\lambda_1}\right) \cdot I_{\nu_1 f_0/(\nu_2+\nu_1 f_0)}(\tfrac{1}{2}\nu_1 + j; \tfrac{1}{2}\nu_2).$$

Since it is possible (Chapter 24) to express the incomplete beta function ratio in different ways, there is a corresponding range of different expressions for the cumulative distribution function of noncentral F. For the special case when ν_2 is an even integer, there are some quite simple expressions in finite terms. Sibuya [25] has pointed out that these all can be obtained by using the formal identity

(8)
$$\sum_{j=0}^{\infty} \frac{(\frac{1}{2}\lambda_1)^j}{j!} e^{-\frac{1}{2}\lambda_1} f(j) = \sum_{j=0}^{\infty} \frac{(\frac{1}{2}\lambda_1)^j}{j!} \Delta^j f(0) \qquad \text{(cf. equation (9), Chapter 28)}$$

with $f(\cdot)$ an incomplete beta function *ratio*, together with recurrence relationships satisfied by this function.

In particular, Sibuya [25] shows that (if ν_2 is an even integer)

(9)
$$\Pr[F'_{\nu_1,\nu_2}(\lambda_1) \leq f_0] = e^{-\frac{1}{2}\lambda_1(1-Y)} \sum_{i=0}^{\frac{1}{2}\nu_2-1} \frac{[\frac{1}{2}\lambda_1(1-Y)]^i}{i!} I_Y(\tfrac{1}{2}\nu_1 + i, \tfrac{1}{2}\nu_2 - i)$$

where $Y = \nu_1 f_0/(\nu_2 + \nu_1 f_0)$.

Replacing $I_Y(\frac{1}{2}\nu_1 + i, \frac{1}{2}\nu_2 - i)$ by a polynomial, we obtain a formula that was given in 1963 by Seber [23]:

(10)
$$Y^{\frac{1}{2}\nu_1}\{\exp[-\tfrac{1}{2}\lambda_1(1-Y)]\} \sum_{i=0}^{\frac{1}{2}\nu_2-1} (1-Y)^i \sum_{j=0}^{i} \left\{\binom{i+\frac{1}{2}\nu_1-1}{i-j} [\tfrac{1}{2}\lambda_1 Y]^j/j!\right\}$$
$$= Y^{\frac{1}{2}\nu_1}\{\exp[-\tfrac{1}{2}\lambda_1(1-Y)]\} \sum_{i=0}^{\frac{1}{2}\nu_2-1} T_i$$

where

$$T_{-1} = 0; \quad T_0 = 1$$

and

$$T_i = i^{-1}(1-Y)\{(2i-2+\tfrac{1}{2}\nu_1+\tfrac{1}{2}\lambda_1 Y)T_{i-1} - (i+\tfrac{1}{2}\nu_1-2)(1-Y)T_{i-2}\}$$
$$(i = 1,2,\ldots,\tfrac{1}{2}\nu_2 - 1).$$

This formula was obtained (in slightly different form) in 1954 by Nicholson [16] and Hodges [8], though these authors did not give the recurrence formula for T_i. An expression of similar type, given by Wishart [34] and Tang [27] is:

(11) $$Y^{\frac{1}{2}(\nu_1+\nu_2)-1} e^{-\frac{1}{2}\lambda_1(1-Y)} \sum_{i=0}^{\frac{1}{2}\nu_2-1} T'_i$$

where

$$T'_{-1} = 0; \quad T'_0 = 1;$$

and

$$T'_i = i^{-1}(Y^{-1} - 1)\{[\tfrac{1}{2}(\nu_1 + \nu_2) - i + \tfrac{1}{2}\lambda_1 Y]T'_{i-1} + \tfrac{1}{2}\lambda_1(1 - Y)T'_{i-2}\}$$
$$(i = 1,2,\ldots,\tfrac{1}{2}\nu_2 - 1).$$

Each of the expressions (9)–(11) applies only where ν_2 is an even integer. Price [21] has obtained some finite expressions which are applicable when ν_2 is an odd integer. These are rather complicated, and are not reproduced here. If formula (8) be applied directly to (7), the following *infinite* series expansion, valid for general ν_2, is obtained:

(12)
$$\Pr[F'_{\nu_1,\nu_2}(\lambda_1) \leq f_0] = I_Y(\tfrac{1}{2}\nu_1,\tfrac{1}{2}\nu_2) - \frac{\Gamma(\frac{1}{2}(\nu_1 + \nu_2))}{\Gamma(\frac{1}{2}\nu_1 + 1)} Y^{\frac{1}{2}\nu_1}(1 - Y)^{\frac{1}{2}\nu_2} \sum_{j=0}^{\infty} \frac{\lambda^j}{j!} \Delta^{j-1} t_j$$

where

$$t_1 = 1; \quad t_{j+1} = (\tfrac{1}{2}\nu_1 + j)^{-1}[\tfrac{1}{2}(\nu_1 + \nu_2) + j - 1]Yt_j.$$

This can be expressed in terms of generalized Laguerre polynomials (see Tiku [28]). Tiku also obtained (13), which is a more complicated but more rapidly convergent expansion than (12):

(13)
$$\Pr[F'_{\nu_1,\nu_2}(\lambda_1) \leq f_0] = I_{Y'}(\tfrac{1}{2}a,\tfrac{1}{2}\nu_2) + \sum_{j=3}^{\infty} \frac{(-1)^j}{j!} b_j \frac{Y'^{\frac{1}{2}a}(1 - Y')^{\frac{1}{2}\nu_2}}{B(\frac{1}{2}a,\frac{1}{2}\nu_2)} \Delta^{j-1} t_j$$

where the t_j's are as defined in (12),

$$a = (\nu_1 + \lambda_1)^2(\nu_1 + 2\lambda_1)^{-1}; \quad Y' = 1 - \left[1 + \frac{\nu_1(\nu_1 + \lambda_1)}{\nu_2(\nu_1 + 2\lambda_2)} f_0\right]^{-1};$$

$$b_3 = 2\lambda_1^2(\nu_1 + 2\lambda_1)^{-2}; \quad b_4 = 6\lambda_1^2(\nu_1 + 4\lambda_1)(\nu_1 + 2\lambda_1)^{-3};$$

$$b_5 = 24\lambda_1^2(\nu_1 + 6\nu_1\lambda_1 + 11\lambda_1^2)(\nu_1 + 2\lambda_1)^{-4}; \quad \ldots$$

(Note that the b's do not depend on ν_2.)

It can be shown that (as would intuitively be expected) $\Pr[F'_{\nu_1,\nu_2}(\lambda_1) \leq f_0]$ is a decreasing function of λ_1. The probability density function is unimodal.

As ν_2 tends to infinity the distribution of $F'_{\nu_1,\nu_2}(\lambda_1)$ approaches that of ν_1^{-1} (noncentral χ^2 with ν_1 degrees of freedom and noncentrality parameter λ_1). Also, of course, as λ_1 tends to zero the distribution tends to the (central) F_{ν_1,ν_2} distribution.

4. Tables

The earliest tables to be published (1938) were those of Tang [27]. These were motivated by the calculation of power functions of variance ratio tests, and give values of $\Pr[F'_{\nu_1,\nu_2}(\lambda_1) > F_{\nu_1,\nu_2,\alpha}]$ to 3 decimal places for $\alpha = 0.95, 0.99$; $\nu_1 = 1(1)8$; $\nu_2 = 2(2)6(1)30$, 60, ∞ and $\sqrt{\lambda_1/(\nu_1 + 1)} = 1.0(0.5)3.0(1)8$.

This table has been reproduced in a number of text books. It has been extended by Lachenbruch [10] who gives values of the probability to 4 decimal places for the same values of α and $\nu_1 = 1(1)12(2)16(4)24, 30(10)50, 75$; $\nu_2 = 2(2)20(4)40(10)80$; and $\sqrt{\lambda_1/(\nu_1+1)} = 1.0(0.5)3.0(1)8$. Lachenbruch also gives tables of percentiles $F'_{\nu_1,\nu_2}(\lambda_1)$ of the noncentral F distribution for $\lambda_1 = 2(2)20$, $\alpha = 0.01, 0.025, 0.05, 0.1, 0.5, 0.9, 0.95, 0.975, 0.99$; $\nu_1 = 1(1)10, 15, 20, 30, 50, 60, 120$; $\nu_2 = 2(2)10(10)40, 60$. Values are generally to 4 decimal places, except for $\nu_1 = 1$ or $\nu_2 \geq 30$, when they are only to three significant figures. (Three significant figures, only, are also given for $\nu_1 = 2, 3, 4$ with $\nu_2 = 2$; $\nu_1 = 2, 3$ with $\nu_2 = 4$; $\nu_1 = 2$ with $\nu_2 = 6$ and $\nu_1 = 120$ with $\nu_2 = 30$).

Pearson and Hartley [20] have given graphical representations of Tang's tables, in the form of power functions of analysis of variance tests. Patnaik [18] has published a chart showing the relations among λ_1, ν_1 and ν_2 implied by the constraints

$$\Pr[F'_{\nu_1,\nu_2}(\lambda_1) \geq F_{\nu_1,\nu_2,\alpha}] = \beta \tag{14}$$

for $\alpha = 0.95$; $\beta = 0.5$ or 0.9. Fox [7] has given charts showing contours of $\phi = \sqrt{\lambda/(\nu_1+1)}$ in the (ν_1,ν_2) plane when (14) is satisfied, for $\alpha = 0.95, 0.99$; $\beta = 0.5(0.1)0.9$.

Lehmer [13] has given values of ϕ to 3 decimal places for $\alpha = 0.95, 0.99$; $\beta = 0.2, 0.3$ with $\nu_1 = 1(1)9$; and $120/\nu_1 = 1(1)6(2)12$; $\nu_2 = 2(2)18$ and $240/\nu_2 = 1(1)4(2)12$. There is also a table of this kind published by Ura [32]. It gives $\sqrt{\lambda/\nu_1}$ to *two* decimal places for $\alpha = 0.95$, $\beta = 0.90$ with $\nu_1 = 1(1)9$ and $120/\nu_1 = 0(1)6(2)12$, $\nu_2 = 2(2)18$ and $120/\nu_2 = 0(1)6$.

Tiku [31] has given values of $\Pr[F'_{\nu_1,\nu_2}(\lambda_1) > F_{\nu_1,\nu_2,1-\alpha}]$ to 4 decimal places for $\alpha = 0.005, 0.01, 0.025, 0.05$; $\nu_1 = 1(1)10, 12$; $\nu_2 = 2(2)30, 40, 60, 120, \infty$; $[\lambda_1/(\nu_1+1)]^{\frac{1}{2}} = 0.5(0.5)3.0$.

Bargmann and Ghosh [1] have reported a Fortran program which computes the probability density and cumulative distribution functions of the noncentral F distribution.

5. Approximations

From formula (2) it can be seen that approximations to noncentral χ^2 distributions can be used to give approximations to noncentral F distributions. Thus the simple approximation to the distribution of $\chi'^2_{\nu_1}(\lambda_1)$ by that of $c\chi^2_\nu$, with $c = (\nu_1 + 2\lambda_1)(\nu_1 + \lambda_1)^{-1}$; $\nu = (\nu_1 + \lambda_1)^2(\nu_1 + 2\lambda_1)^{-1}$ gives an approximation to the distribution of $F'_{\nu_1,\nu_2}(\lambda_1)$ by that of $(c\nu/\nu_1)F_{\nu,\nu_2} = (1 + \lambda_1\nu_1^{-1})F_{\nu,\nu_2}$. (Note the need for the factor ν/ν_1.) The accuracy of this approximation has been studied by Patnaik [18]. Of course, the distribution of F_{ν,ν_2} itself may also be approximated by one of the methods described in Chapter 26, leading to a composite approximation to the distribution of $F'_{\nu_1,\nu_2}(\lambda_1)$. Thus, using Paulson's approximation, Severo and Zelen [24] were led to suggest

$$\text{(15)} \qquad \frac{(1 - \frac{2}{9}\nu_2^{-1})\left(\dfrac{\nu_1 F'}{\nu_1 + \lambda_1}\right)^{\frac{1}{3}} - [1 - \frac{2}{9}(\nu_1 + 2\lambda_1)(\nu_1 + \lambda_1)^{-2}]}{\left[\frac{2}{9}(\nu_1 + 2\lambda_1)(\nu_1 + \lambda_1)^{-2} + \frac{2}{9}\nu_2^{-1}\left(\dfrac{\nu_1 F'}{\nu_1 + \lambda_1}\right)^{\frac{2}{3}}\right]^{\frac{1}{2}}}$$

as having, approximately, a unit normal distribution. Laubscher [11] compared the accuracy of this approximation with that obtained by using Fisher's approximation to the distribution of the χ^2's and F_{ν_1,ν_2}. The results were markedly favorable to (15). Laubscher also considered, as a possible normalizing transformation,

$$(\tfrac{1}{2}\nu_2 - 2)^{\frac{1}{2}} \cosh^{-1}\left[\frac{\nu_1(\nu_2 - 2)^{\frac{1}{2}}(F' + \nu_2/\nu_1)}{\nu_2(\nu_1 + \nu_2 - 2)^{\frac{1}{2}}}\right],$$

suggested as an approximately variance-equalizing transformation, but this gave very poor results. Tiku [30] obtained quite good results by fitting the distribution of $F'_{\nu_1,\nu_2}(\lambda_1)$ by that of $(b + cF_{\nu_1',\nu_2})$, choosing b, c, and ν_1', so as to make the first three moments agree. The values which do this are

$$\begin{aligned} \nu_1' &= \tfrac{1}{2}(\nu_2 - 2)\left[\sqrt{\frac{H^2}{H^2 - 4K^3}} - 1\right] \\ c &= (\nu_1'/\nu_1)(2\nu_1' + \nu_2 - 2)^{-1}(H/K) \\ b &= -\nu_2(\nu_2 - 2)^{-1}(c - 1 - \lambda_1\nu_1^{-1}) \end{aligned}$$

where

$$\begin{aligned} H &= 2(\nu_1 + \lambda_1)^3 + 3(\nu_1 + \lambda_1)(\nu_1 + 2\lambda_1)(\nu_2 - 2) + (\nu_1 + 3\lambda_1)(\nu_2 - 2)^2 \\ K &= (\nu_1 + \lambda_1)^2 + (\nu_2 - 2)(\nu_1 + 2\lambda_1). \end{aligned}$$

Comparisons between various approximations were carried out by Tiku. His conclusions were: the 3-moment central F approximation is the most accurate, Severo and Zelen's normal approximation seems to be the easiest to compute although slightly less accurate than Patnaik's approximation for large ν_2.

By analogy with the central F distribution, one might expect to obtain useful approximations by considering the distribution of $z'_{\nu_1,\nu_2}(\lambda_1) = \frac{1}{2}\log F'_{\nu_1,\nu_2}(\lambda_1)$.* Since $z' = z'_{\nu_1,\nu_2}(\lambda_1) = \frac{1}{2}\log(\nu_2/\nu_1) + \frac{1}{2}\log \chi'^2_{\nu_1}(\lambda_1) - \frac{1}{2}\log \chi^2_{\nu_2}$, the cumulants of z' are

$$\text{(16)} \qquad \begin{aligned} \kappa_1(z') &= \tfrac{1}{2}[\log(\nu_2/\nu_1) + \kappa_1(\chi'^2_{\nu_1}(\lambda_1)) - \kappa_1(\chi^2_{\nu_2})] \\ \kappa_r(z') &= 2^{-r}[\kappa_r(\chi'^2_{\nu_1}(\lambda_1)) + (-1)^r\kappa_r(\chi^2_{\nu_2})] \qquad (r \geq 2). \end{aligned}$$

Barton *et al.* [2] have given formulas to aid in the computation of $\kappa_r(\chi'^2_{\nu_1}(\lambda_1))$, and have utilized them in calculating the power function of the F-test, by fitting Edgeworth series to the distribution of z'. We reproduce here from [2] a table

*This distribution is called the *noncentral z distribution.*

TABLE 1

Values of $\sqrt{\beta_1}$ and β_2 for the Distribution of z'.

		$\sqrt{\beta_1}$				β_2			
		ν_2				ν_2			
λ_1	ν_1	2	4	10	20	2	4	10	20
0	2	0	−0.5772	−0.9238	−1.0336	4.2	4.3327	4.8707	5.1210
$\frac{1}{2}$		0.0014	−0.5850	−0.9406	−1.0540	4.2192	4.3634	4.9239	5.1855
1		0.0088	−0.5977	−0.9741	−1.0958	4.2647	4.4357	5.0526	5.3430
3		0.1165	−0.5835	−1.0787	−1.2533	4.4884	4.7686	5.7550	6.2616
0	4	0.5772	0	−0.4407	−0.6050	4.3327	3.5938	3.6868	3.8820
$\frac{1}{2}$		0.5790	0.0004	−0.4436	−0.6098	4.3380	3.5980	3.6940	3.8928
1		0.5845	0.0025	−0.4490	−0.6198	4.3517	3.6084	3.7118	3.9195
5		0.6901	0.0985	−0.4223	−0.6453	4.5420	3.7025	3.8343	4.1501
0	10	0.9238	0.4407	0	−0.2023	4.8707	3.6868	3.2187	3.2228
$\frac{1}{2}$		0.9241	0.4410	0.0000	−0.2026	4.8714	3.6874	3.2191	3.2233
1		0.9249	0.4419	0.0003	−0.2033	4.8736	3.6891	3.2202	3.2248
0	20	1.0336	0.6050	0.2023	0	5.1210	3.8820	3.2228	3.1049
$\frac{1}{2}$		1.0337	0.6051	0.2024	0.0000	5.1211	3.8821	3.2229	3.1049
1		1.0338	0.6053	0.2025	0.0000	5.1215	3.8825	3.2230	3.1050

of values of $\sqrt{\beta_1}$ and β_2 for the distribution of z'. (This paper also contains a useful table of $\psi^{(r)}(x)$ (defined in Chapter 1, Equation (34)) to 8 decimal places for $r = 0(1)3$, $x = 1.0(0.5)15.5$).

Pearson [19] has obtained good results by fitting the distribution of z' by S_U distributions (see Chapter 12). (It should be noted that Tiku [28] stated, incorrectly, that Pearson fitted F' itself by an S_U distribution.) However, the computation of the cumulants of z' is rather laborious. (In [2], where details are given, the cumulants are expressed in terms of polygamma functions $\psi^{(r)}(x)$ and special $\mathcal{R}$-functions tabulated on pages 426–429 of [2].)

6. Related Distributions

The *doubly noncentral F distribution* defined by (1) has already been noted. Using the representation of each of the noncentral χ^2's as mixtures of central χ^2 distributions we see that

$$G''_{\nu_1,\nu_2}(\lambda_1,\lambda_2) = \chi'^2_{\nu_1}(\lambda_1)/\chi'^2_{\nu_2}(\lambda_2)$$

is distributed as a mixture of G_{ν_1+2j,ν_2+2k} distributions in proportions $[e^{-\frac{1}{2}\lambda_1}(\frac{1}{2}\lambda_1)^j/j!] \times [e^{-\frac{1}{2}\lambda_2}(\frac{1}{2}\lambda_2)^k/k!]$. Hence (using the contracted forms F'', G'' for the variables) the probability density function of G'' is

(17) $$p_{G''}(g) = \sum_{j=0}^{\infty}\sum_{k=0}^{\infty} [e^{-\frac{1}{2}\lambda_1}(\tfrac{1}{2}\lambda_1)^j/j!][e^{-\frac{1}{2}\lambda_2}(\tfrac{1}{2}\lambda_2)^k/k!] \times [B(\tfrac{1}{2}\nu_1 + j, \tfrac{1}{2}\nu_2 + k)]^{-1} g^{\frac{1}{2}\nu_1+j-1}(1+g)^{-\frac{1}{2}(\nu_1+\nu_2)-j-k}$$

and that of F'' is

(18)
$$p_{F''}(f) = \sum_{j=0}^{\infty}\sum_{k=0}^{\infty} [e^{-\frac{1}{2}\lambda_1}(\tfrac{1}{2}\lambda_1)^j/j!][e^{-\frac{1}{2}\lambda_2}(\tfrac{1}{2}\lambda_2)^k/k!]\nu_1^{\frac{1}{2}\nu_1+j}\nu_2^{\frac{1}{2}\nu_2+k} f^{\frac{1}{2}\nu_1+j-1} \times (\nu_2+\nu_1 f)^{-\frac{1}{2}(\nu_1+\nu_2)-j-k}[B(\tfrac{1}{2}\nu_1+j, \tfrac{1}{2}\nu_2+k]^{-1}$$

$$= p_{F_{\nu_1,\nu_2}}(f)\cdot \sum_{j=0}^{\infty}\sum_{k=0}^{\infty}\left[e^{-\frac{1}{2}\lambda_1}\left(\frac{\frac{1}{2}\lambda_1\nu_1 f}{\nu_2+\nu_1 f}\right)^j \Big/ j!\right] \times \left[e^{-\frac{1}{2}\lambda_2}\left(\frac{\frac{1}{2}\lambda_2\nu_2}{\nu_2+\nu_1 f}\right)^k \Big/ k!\right]\frac{B(\frac{1}{2}\nu_1,\frac{1}{2}\nu_2)}{B(\frac{1}{2}\nu_1+j,\frac{1}{2}\nu_2+k)}.$$

Approximations to the distributions of the noncentral χ^2 distributions can be used to derive approximations to the doubly noncentral F distributions. It is, of course, possible to approximate the distribution of just one of the two noncentral χ^2's. Thus if the distribution of $\chi'^2_{\nu_2}(\lambda_2)$ be approximated by that of $c'\chi^2_{\nu'}$, with $c' = (\nu_2 + 2\lambda_2)(\nu_2+\lambda_2)^{-1}$ and $\nu' = (\nu_2+\lambda_2)^2(\nu_2+2\lambda_2)^{-1}$ then the corresponding approximation to the distribution of $F''_{\nu_1,\nu_2}(\lambda_1,\lambda_2)$ is that of $\left(\frac{\nu_2}{c'\nu'}\right)F'_{\nu_1,\nu'}(\lambda_1) = (1+\lambda_2\nu_2{}^{-1})^{-1}F'_{\nu_1,\nu'}(\lambda_1)$. If both numerator and denominator are approximated, the approximating distribution is that of

(19) $$\frac{1+\lambda_1\nu_1^{-1}}{1+\lambda_2\nu_2^{-1}}F_{\nu,\nu'}$$

with $\nu = (\nu_1+\lambda_1)^2(\nu_1+2\lambda_1)^{-1}$; $\nu' = (\nu_2+\lambda_2)^2(\nu_2+2\lambda_2)^{-1}$.

The doubly noncentral F distribution is used when estimating the effect, on the power function of analysis of variance tests, of non-random effects in the residual variation. For example, in a standard one-way classification if each individual in a group has a departure from the group mean, depending on the order of observation, this makes the residual (within group) sum of squares distributed as a multiple of a noncentral, rather than a central χ^2 variable ([22], pp. 134–135).

If $\chi^2_{\nu_1}$ and $\chi^2_{\nu_2}$ are mutually independent then it is known (Chapter 26) that $\chi^2_{\nu_1}(\chi^2_{\nu_1}+\chi^2_{\nu_2})^{-1}$ has a standard beta distribution with parameters $\frac{1}{2}\nu_1$, $\frac{1}{2}\nu_2$. If $\chi^2_{\nu_1}$ is replaced by the noncentral $\chi'^2_{\nu_1}(\lambda_1)$ the resultant distribution is called the *noncentral beta distribution with (shape) parameters* $\frac{1}{2}\nu_1$, $\frac{1}{2}\nu_2$ and noncentrality parameter λ_1. If both χ^2's are replaced by noncentral χ^2's, giving

(20) $$\chi'^2_{\nu_1}(\lambda_1)\{\chi'_{\nu_1}(\lambda_1)+\chi'_{\nu_2}(\lambda_2)\}^{-1},$$

the corresponding distribution is the *doubly noncentral beta distribution with (shape) parameters* $\frac{1}{2}\nu_1$, $\frac{1}{2}\nu_2$ *and noncentrality parameters* λ_1, λ_2. Noncentral beta distributions can be represented as mixtures of central beta distributions in the same way as noncentral F's can be represented as mixtures of central F's.

Each noncentral χ^2 may be approximated by Patnaik's approximation (Chapter 28). This leads to approximating the distribution by that of

$$\frac{(\nu_1 + 2\lambda_1)(\nu_2 + \lambda_2)}{(\nu_1 + \lambda_1)(\nu_2 + 2\lambda_2)} \times \text{(beta variable with parameters } f_1,\ f_2)$$

where $f_j = (\nu_j + \lambda_j)^2(\nu_j + 2\lambda_j)^{-1}$, $(j = 1,2)$.

DasGupta [4] has compared this approximation with

(a) an expansion using Jacobi's polynomials (see Chapter 1, Section 3) with initial beta distribution having correct first and second moments, and

(b) using Laguerre series expansions (see Chapter 29) for each of the noncentral χ^2 distributions.

He found that the Patnaik approximation was in general sufficiently accurate for practical purposes. Although approximations (a) and (b) are rather more accurate, they are considerably more troublesome to compute.

It will be recalled that the distribution of $\chi_\nu'^2(\lambda)$ is related to that of the difference between two independent Poisson variables (see Chapter 28). By a similar argument (Johnson [9]) it can be shown that, if ν_1 is even,

$$\Pr[F'_{\nu_1,\nu_2}(\lambda_1) < f_0/\nu_1] = \Pr[Y - Z \geq \tfrac{1}{2}\nu_1] \tag{21}$$

where Y and Z are mutually independent, Y has a negative binomial distribution (Chapter 5, Section 1) with parameters $\frac{1}{2}\nu_2$, f_0/ν_2 and Z has a Poisson distribution with parameter $\frac{1}{2}\lambda_1$. By a direct extension of this argument we obtain for the doubly noncentral F distribution, the relation

$$\Pr[F''_{\nu_1,\nu_2}(\lambda_1,\lambda_2) < f_0/\nu_1] = \sum_{j=0}^{\infty} e^{-\frac{1}{2}\lambda_2} \frac{(\frac{1}{2}\lambda_2)^j}{j!} \Pr[Y_j - Z \geq \tfrac{1}{2}\nu_1] \tag{22}$$

where Y_j and Z are independent, Y_j has a negative binomial distribution with parameters $\frac{1}{2}\nu_2 + j$, f_0/ν_2, and Z is distributed as in (21).

REFERENCES

[1] Bargmann, R. E. and Ghosh, S. P. (1964). *Noncentral statistical distribution programs for a computer language*, Report No. RC-1231, IBM Watson Research Center, Yorktown Heights, New York.

[2] Barton, D. E., David, F. N. and O'Neill, A. F. (1960). Some properties of the distribution of the logarithm of non-central F, *Biometrika*, **47,** 417–429.

[3] Bennett, B. M. (1955). Note on the moments of the logarithmic noncentral χ^2 and z distributions, *Annals of the Institute of Statistical Mathematics, Tokyo*, **7,** 57–61.

[4] DasGupta, P. (1968). Two approximations for the distribution of double non-central beta, *Sankhyā, Series B*, **30,** 83–88.

[5] Dixon, W. J. (1962). Rejection of observations, (In *Contributions to Order Statistics*, (Ed. A. E. Sarhan and B. G. Greenberg), pp. 299–342, New York: John Wiley & Sons, Inc.).

[6] Fisher, R. A. (1928). The general sampling distribution of the multiple correlation coefficient, *Proceedings of the Royal Society of London, Series A*, **121,** 654–673.

[7] Fox, M. (1956). Charts of the power of the F-test, *Annals of Mathematical Statistics*, **27,** 484–497.

[8] Hodges, J. L. (1955). On the noncentral beta-distribution, *Annals of Mathematical Statistics*, **26,** 648–653.

[9] Johnson, N. L. (1959). On an extension of the connexion between the Poisson and χ^2 distributions, *Biometrika*, **46,** 352–363.

[10] Lachenbruch, P. A. (1966). The noncentral F distribution — extensions of Tang's tables, *Annals of Mathematical Statistics*, **37,** 774. (Abstract). (Tables in University of North Carolina Mimeo Series No. 531 (1967.))

[11] Laubscher, N. H. (1960). Normalizing the noncentral t and F distributions, *Annals of Mathematical Statistics*, **31,** 1105–1112.

[12] Lehmann, E. L. (1959). *Testing Statistical Hypotheses*, New York: John Wiley & Sons, Inc.

[13] Lehmer, E. (1944). Inverse tables of probabilities of errors of the second kind, *Annals of Mathematical Statistics*, **15,** 388–398.

[14] Madow, W. G. (1948). On a source of downward bias in the analysis of variance and covariance, *Annals of Mathematical Statistics*, **19,** 351–359.

[15] Marakathavalli, N. (1955). Unbiased test for a specified value of the parameter in the non-central F distribution, *Sankhyā*, **15,** 321–330.

[16] Nicholson, W. L. (1954). A computing formula for the power of the analysis of variance test, *Annals of Mathematical Statistics*, **25,** 607–610.

[17] Park, J. H. (1964). Variations of the noncentral t and beta distributions, *Annals of Mathematical Statistics*, **35,** 1583–1593.

[18] Patnaik, P. B. (1949). The noncentral χ^2 and F-distributions and their applications, *Biometrika*, **36,** 202–232.

[19] Pearson, E. S. (1960). Editorial note, *Biometrika*, **47,** 430–431.

[20] Pearson, E. S. and Hartley, H. O. (1951). Charts of the power function for analysis of variance tests, derived from the noncentral *F*-distribution, *Biometrika*, **38,** 112–130.

[21] Price, R. (1964). Some noncentral *F*-distributions expressed in closed form, *Biometrika*, **51,** 107–122.

[22] Scheffé, H. (1959). *The Analysis of Variance*, New York: John Wiley & Sons, Inc.

[23] Seber, G. A. F. (1963). The noncentral chi-squared and beta distributions, *Biometrika*, **50,** 542–544.

[24] Severo, N. and Zelen, M. (1960). Normal approximation to the chi-square and noncentral *F* probability functions, *Biometrika*, **47,** 411–416.

[25] Sibuya, M. (1967). On the noncentral beta distribution function, Unpublished manuscript.

[26] Steffens, F. E. (1968). *Probability integrals of doubly noncentral F- and t-distributions with regression applications*, Research Report No. 267, Council for Scientific and Industrial Research, Pretoria, South Africa.

[27] Tang, P. C. (1938). The power function of the analysis of variance tests with tables and illustrations of their use, *Statistical Research Memoirs*, **2,** 126–150.

[28] Tiku, M. L. (1965). Laguerre series forms of noncentral χ^2 and *F* distributions, *Biometrika*, **52,** 415–427.

[29] Tiku, M. L. (1965). Series expansions for the doubly noncentral *F*-distribution, *Australian Journal of Statistics*, **7,** 78–89.

[30] Tiku, M. L. (1966). A note on approximating to the noncentral *F*-distribution, *Biometrika*, **53,** 606–610.

[31] Tiku, M. L. (1967). Tables of the power of the *F*-test, *Journal of the American Statistical Association*, **62,** 525–539.

[32] Ura, S. (1954). A table of the power function of the analysis of variance test, *Reports of Statistical Application Research, JUSE*, **3,** 23–28.

[33] Weibull, M. (1953). The distributions of *t*- and *F*-statistics and of correlation and regression coefficients in stratified samples from normal populations with different means, *Skandinavisk Aktuarietidskrift*, **1–2,** Supplement.

[34] Wishart, J. (1932). A note on the distribution of the correlation ratio, *Biometrika*, **24,** 441–456.

31

Noncentral t Distribution

1. Definition

The distribution of the ratio

$$t'_\nu(\delta) = (U + \delta)/(\chi_\nu \nu^{-\frac{1}{2}}),$$

where U and χ_ν are independent random variables distributed as standard normal ($N(0,1)$) and chi with ν degrees of freedom respectively, and δ is a constant, is called the *noncentral t distribution with ν degrees of freedom and noncentrality parameter* δ. Sometimes δ^2 (or even $\frac{1}{2}\delta^2$), rather than δ, is termed the noncentrality parameter. If δ is equal to zero, the distribution is that of (central) t with ν degrees of freedom, which has already been discussed.

When there is no fear of confusion, δ may be omitted and t'_ν used instead of $t'_\nu(\delta)$. Occasionally, even the ν may be omitted, and t' used. However, whenever there is possibility of confusion — for example, when two or more values of the noncentrality parameter are under discussion — the full symbol, $t'_\nu(\delta)$, should be used.

2. Historical Remarks

The noncentral t distribution was derived (though not under that name) by Fisher [16], who showed how tables of repeated partial integrals of the standard normal distribution could be used in connection with this distribution. Tables given by Neyman *et al.* [41] and Neyman and Tokarska [42] were based on evaluation of probability integrals of certain noncentral t distributions.

Tables from which percentage points of noncentral t distributions could be obtained were given by Johnson and Welch [30]. Later tables (Resnikoff and Lieberman [53], Owen [43], Locks *et al.* [35]) are fuller and require less calculation. Charts based on the probability integral are given by Pearson and

Hartley [52] (see also Section 7).

Formal expressions for the distribution function are rather more complicated than those for noncentral χ^2 and F. A number of different formulas can be found in Amos [2].

A number of approximations for probability integrals, and for percentage points, of noncentral t distributions have been proposed. Comparisons of various approximations for percentage points have been made by van Eeden [15] and by Owen [43].

Computer programs for calculating percentage points have been described by Owen and Amos [46] and by Bargmann and Ghosh [5]. Amos [2] has reported on comparisons of two such computer programs.

3. Genesis and Applications

The statistic $\sqrt{n}(\bar{X} - \xi_0)/S$ is used in testing the hypothesis that the mean of a normal population is equal to ξ_0. If $\bar{X}(= n^{-1} \sum_{i=1}^{n} X_i)$ and

$$S\left(= \sqrt{(n-1)^{-1} \sum_{i=1}^{n} (X_i - \bar{X})^2}\right)$$

are calculated from a random sample of size n, and the population mean is equal to ξ_0, then $\sqrt{n}\,(\bar{X} - \xi_0)/S$ should be distributed as (central) t with $(n - 1)$ degrees of freedom. If however the population mean (ξ) is not equal to ξ_0, then $\sqrt{n}\,(\bar{X} - \xi_0)/S$ is distributed as $t'_{n-1}(\sqrt{n}\,(\xi - \xi_0)/\sigma)$ where σ is the population standard deviation. The power of the test is calculated as a partial integral of the probability density function of this noncentral t distribution.

Similarly, a statistic used in testing equality of means of two normal populations, (I) and (II) (with common, though unknown, variance, σ^2), using random samples of sizes n_1, n_2 respectively, is, in an obvious notation

$$\sqrt{n_1 n_2 (n_1 + n_2)^{-1}}(\bar{X}_1 - \bar{X}_2)/\sqrt{(n_1 + n_2 - 2)^{-1}[(n_1 - 1)S_1^2 + (n_2 - 1)S_2^2]}.$$

If the two population means are indeed equal, this statistic should be distributed as (central) t with $(n_1 + n_2 - 2)$ degrees of freedom. If, however, (mean of population (I)) $-$ (mean of population (II)) $= \theta$ then the statistic is distributed as $t'_{n_1+n_2-2}(\theta\sigma^{-1}\sqrt{n_1 n_2 (n_1 + n_2)^{-1}})$. Here, again, the power of the test can be calculated as a partial integral of the appropriate noncentral t distribution.

Charts giving powers of t-tests have been published in Pearson and Hartley [52] and Croarkin [11]. Tables of the power function are also available in Neyman *et al.* [41] and Davies [13].

It is sometimes desired to calculate confidence intervals for the ratio of population mean to standard deviation (reciprocal of the coefficient of variation). Such intervals may be computed in suitable cases by noting that if $\bar{X}$ and

$S^2 = (n-1)^{-1} \sum_{j=1}^{n} (X_j - \bar{X})^2$ be calculated from a random sample of values $X_1, X_2, \ldots, X_n$ from a normal population with expected value ξ and standard deviation σ, then $\sqrt{n}\,\bar{X}/S$ is distributed as $t'_{n-1}(\sqrt{n}\,\xi/\sigma)$. Symmetrical $100(1-\alpha)\%$ confidence limits for ξ/σ are obtained as solutions for θ of the equations

$$t'_{n-1,\alpha/2}(\sqrt{n}\,\theta) = \sqrt{n}\,\bar{X}/S$$

and

$$t'_{n-1,1-\alpha/2}(\sqrt{n}\,\theta) = \sqrt{n}\,\bar{X}/S.$$

(Approximations to the sample coefficient of variation in relation to the noncentral t distribution were studied by McKay [39] and an accurate (to four decimal places) approximation to the percentage points in terms of the χ^2 percentage points was derived by Iglewicz *et al.* [27].)

Belobragina and Eliseyev [7] have constructed a nomogram which indirectly gives the lower $(100-\alpha)\%$ bounds for ξ/σ. Actually their charts show upper bounds for $\Phi(-\xi/\sigma)$, given $\bar{X}/S$. Sample sizes included are 5, 10, 20, 50 and 1000; and confidence coefficients 90%, 95%, 97%, 99%, and 99.9%.

Finally consider the problem of constructing a 'tolerance limit', using $\bar{X}$ and S as defined above. Suppose we wish to find a number k such that

$$\Pr[\bar{X} + kS < \xi + U_\alpha \sigma] = \gamma$$

(ξ and σ defined as above). The inequality can be re-written

$$\sqrt{n}\,(\bar{X} - \xi - U_\alpha\sigma)/S < -k\sqrt{n}.$$

and so $(-k\sqrt{n})$ is the lower $100\gamma\%$ point of the $t'_{n-1}(-U_\alpha\sqrt{n})$ distribution (see e.g. [58]). A more detailed survey of these and some other applications (with an extensive bibliography of over 100 references) of the noncentral t-distribution is given in a recent paper by Owen [45].

4. Moments

The rth moment of t'_ν about zero is

$$E(t'^r_\nu) = \nu^{\frac{1}{2}r} E(\chi_\nu^{-r}) E[(U+\delta)^r] \tag{1}$$

$$= (\tfrac{1}{2}\nu)^{\frac{1}{2}r} \frac{\Gamma(\frac{1}{2}(\nu - r))}{\Gamma(\frac{1}{2}\nu)} \sum_{0 \le j \le r/2} \binom{r}{2j} \frac{(2j)!}{2^j j!} \delta^{r-2j}.$$

Hogben *et al.* [25] give an alternative form, in which the sum is replaced by the expression $e^{-\frac{1}{2}\delta^2}(d/d\delta)^r(e^{\frac{1}{2}\delta^2})$. This paper also gives tables of the coefficients in the polynomials in δ which represent the central moments of t'_ν. Merrington and Pearson [40] give the following expressions:

$$\mu'_1 = (\tfrac{1}{2}\nu)^{\frac{1}{2}} \frac{\Gamma(\frac{1}{2}(\nu-1))}{\Gamma(\frac{1}{2}\nu)}\,\delta; \tag{2.1}$$

$$\text{(2.2)} \qquad \operatorname{var}(t'_\nu) = \mu_2 = \frac{\nu}{\nu-2}(1+\delta^2) - \mu_1'^2;$$

$$\text{(2.3)} \qquad \mu_3 = \mu_1'\left\{\frac{\nu(2\nu-3+\delta^2)}{(\nu-2)(\nu-3)} - 2\mu_2\right\};$$

$$\text{(2.4)} \qquad \mu_4 = \frac{\nu^2}{(\nu-2)(\nu-4)}(3+6\delta^2+\delta^4) - \mu_1'^2\left\{\frac{\nu[(\nu+1)\delta^2+3(3\nu-5)]}{(\nu-2)(\nu-3)} - 3\mu_2\right\}.$$

For ν large (δ being fixed),

$$\text{(3)} \qquad \mu_1' \doteqdot \delta; \quad \operatorname{var}(t'_\nu) \doteqdot 1 + \tfrac{1}{2}\delta^2\nu^{-1}; \quad \mu_3 \doteqdot \nu^{-1}\delta[3 + \tfrac{5}{4}\delta^2\nu^{-1}]$$

and the skewness moment-ratio, $\sqrt{\beta_1}$, is approximately $\nu^{-1}\delta(3-\delta^2\nu^{-1})$. Note that the skewness has the same sign as δ; this is also true of the expected value. Further, the distribution of $t'_\nu(-\delta)$ is a mirror-image (reflected at $t'_\nu = 0$) of that of $t'_\nu(\delta)$. Thus, for example, in the last paragraph of Section 3, $k\sqrt{n}$ is the upper $100\gamma\%$ point of the distribution of $t'_{n-1}(U_\alpha\sqrt{n})$.

The (β_1,β_2) values of the t'_ν distribution fall in the Type IV region of the Pearson system of frequency distributions. Merrington and Pearson [40] found the interesting approximate relation:

$$\text{(4)} \qquad \beta_2 \doteqdot \frac{1.406(\nu-3.2)}{\nu-4}\beta_1 + \frac{3(\nu-2)}{\nu-4}.$$

5. Distribution Function

The relationship

$$\Pr[t'_\nu \le t] = \Pr[U + \delta \le t\chi_\nu/\sqrt{\nu}]$$

leads to

(5)

$$\Pr[t'_\nu \le t] = \frac{1}{2^{\frac{1}{2}\nu-1}\Gamma(\frac{1}{2}\nu)}\int_0^\infty x^{\nu-1}e^{\frac{1}{2}x^2}\frac{1}{\sqrt{2\pi}}\int_0^{tx/\sqrt{\nu}} \exp[-\tfrac{1}{2}(u-\delta)^2]\,du\,dx.$$

Differentiating with respect to t gives the probability density function

(6)

$$p_{t'_\nu}(t) = \frac{e^{\frac{1}{2}\delta^2}}{2^{\frac{1}{2}(\nu-1)}\sqrt{\pi\nu}\,\Gamma(\frac{1}{2}\nu)}\int_0^\infty x^\nu \exp[-\tfrac{1}{2}\{(1+t^2\nu^{-1})x^2 - 2(t\nu^{\frac{1}{2}})x\}]\,dx$$

$$= \frac{\nu!}{2^{\frac{1}{2}(\nu-1)}\sqrt{\pi\nu}\,\Gamma(\frac{1}{2}\nu)}\left\{\exp\left[-\frac{\nu\delta^2}{\nu+t^2}\right]\right\} \times \left(\frac{\nu}{\nu+t^2}\right)^{\frac{1}{2}(\nu+1)} Hh_\nu\left[-\frac{\delta t}{\sqrt{\nu+t^2}}\right]$$

where $Hh_\nu(x) = (\nu!)^{-1}\int_0^\infty v^\nu \exp[-\tfrac{1}{2}(v + x)^2]\,dv$.

This form was given by Fisher [16], in the introduction to Airey's [1] tables of the Hh functions. Note that $(\sqrt{2\pi})^{-1}Hh_\nu(x)$ is the νth repeated partial integral of the standard normal probability density function.

$$(7)\quad (\sqrt{2\pi})^{-1}Hh_\nu(x) = (\sqrt{2\pi})^{-1}\int_x^\infty\int_{u_\nu}^\infty\cdots\int_{u_3}^\infty\int_{u_2}^\infty e^{-\frac{1}{2}u_1^2}\,du_1\,du_2\ldots du_\nu.$$

There are alternative formulas for the distribution of t'_ν. Firstly we have

$$(8)\qquad p_{t'_\nu}(t) = \frac{e^{-\frac{1}{2}\delta^2}\Gamma(\frac{1}{2}(\nu+1))}{\sqrt{\pi\nu}\,\Gamma(\frac{1}{2}\nu)}\left(\frac{\nu}{\nu+t^2}\right)^{\frac{1}{2}(\nu+1)} \times \sum_{j=0}^\infty \frac{\Gamma(\frac{1}{2}(\nu+j+1))}{j!\Gamma(\frac{1}{2}(\nu+1))}\left[\frac{t\,\delta\sqrt{2}}{\sqrt{\nu+t^2}}\right]^j.$$

This is of form $p_{t'_\nu}(t) = \sum_{j=0}^\infty c_j[\theta(t)]^j$ where the c_j's are constants and $\theta(t) = t\,\delta\sqrt{2}\,(\nu+t^2)^{-\frac{1}{2}}$. The expression $[\theta(t)]^0$ is to be interpreted as 1, for all values (including 0) of $\theta(t)$. If $\delta = 0$, the expression for $p_{t'_\nu}(t)$ reduces to that of a (central) t_ν density. Note that if δ and t are of opposite signs the series alternates in sign. The series in (8) can be evaluated term by term, to give values for probabilities in terms of incomplete beta function ratios. If the range of integration be taken from 0 to t (> 0), each term is positive, and

$$(9)\qquad \Pr[0 < t'_\nu \le t] = \tfrac{1}{2}e^{-\frac{1}{2}\delta^2}\sum_{j=0}^\infty \frac{(\frac{1}{2}\delta^2)^{\frac{1}{2}j}}{\Gamma(\frac{1}{2}j+1)} I_{t^2/(\nu+t^2)}(\tfrac{1}{2}(j+1),\tfrac{1}{2}\nu).$$

It is possible to obtain a similar expansion for $\Pr[-t < t'_\nu \le 0]$ with $-t < 0$, but the terms now alternate in sign. However probabilities of the latter kind can be evaluated from (9) and the value of $\Pr[-t < t'_\nu \le t]$. This last is equal to $\Pr[t'^2_\nu \le t^2]$, and we note that t'^2_ν has the noncentral F distribution $F'_{1,\nu}(\delta)$ (Chapter 30, Section 4). Hence

$$(10)\qquad \Pr[-t < t'_\nu \le t] = e^{-\frac{1}{2}\delta^2}\sum_{j=0}^\infty \frac{(\frac{1}{2}\delta^2)^j}{j!}\Pr[F_{1+2j,\nu} < (1+2j)^{-1}t^2]$$
$$= e^{-\frac{1}{2}\delta^2}\sum_{j=0}^\infty \frac{(\frac{1}{2}\delta^2)^j}{j!} I_{t^2/(\nu+t^2)}(j+\tfrac{1}{2},\tfrac{1}{2}\nu).$$

If $\Pr[t'_\nu \le 0]$ is required, the formula

$$\Pr[t'_\nu \le 0] = \Pr[U + \delta \le 0] = (\sqrt{2\pi})^{-1}\int_{-\infty}^{-\delta} e^{-\frac{1}{2}u^2}\,du$$

can be used.

If ν is even, and $t > 0$, $\Pr[0 < t'_\nu < t]$ can be expressed as a finite sum in terms of Hh functions:

$$\Pr[0 < t'_\nu < t] = \frac{1}{\sqrt{2\pi}} \exp\left[-\frac{\delta^2\nu}{2(\nu+t^2)}\right] \times \sum_{j=0}^{\frac{1}{2}(\nu-2)} \frac{(2j)!}{2^j j!}\left[\frac{\nu}{2(\nu+t^2)}\right]^j Hh_{2j}\left[-\frac{\delta t}{\sqrt{\nu+t^2}}\right]. \tag{11}$$

Use of this formula for calculation would require extensive tables of Hh functions.

Among the formulas given by Amos [2], one to which he gave special attention, in regard to its computing use, expresses the cumulative distribution function of t'_ν in terms of confluent hypergeometric functions

$$M(a,b;x) = \sum_{j=0}^{\infty} \frac{\Gamma(a+j)\Gamma(b)}{\Gamma(a)\Gamma(b+j)} \cdot \frac{x^j}{j!};$$

(b cannot be zero, or a negative integer).

The formula (for $\nu > 2$) is

$$\begin{aligned}
(12)\quad \Pr[t'_\nu \le t] &= 1 - (\sqrt{2\pi})^{-1}\int_{-\infty}^{\delta\nu^{\frac{1}{2}}/(\nu+t^2)^{\frac{1}{2}}} \exp\left(-\tfrac{1}{2}u^2\right) du \\
&+ (\sqrt{\pi})^{-1}\Bigg[\frac{t}{\sqrt{\nu+t^2}} \cdot \frac{\Gamma(\frac{1}{2}(\nu+1))}{\Gamma(\frac{1}{2}\nu)} \\
&\times \sum_{j=0}^{\infty} \frac{\Gamma(1-\frac{1}{2}\nu+j)}{j!(2j+1)\Gamma(1-\frac{1}{2}\nu)}\left(\frac{t^2}{\nu+t^2}\right)^j M\left(j+\tfrac{1}{2},\tfrac{1}{2};\frac{-\delta^2\nu}{2(\nu+t^2)}\right) \\
&- \frac{\delta\sqrt{\nu}}{\sqrt{2(\nu+t^2)}} \\
&\times \sum_{j=1}^{\infty} \frac{\Gamma(\frac{1}{2}(1-\nu)+j)}{j!\Gamma(\frac{1}{2}(1-\nu))}\left(\frac{t^2}{\nu+t^2}\right)^j M\left(j+\tfrac{1}{2},\tfrac{3}{2};\frac{-\delta^2\nu}{2(\nu+t^2)}\right)\Bigg].
\end{aligned}$$

If ν is even, the first of the two summations contains only a finite number $(\frac{1}{2}\nu + 1)$ of terms; if ν is odd, the second summation contains only a finite number $(\frac{1}{2}(\nu + 3))$ of terms.

Hodges and Lehmann [23] derived an asymptotic (with $\nu \to \infty$) series for the power of the t test (see Section 3) with ν degrees of freedom in terms of the central moments of $\chi_\nu/\sqrt{\nu}$.

They found that using this series for ν "not too small" (the case $\nu = 40$ is investigated in detail) determination of the power with sufficient precision is possible in many cases; moreover, the series is useful as an indicator of the proper interpolation procedures with respect to δ in the noncentral t tables (see Section 6).

6. Approximations

If δ is kept at a fixed value, and ν is increased without limit, the t'_ν distribution tends to the normal distribution $N(\delta,1)$. If ν (> 2) is kept fixed, and δ increased without limit, the standardized t'_ν distribution tends to the standardized χ_ν^{-1} distribution.

Earlier approximations were based on an indirect approach. Jennett and Welch [28] used the approximate normality of $(u - t'_0\chi_\nu\nu^{-\frac{1}{2}})$ in the equation

$$\Pr[t'_\nu \leq t] = \Pr[u - t\chi_\nu\nu^{-\frac{1}{2}} \leq -\delta] \tag{13}$$

to obtain

$$\Pr[t'_\nu \leq t] \doteqdot (\sqrt{2\pi})^{-1}\int_{-\infty}^{X} e^{-\frac{1}{2}u^2}\,du$$

with

$$X = [1 + t^2\nu^{-1}\operatorname{var}(\chi_\nu)]^{-\frac{1}{2}}[-\delta + t\nu^{-\frac{1}{2}}E(\chi_\nu)].$$

An approximation to the percentage point $t'_{\nu,\alpha}(\delta)$, defined by

$$\Pr[t'_\nu(\delta) \leq t'_{\nu,\alpha}(\delta)] = \alpha,$$

is found by putting $X = U_\alpha$ and solving for t. The resulting approximation is

$$t'_{\nu,\alpha} \doteqdot \frac{\delta b_\nu + U_\alpha\sqrt{b_\nu^2 + (1 - b_\nu^2)(\delta^2 - U_\alpha^2)}}{b_\nu^2 - U_\alpha^2(1 - b_\nu^2)} \tag{14.1}$$

where

$$b_\nu = \nu^{-\frac{1}{2}}E(\chi_\nu) = \left(\frac{2}{\nu}\right)^{\frac{1}{2}}\Gamma(\tfrac{1}{2}(\nu + 1))/\Gamma(\tfrac{1}{2}\nu), \quad (\text{and } \operatorname{var}(\chi_\nu) = \nu(1 - b_\nu^2)).$$

Values of b_ν are given in Table 35 of Pearson and Hartley [52] for $\nu = 1(1)$ $20(5)50(10)100$, and also in van Eeden [14].

Johnson and Welch [30] introduced the further approximations $\operatorname{var}(\chi_\nu) \doteqdot \frac{1}{2}$, and $E(\chi_\nu) \doteqdot \sqrt{\nu}$, giving the value $(1 + \frac{1}{2}t^2\nu^{-1})^{-\frac{1}{2}}(t - \delta)$ for X, and leading to the approximation

$$t'_{\nu,\alpha} \doteqdot [\delta + U_\alpha\sqrt{1 + \tfrac{1}{2}(\delta^2 - U_\alpha^2)\nu^{-1}}\,][1 - \tfrac{1}{2}U_\alpha^2\nu^{-1}]^{-1}. \tag{14.2}$$

(Masuyama [38] has shown how rough values of this approximation may be obtained using a special type of "improved binomial probability paper".)

An approximation intermediate between (14.1) and (14.2) is obtained by using the correct value of $E(\chi_\nu)$, but replacing $\operatorname{var}(\chi_\nu)$ by $\frac{1}{2}$. This is given by van Eeden [14] as

$$t'_{\nu,\alpha} \doteqdot \frac{\delta b_\nu + U_\alpha\sqrt{b_\nu^2 + \frac{1}{2}(\delta^2 - U_\alpha^2)\nu^{-1}}}{b_\nu^2 - \frac{1}{2}U_\alpha^2\nu^{-1}}. \tag{14.3}$$

These three approximations give real values for $t'_{\nu,\alpha}$ only for limited ranges of

δ and U_α:
For (14.1), we must have $b_\nu^2 + (1 - b_\nu^2)(\delta^2 - U_\alpha^2) > 0$;

$$U_\alpha^2 < b_\nu^2(1 - b_\nu^2)^{-1} + \delta^2.$$

For (14.2), we must have $1 + \frac{1}{2}(\delta^2 - U_\alpha^2)\nu^{-1} > 0$;

$$U_\alpha^2 < 2\nu + \delta^2.$$

For (14.3), we must have $b_\nu^2 + \frac{1}{2}(\delta^2 - U_\alpha^2)\nu^{-1} > 0$;

$$U_\alpha^2 < 2\nu b_\nu^2 + \delta^2.$$

Since $2\nu b_\nu^2 < b_\nu^2(1 - b_\nu^2)^{-1} < 2\nu$, it follows that (14.2) can be used over a wider range of values of α (for given δ) than (14.1), and (14.1) over a wider range than (14.3). However, it should be borne in mind that when approaching limiting possible values of α, the formulas may become less reliable. Also, the wider range of (14.2) is offset by its lower accuracy (van Eeden [14]).

Direct approximations to the distribution of noncentral t are of rather later date. For small values of δ and large (> 20) values of ν, the simple approximation of the standardized t_ν' variable by a standard normal variable gives fair results. This is equivalent to using the formula

$$(15) \qquad t_{\nu,\alpha}'(\delta) \doteqdot \frac{\nu}{\nu - 1}\,\delta b_\nu + U_\alpha \sqrt{\frac{\nu}{\nu - 2}(1 + \delta^2) - \frac{\nu^2}{(\nu - 1)^2}\,\delta^2 b_\nu^2}.$$

Since $\mathrm{var}(t_\nu')$ is not finite if $\nu \le 2$, this formula cannot be used for $\nu \le 2$. In fact, as implied above, it is unlikely to be useful unless ν is fairly large, and also δ is fairly small. When $\delta = 0$, the approximation becomes

$$t_{\nu,\alpha} = U_\alpha\sqrt{\nu/(\nu - 2)}$$

which has already been noted as a fair approximation in Section 4 of Chapter 27.

A better approximation is obtained, as is to be expected, if a Pearson Type IV distribution is fitted, making the first four moments agree with those of the noncentral t distribution. Merrington and Pearson [40] found that this gives upper and lower 5%, 1% and 0.5% points with an error no greater than 0.01, for a considerable range of values of δ and ν (including ν as small as 8). Further investigations, by Pearson [50] have confirmed the closeness of the two systems of distributions.

If a Cornish-Fisher expansion be applied to the distribution of $t_\nu'(\delta)$, the following approximate expansion (up to and including terms in ν^{-2}) is obtained:

$$\begin{aligned}(16.1) \quad t_{\nu,\alpha}'(\delta) \doteqdot\ & U_\alpha + \delta + \tfrac{1}{4}[U_\alpha^3 + U_\alpha + (2U_\alpha^2 + 1)\delta + U_\alpha\delta^2]\nu^{-1} \\ & + \tfrac{1}{96}[5U_\alpha^5 + 16U_\alpha^3 + 3U_\alpha + 3(4U_\alpha^4 + 12U_\alpha^4 + 1)\delta \\ & + 6(U_\alpha^3 + 4U_\alpha)\delta^2 - 4(U_\alpha^2 - 1)\delta^3 - 3U_\alpha\delta^4]\nu^{-2}.\end{aligned}$$

Putting $\delta = 0$ in (16.1) one obtains the approximation (see Chapter 27, Equation (11)) for central t_ν percentage points

$$t_{\nu,\alpha} \doteqdot U_\alpha + \tfrac{1}{4}(U_\alpha^3 + U_\alpha)\nu^{-1} + \tfrac{1}{96}(5U_\alpha^5 + 16U_\alpha^3 + 3U_\alpha)\nu^{-2}.$$

If these terms in (16.1) be replaced by the exact value of $t_{\nu,\alpha}$ then the approximation becomes

$$\begin{aligned}(16.2)\qquad t'_{\nu,\alpha}(\delta) \doteqdot t_{\nu,\alpha} &+ \delta + \tfrac{1}{4}\,\delta(1 + 2U_\alpha^2 + U_\alpha\delta)\nu^{-1}\\ &+ \tfrac{1}{96}\,\delta[3(4U_\alpha^4 + 12U_\alpha^2 + 1) + 6(U_\alpha^3 + 4U_\alpha)\delta\\ &- 4(U_\alpha^2 - 1)\delta^2 - 3U_\alpha\delta^3]\nu^{-2}.\end{aligned}$$

Extensive numerical comparisons for $\nu = 2(1)9$, given by van Eeden [14] indicate that, for $\delta > 0$, formula (16.1) gives the better results for lower percentage points ($\alpha < \frac{1}{2}$), while (16.2) is better for $\alpha > \frac{1}{2}$.

Azorín [4] obtained another type of approximation by constructing an approximate variance equalizing transformation. Starting from the relationship

$$\operatorname{var}(t'_\nu) = a^2 + b^2[E(t'_\nu)]^2$$

with

$$a = \sqrt{\nu/(\nu - 2)}; \quad b = \Gamma(\tfrac{1}{2}\nu)\sqrt{2\{(\nu - 2)[\Gamma(\tfrac{1}{2}(\nu - 1))]^2 - 1\}^{-1}}$$

one obtains the transformation

$$(17.1)\qquad \frac{1}{b}\sinh^{-1}\frac{bt'_\nu}{a} - \sinh^{-1}\frac{bE(t'_\nu)}{a}$$

which is to be approximated as a standard normal variable. This transformation was studied by Laubscher [34]. Azorín suggested two similar transformations of simpler form (each to be approximated as a standard normal variable):

$$(17.2)\qquad \sqrt{\nu}\,\sinh^{-1}(t'_\nu/\sqrt{\nu}) - \delta;$$

$$(17.3)\qquad \sqrt{\tfrac{2}{3}\nu}\,\sinh^{-1}(t'_\nu/\sqrt{\tfrac{2}{3}\nu}) - \delta.$$

Modified versions of (17.2) and (17.3), correcting mean and standard deviation of the transformed variables to terms of order ν^{-1}, are

$$(17.4)\qquad \frac{\sqrt{\nu}\,\sinh^{-1}(t'_\nu/\sqrt{\nu}) - \delta - \frac{1}{2}\,\delta^2\nu^{-\frac{1}{2}} - \frac{1}{4}\,\delta\nu^{-1}}{1 + \frac{1}{4}(2 - \delta^2)\nu^{-1}};$$

$$(17.5)\qquad \frac{\sqrt{\frac{2}{3}\nu}\,\sinh^{-1}(t'_\nu/\sqrt{\frac{2}{3}\nu}) - \delta - \frac{1}{2}\delta^2/\sqrt{\frac{2}{3}\nu}}{1 + \frac{1}{4}(1 - \delta^2)\nu^{-1}}.$$

(Note that there is no term in ν^{-1} in the numerator of (17.5).)

Transformations of type (17) would be expected to give accuracy comparable to the Type IV approximation, in view of the close similarity between Type IV and S_U distributions (see Chapter 12).

A remarkably accurate transformation was suggested by Harley [22]. She suggested that the distribution of $t'_\nu(\delta)$ be approximated by that of a function of the sample correlation coefficient, (R), (Chapter 32) in random samples of size $(\nu + 2)$ from bivariate normal population with the population correlation coefficient

$$\rho = \delta\sqrt{2/(2\nu + 1 + \delta^2)}.$$

The proposed function is

$$\frac{R}{\sqrt{1 - R^2}}\sqrt{\frac{\nu(2\nu + 1)}{2\nu + 1 + \delta^2}}. \tag{18}$$

While percentage points of t'_ν can be approximated from those of R, using (18), it is also possible to approximate percentage points of R, using those of t'_ν. It is this latter use which appears to be the more valuable, in the opinion of van Eeden [14] and Owen [43].

Hogben *et al.* [26] approximate the distribution of $Q = t'_\nu(\nu + t'^2_\nu)^{-\frac{1}{2}}$ and hence the distribution of t'_ν. Fitting a Type I (beta) distribution (with range -1, 1) to Q is, of course, equivalent to approximating the distribution of t'_ν. This approximation is claimed to be especially useful for small values of δ.

Finally, mention may be made of approximations proposed by Halperin [21]. These are not of great accuracy, but appear to provide a bound for percentage points (though this is not established completely). They are also simple to compute, using tables of percentage points of central $t_\nu(t_{\nu,\alpha})$ and $\chi^2_\nu(\chi^2_{\nu,\alpha})$. The suggested inequalities are (assuming $\delta > 0$):

$$t'_{\nu,\alpha}(\delta) \le \frac{\delta\sqrt{\nu}}{\chi_{\nu,1-\alpha}} + t_{\nu,\alpha} \qquad (\alpha \ge \tfrac{1}{2}), \tag{19}$$

$$t'_{\nu,\alpha}(\delta) \ge \frac{\delta\sqrt{\nu}}{\chi_{\nu,1-\alpha}} + t_{\nu,\alpha} \qquad (\alpha \le 0.43).$$

7. Tables and Charts

The tables of Neyman *et al.* [41] and Neyman and Tokarska [42] were calculated to give the power of the t-test. The first paper (Table III) gives the operating characteristic (= 1 − Power) of a 5% significance limit t-test at values $\delta = 1(1)9$ for $\nu = 1(1)30$ — (i.e. values of $\Pr[t'_\nu(\delta) \le t_{\nu,0.95}]$) — and also values of δ satisfying the equation $\Pr[t'_\nu(\delta) \le t_{\nu,0.95}] = 0.05$. The second paper gives more extensive tables of the same kind. Owen [44] has given tables, to 5 decimal places, of values of δ such that $\Pr[t'_\nu(\delta) \le t_{\nu,1-\alpha}] = \beta$ for

$$\nu = 1(1)30(5)100(10)200, \infty;$$
$$\alpha = 0.005, 0.01, 0.025, 0.05;$$
$$\beta = 0.01, 0.05, 0.1(0.1)0.9.$$

The tables of Johnson and Welch [30] give values of a quantity λ such that

$$(20) \qquad t'_{\nu,\alpha}(\delta) = \frac{\delta + \lambda[1 + \frac{1}{2}(\delta^2 - \lambda^2)\nu^{-1}]^{\frac{1}{2}}}{1 - \frac{1}{2}\lambda^2\nu^{-1}}.$$

Comparing with approximation (14.2), it is seen that one might expect $\lambda \doteqdot U_\alpha$, so that values of λ should not vary too much as δ and ν vary, thus making interpolation simpler. Values of λ are given for $\nu = 4(1)9, 16, 36, 144, \infty$ (i.e. U_α); above $\nu = 9$, interpolation with respect to $12\nu^{-\frac{1}{2}}$ is suggested. The argument used is $y = (1 + \frac{1}{2}t'^2_\nu\nu^{-1})^{-\frac{1}{2}}$ for $0.6 \leq |y| \leq 1$; and $y' = yt'_\nu/\sqrt{2\nu}$ for $|y| \leq 0.6$. Values of α ($(1 - \epsilon)$ in the original) are 0.005, 0.01, 0.025, 0.05, 0.1, 0.2, 0.3, 0.4, 0.5, 0.6, 0.7, 0.8, 0.9, 0.95, 0.975, 0.99 and 0.995. As the argument is a function of t', direct entry into the tables leads to a value

$$(21) \qquad \delta' = \delta(\nu,t',\alpha) = t' - \lambda/y$$

such that $t'_{\nu,\alpha}(\delta') = t'$. In order to obtain $t'_{\nu,\alpha}(\delta)$ for a given δ, an iterative (or inverse interpolation) procedure is necessary. A table giving the results of such procedures, for the case $\alpha = 0.95$ only, is also provided. This table gives λ to 3 or 4 decimal places as a function of $\eta = [\delta/\sqrt{2\nu}](1 + \frac{1}{2}\delta^2\nu^{-1})^{-\frac{1}{2}}$. The later tables of Owen [43] include a substantial extension of this last table. The argument η is tabulated at intervals of 0.01, instead of 0.1; λ is given to 5 decimal places, and values $\nu = 1, 2, 3$ are included, in addition to those in the Johnson and Welch tables. Values of α, however, are now 0.005, 0.01, 0.025, 0.05, 0.1, 0.25, 0.5, 0.75, 0.9, 0.95, 0.975, 0.99, 0.995. Owen also gives tables of λ (to 5 decimal places) as a function of y and y' (in the same form as Johnson and Welch) for these new sets of values of ν and α. There are also extensive tables of quantity k (to 3 decimal places) such that

$$(22) \qquad t'_{\nu,\alpha}(U_p\sqrt{\nu + 1}) = k\sqrt{\nu + 1}$$

for $p = 0.75, 0.9, 0.95, 0.975, 0.99, 0.999, 0.9999, 0.99999$; $\nu + 1 = 2(1)200(5)400(25)1000(500)2000, 3000, 5000, 10000, \infty$; α has the same values as in Owen's other tables. The choice of $U_p\sqrt{\nu + 1}$ as noncentrality parameter makes it convenient to use the tables in calculating confidence intervals for percentage points of normal distributions. Thus, since the inequality $\bar{X} - kS < \xi - U_p\sigma$ is equivalent to

$$\frac{\sqrt{n}\,(\bar{X} - \xi)\sigma^{-1} + \sqrt{n}U_p}{S\sigma^{-1}} < k\sqrt{n}$$

it follows that

$$(23) \qquad \Pr[\bar{X} - kS < \xi - U_p\sigma] = \Pr[t'_{n-1}(U_p\sqrt{n}) < k\sqrt{n}].$$

This probability is equal to α if

$$t'_{n-1,\alpha}(U_p\sqrt{n}) = k\sqrt{n}.$$

Putting $n = \nu + 1$, Equation (22) is obtained.

Owen also gives values of k satisfying

(24) $$t'_{\nu,\alpha}(U_p\sqrt{n}) = k\sqrt{n}$$

for $n = 1$, $\nu = 1, 2$ (and ∞); $\alpha = 0.90, 0.95, 0.99$; for $p = 0.50(0.01)0.90$ $(0.005)0.990(0.001)0.999(0.0001)0.9999$ and some even higher values of p. Further tables give values of k satisfying (24) for

$$n = 1(1)10, 17, 37, 145, 500, \infty;$$
$$\alpha = 0.90, 0.95, 0.99;$$
$$p = 0.75, 0.9, 0.95, 0.975, 0.99, 0.999, 0.9999, 0.99999;$$

and

$$\nu = 1(1)75(5)100(10)150, 200, 300, 500, 1000.$$

Some extracts from Owen's tables are included in the survey [45].

The tables of Resnikoff and Lieberman [53] are also based on Equation (22). They give $t'_{\nu,\alpha}(U_p\sqrt{\nu+1})/\sqrt{\nu}$ for

$$p = 0.001, 0.0025, 0.004, 0.01, 0.025, 0.04, 0.065, 0.1, 0.15, 0.25;$$
$$\alpha = 0.005, 0.01, 0.05, 0.1, 0.25, 0.5, 0.75, 0.9, 0.95, 0.99, 0.995;$$
$$\nu = 2(1)24(5)49.$$

(For $\nu = 2$, 3, and 4, values are not given corresponding to $\alpha = 0.99, 0.995$.)

These tables also give probability integrals of the noncentral t distribution (to 4 decimal places)

$$\Pr[t'_\nu(U_p\sqrt{\nu+1}) \le x\sqrt{\nu}]$$

for the same values of p and ν, and for x at intervals of 0.05. There is also a table of values of the probability density function, for the same values of p, ν and x.

Locks *et al.* [35] have produced a similar, but more extensive set of tables of the probability integral and the probability density function. In order to facilitate use in connection with multiple regression, values of the noncentrality parameter equal to $U_p\sqrt{\nu+2}$, as well as $U_p\sqrt{\nu+1}$, are used. However the tables do not use specified values of p, but rather take $U_p = 0.0(0.025)3.0$.

Hodges and Lehmann [24] have noted that if normal equivalent deviates are used, problems of interpolation are much less troublesome. Thus if

$$\Pr[t'_\nu(\delta) \ge t_{\nu,1-\alpha}] = \beta$$

then, they table a quantity A satisfying the equation

$$U_{1-\alpha} + U_\beta = \delta(1 - \tfrac{1}{4}U_{1-\alpha}^2\nu^{-1}) - AU_{1-\alpha}^2(U_{1-\alpha} + U_\beta)\nu^{-2}.$$

Values of A are given to four places of decimals for $\alpha = 0.005, 0.01, 0.025, 0.05, 0.1$; $\beta = 0.5(0.1)0.9, 0.95, 0.99$; and $\nu = 3(1)6, 8, 12, 24, \infty$.

For given α and β, A is a rather smooth function of ν. For $\nu > 6$, $\alpha \ge 0.01$ and $\beta \le 0.9$, linear harmonic interpolation will give good results, and practically useful values can be obtained well beyond these limits.

Among shorter, but useful, tables, mention must be made of those of van Eeden [15] and Scheuer and Spurgeon [54]. Van Eeden gives $t'_{\nu,\alpha}(U_p\sqrt{\nu+1})$ directly (to 3 decimal places (2 for $\alpha = 0.99, 0.01$)) for $\nu = 4(1)9$; $\alpha = 0.01, 0.025, 0.05, 0.95, 0.975, 0.99$ and $p = 0.125, 0.15(0.05)0.45$. Scheuer and Spurgeon give values of the same function (to 3 decimal places) for the values of p and ν used by Resnikoff and Lieberman, but only for $\alpha = 0.025, 0.975$.

Bruscantini [8] has made a detailed study of the distribution of $Y = U + \theta\chi_2$ He refers to, and gives a short extract from, unpublished tables giving values, to 5 decimal places, of the cumulative distribution function of the standardized variable

$$Y' = (Y - \theta\sqrt{\pi/2})[1 + \theta^2(2 - \tfrac{1}{2}\pi)]^{-\frac{1}{2}}$$

for argument y' at intervals of 0.5 and $\theta = 2.00(0.05)7.20$. These are, in fact values of $\Pr[t'_2(y'') > \theta]$ with

$$y'' = \theta\sqrt{\pi/2} + y'\sqrt{1 + \theta^2(2 - \tfrac{1}{2}\pi)}.$$

8. Related Distributions

8.1 *Noncentral Beta Distribution*

The *noncentral beta distribution* is defined as the distribution of the ratio

$$b'_{\nu_1,\nu_2}(\lambda) = \chi'^2_{\nu_1}(\lambda)/(\chi^2_{\nu_2} + \chi'^2_{\nu_1}(\lambda)).$$

It can be seen that $[t'_\nu(\delta)]^2/[\nu + \{t'_\nu(\delta)\}^2]$ is distributed as $b'_{1,\nu}(\delta^2)$. This is also the distribution of Q^2, where Q is the variable (mentioned in Section 6) studied by Hogben *et al.* [26]. (See also page 434 of David and Paulson [12].)

8.2 *Double Noncentral t Distribution*

If the χ_ν in the denominator of $t'_\nu(\delta)$ is replaced by a noncentral χ_ν (noncentrality parameter λ), the distribution of the modified variable is called a *double noncentral t distribution with noncentrality parameters* (δ,λ) and ν degrees of freedom.

Symbolically

$$t''_\nu(\delta,\lambda) = \frac{U + \delta}{\chi'_\nu(\lambda)/\sqrt{\nu}}\cdot$$

Since $\chi'_\nu(\lambda)$ is distributed as a mixture of $\chi_{\nu+2j}$ distributions in proportions $e^{-\frac{1}{2}\lambda}(\tfrac{1}{2}\lambda)^j/j!$ $(j = 0,1,2,\ldots)$, the distribution of $t''_\nu(\delta,\lambda)$ is a mixture of

$$\sqrt{\nu(\nu + 2j)^{-1}}\, t'_{\nu+2j}(\delta)$$

distributions in these same proportions. Hence all formulas, approximations, tables, etc. for the noncentral t distribution can be applied to the double noncentral t distribution. Thus, for example, the rth moment of t_ν'' about zero is

$$(25) \qquad E(t_\nu''^r) = (\tfrac{1}{2}\nu)^{\frac{1}{2}r} e^{-\frac{1}{2}\lambda}\left[\sum_{j=0}^{\infty} \frac{(\frac{1}{2}\lambda)^j}{j!} \cdot \frac{\Gamma(\frac{1}{2}\nu)\Gamma(\frac{1}{2}(\nu-r)+j)}{\Gamma(\frac{1}{2}(\nu-r))\Gamma(\frac{1}{2}\nu+j)}\right] \times E[(U+\delta)^r]\frac{\Gamma(\frac{1}{2}(\nu-r))}{\Gamma(\frac{1}{2}\nu)}.$$

Krishnan [31] has pointed out that the summation in the above formula can be expressed as the confluent hypergeometric function $M(\frac{1}{2}(\nu-r);\frac{1}{2}\nu;\frac{1}{2}\lambda)$, and that an even simpler form can be obtained using Kummer's formula $e^{-\frac{1}{2}\lambda}M(\frac{1}{2}(\nu-r),\frac{1}{2}\nu,\frac{1}{2}\lambda) = M(\frac{1}{2}r,\frac{1}{2}\nu,-\frac{1}{2}\lambda)$. She has obtained some recurrence relations between moments about zero of $t_\nu''(\delta,\lambda)$, $t_{\nu-2}''(\delta,\lambda)$ and $t_{\nu-4}''(\delta,\lambda)$.

These formulas are most conveniently given in terms of the quantities

$$\mu'_{r,\nu} = \mu'_r(t_\nu'')/\nu^{\frac{1}{2}r}.$$

They are:

$$(26.1) \quad \mu'_{1,\nu} = \left(1 - \frac{\nu-4}{\lambda}\right)\mu'_{1,\nu-2} + \frac{\nu-5}{\lambda}\mu'_{1,\nu-4} \qquad (\nu > 5)$$

$$(26.2) \quad \mu'_{2,\nu} = \lambda^{-1}[\delta^2 + 1 - \mu'_{2,\nu-2}] \qquad (\nu > 4)$$

$$(26.3) \quad \mu'_{3,\nu} = (\delta^2 + 3)(\mu'_{1,\nu-2} - \mu'_{1,\nu}) \qquad (\nu > 3)$$

$$(26.4) \quad \mu'_{4,\nu} = \tfrac{1}{2}(\delta^4 + 6\delta^2 + 3)(\delta^2 + 1)^{-1}(\mu'_{2,\nu-2} - \mu'_{2,\nu}) \qquad (\nu > 4).$$

For ν large, δ and λ remaining fixed,

$$(27.1) \qquad \mu_1'(t_\nu'') = \delta[1 + (\tfrac{3}{4} - \tfrac{1}{2}\lambda)\nu^{-1} + O(\nu^{-2})]$$

$$(27.2) \qquad \mu_2'(t_\nu'') = (\delta^2 + 1)[1 + (2 - \lambda)\nu^{-1} + O(\nu^{-2})]$$

$$(27.3) \qquad \mu_3'(t_\nu'') = \delta(\delta^2 + 3)[1 + 3(\tfrac{5}{4} - \tfrac{1}{2}\lambda)\nu^{-1} + O(\nu^{-2})]$$

$$(27.4) \qquad \mu_4'(t_\nu'') = (\delta^4 + 6\delta^2 + 3)[1 + (6 - 2\lambda)\nu^{-1} + O(\nu^{-2})].$$

Since the available tables of hypergeometric and gamma functions are inadequate to be used for computing moments of $t_\nu''(\delta,\lambda)$, Krishnan [31] presents tables, to six decimal places, and $\lambda = 2(2)8(4)20$ of

$$c_1 = \mu_1'(t_\nu'')/\delta \qquad \text{for } \nu = 2(1)20,$$
$$c_2 = \mu_2'(t_\nu'')/(\delta^2 + 1) \qquad \text{for } \nu = 3(1)20,$$
$$c_3 = \mu_3'(t_\nu'')/[\delta(\delta^2 + 3)] \qquad \text{for } \nu = 4(1)20,$$
$$c_4 = \mu_4'(t_\nu'')/(\delta^4 + 6\delta^2 + 3) \text{ for } \nu = 5(1)20.$$

(Note that c's are independent of δ.)

In the same paper Krishnan also considered two approximations to the distribution of $t_\nu''(\delta,\lambda)$. In three special cases, good results were obtained using a method suggested by Patnaik [48] in which the distribution is approximated by that of $ct_f'(\delta)$, c and f being chosen to give correct values for the first two moments. The other method, an extension of Harley's [22] method (see Section 6), also gave useful, though not quite as accurate, results. For this approximation one calculates $L = [(\nu - 3)\mu_{3,\nu}'][\nu\mu_{1,\nu}']^{-1}$; $K = (1 - 2\nu^{-1})\mu_{2,\nu}'$ and

$$p = [(3K - L)\{\nu L - (\nu - 1)K\}^{-1}]^{\frac{1}{2}}.$$

Then the distribution of $t_\nu''(\delta,\lambda)$ is approximately that of

$$[\nu K\{1 + (\nu + 1)\rho^2/(1 - \rho^2)\}^{-1}]^{\frac{1}{2}} R/\sqrt{1 - R^2}$$

where R is distributed as the product moment sample correlation in a sample of size $(\nu + 2)$ from bivariate normal population with correlation ρ (see Chapter 32, Section 2).

Krishnan [32] gives tables of $\Pr[t_\nu''(\delta,\lambda) \leq 1]$ to 4 decimal places for $\nu = 2(1)20$; $\delta = -5(1)5$; $\lambda = 0(2)8$; and reports availability of similar tables of $\Pr[t_\nu''(\delta,\lambda) \leq 2]$. (Note that, for $t_0 < 0$, $\Pr[t_\nu''(\delta,\lambda) \leq t_0]$ can be evaluated as $\Pr[t_\nu''(-\delta,\lambda) \geq -t_0]$.)

Krishnan, and also Bulgren and Amos [9], give the following formula

$$\Pr[t_\nu'' \leq t_0] = 1 - \Phi(\sqrt{\beta}) + Z(\sqrt{\beta}) \sum_{j=0}^{\infty} E(\lambda,j)$$
$$\times \left[\frac{\sqrt{a}\Gamma(\frac{1}{2}(\nu + 1) + j)}{\Gamma(\frac{1}{2}\nu + j)} \sum_{i=0}^{\infty} \frac{(1 - \frac{1}{2}\nu - j)^{[i]}}{i!(2i + 1)} a^i M(-i;\tfrac{1}{2};\tfrac{1}{2}\beta) \right.$$
$$\left. - \sqrt{\tfrac{1}{2}\beta} \sum_{i=0}^{\infty} \frac{(1 - \frac{1}{2}\nu - j)^{[i]}}{i!} a^i M(1 - i;\tfrac{3}{2};\tfrac{1}{2}\beta) \right]$$

where

$$a = t_0^2/(\nu + t_0^2); \quad \beta = \delta^2(1 - a) \qquad (\sqrt{\beta} > 0)$$

and

$$E(\lambda,j) = [\Gamma(j + 1)]^{-1} \cdot (\tfrac{1}{2}\lambda)^j e^{-\frac{1}{2}\lambda}.$$

Bulgren and Amos [9] give some other series representations and also a table of values of $\Pr[t_\nu''(\delta,\lambda) \leq t_0]$ to 6 decimal places for $t_0 = 1$, 2 and $\nu = 2$, 5(5)20; $\delta = -4(2)4$; $\lambda = 0(4)8$. They report availability of more extensive tables covering $\nu = 2(1)20$; $\delta = -5(1)5$; $\lambda = 0(2)8$.

8.3 *Modified Noncentral t Distributions*

The most common modified t statistic is obtained by replacing the S in the denominator (see Section 3) by the sample range W, or by the mean of a number of independent sample ranges (see Lord [36], [37] and Chapter 13). Thus, in place of the χ_ν in the denominator of $t_\nu'(\delta)$ (see Section 1) there is a variable having some other distribution, but still independent of the U variable in the numerator.

The noncentrality of the distribution is associated with the δ in the numerator.

The denominator (in both original and modified forms) has the same distribution for both the central and noncentral cases. So approximations to this latter distribution which have been used for the central case can also be applied to the noncentral case, with reasonable hope of obtaining useful results. For example, if the distribution of W is approximated by that of $\chi_{\nu'}(c'\nu'^{\frac{1}{2}})^{-1}$, then U/W may be approximated by (central) $c't_{\nu'}(\delta)$.

Discussion of approximations to the distribution of noncentral modified t will be found in Lord [37] and Zaludová [59].

8.4 *Distribution of Noncentral t-Statistic when the Population is Non-normal*

These distributions have been studied in order to assess the effect of non-normality on the power of the t-test. We note, in particular, the work of Ghurye [20] and Srivastava [55]. The first of these authors extended some results of Geary [18], [19] who supposed that the population density function could be adequately represented as

(28)

$$p_X(x) = \frac{1}{\sqrt{2\pi}\,\sigma}\left\{\exp\left[-\frac{1}{2}\left(\frac{x-\xi}{\sigma}\right)^2\right]\right\}\left[1+\frac{\sqrt{\beta_1}}{6}\left\{\left(\frac{x-\xi}{\sigma}\right)^3 - 3\left(\frac{x-\xi}{\sigma}\right)\right\}\right].$$

Srivastava, utilizing later results of Gayen [17], obtained formulas for the case when the population density function is as in (28), with additional terms

$$\frac{1}{\sqrt{2\pi}\,\sigma}\left\{\exp\left(-\frac{1}{2}\left(\frac{x-\xi}{\sigma}\right)^2\right)\right\}$$
$$\times\left[\frac{\beta-3}{24}\left\{\left(\frac{x-\xi}{\sigma}\right)^4 - 6\left(\frac{x-\xi}{\sigma}\right)^2 + 3\right\}\right.$$
$$\left. + \frac{\beta_1}{72}\left\{\left(\frac{x-\xi}{\sigma}\right)^6 - 15\left(\frac{x-\xi}{\sigma}\right)^4 + 45\left(\frac{x-\xi}{\sigma}\right)^2 - 15\right\}\right]$$

(i.e. the next terms in the Edgeworth expansion).

The correction to the normal theory power is of the form

$$-\sqrt{\beta_1}\,P_{\sqrt{\beta_1}} - (\beta_2 - 3)P_{\beta_2} - \beta_1 P_{\beta_1},$$

where the P's do *not* depend on the β's but on the noncentrality, the degrees of freedom and the significance level of the test.

REFERENCES

[1] Airey, J. R. (1931). *Table of Hh Functions*, Table XV, pp. 60–72 in British Association Mathematical Tables, Vol. **1**, British Association, London, (Second edition 1946, Third edition 1951).

[2] Amos, D. E. (1964). Representations of the central and noncentral t distributions, *Biometrika*, **51**, 451–458.

[3] Anscombe, F. J. (1950). Table of the hyperbolic transformation $\sinh^{-1}\sqrt{x}$, *Journal of the Royal Statistical Society, Series A*, **113**, 228–229.

[4] Azorín, P. F. (1953). Sobre la distribución t no central. I, II, *Trabajos de Estadística*, **4**, 173–198 and 307–337.

[5] Bargmann, R. E. and Ghosh, S. P. (1964). *Noncentral statistical distribution programs for a computer language*, IBM Research Report, RC-1231.

[6] Bartlett, M. S. (1935). The effect of non-normality on the t-distribution, *Proceedings of the Cambridge Philosophical Society*, **31**, 223–231.

[7] Belobragina, L. S. and Eliseyev, V. K. (1967). Statistical estimation of a recognition error based on experimental data, *Kibernetika*, **4**, 81–89. (In Russian)

[8] Bruscantini, S. (1968). Origin, features and use of the pseudo-normal distribution, *Statistica* (*Bologna*), **28**, 102–124.

[9] Bulgren, W. G. and Amos, D. E. (1968). A note on representations of the doubly noncentral t distribution, *Journal of the American Statistical Association*, **63**, 1013–1019.

[10] Craig, C. C. (1941). Note on the distribution of noncentral t with an application, *Annals of Mathematical Statistics*, **12**, 224–228.

[11] Croarkin, Mary C. (1962). Graphs for determining the power of Student's t-test, *Journal of Research of the National Bureau of Standards*, **66B**, 59–70, (Correction in *Mathematics of Computation*, (1963), **17**, 83 (334)).

[12] David, H. A. and Paulson, A. S. (1965). The performance of several tests for outliers, *Biometrika*, **52**, 429–436.

[13] Davies, O. L. (1954). *The Design and Analysis of Industrial Experiments*, New York: Hafner.

[14] Eeden, Constance van (1958). *Some approximations to the percentage points of the noncentral t-distribution*, Report S242, Statistics Department, Mathematics Center, Amsterdam.

[15] Eeden, Constance van (1961). Ibid., *Revue de l'Institut Internationale de Statistique*, **29**, 4–31.

[16] Fisher, R. A. (1931). Introduction of *Table of Hh Functions*, pp. xxvi–xxxv (see [1]).

[17] Gayen, A. K. (1949). The distribution of "Student's" t in random samples of any size drawn from non-normal universes, *Biometrika*, **36**, 353–369.

[18] Geary, R. C. (1936). The distribution of "Student's" ratio for non-normal samples, *Journal of the Royal Statistical Society*, *Series B*, **3**, 178–184.

[19] Geary, R. C. (1947). Testing for normality, *Biometrika*, **34**, 209–242.

[20] Ghurye, S. G. (1949). On the use of Student's t-test in an asymmetrical population, *Biometrika*, **36**, 426–430.

[21] Halperin, M. (1963). Approximations to the noncentral t, with applications, *Technometrics*, **5**, 295–305.

[22] Harley, B. I. (1957). Relation between the distributions of noncentral t and a transformed correlation coefficient, *Biometrika*, **44**, 219–224.

[23] Hodges, J. L. and Lehmann, E. L. (1965). Moments of chi and power of t, *Proceedings of the Fifth Berkeley Symposium on Mathematical Statistics and Probability*, **1**, 187–201.

[24] Hodges, J. L. and Lehmann, E. L. (1968). A compact table for power of the t-test, *Annals of Mathematical Statistics*, **39**, 1629–1637.

[25] Hogben, D., Pinkham, R. S. and Wilk, M. B. (1961). The moments of the noncentral t distribution, *Biometrika*, **48**, 465–468.

[26] Hogben, D., Pinkham, R. S. and Wilk, M. B. (1964). An approximation to the distribution of Q (a variate related to the noncentral t), *Annals of Mathematical Statistics*, **35**, 315–318.

[27] Iglewicz, B., Myers, R. H. and Howe, R. B. (1968). On the percentage points of the sample coefficient of variation, *Biometrika*, **55**, 580–581.

[28] Jennett, W. J. and Welch, B. L. (1939). The control of proportion defective as judged by a single quality characteristic varying on a continuous scale, *Journal of the Royal Statistical Society, Series B*, **6**, 80–88.

[29] Jílek, M. and Líkař, O. (1959). Coefficients for the determination of one-sided tolerance limits of normal distributions, *Annals of the Institute of Statistical Mathematics, Tokyo*, **11**, 45–48.

[30] Johnson, N. L. and Welch, B. L. (1940). Applications of the noncentral t distribution, *Biometrika*, **31**, 362–389.

[31] Krishnan, M. (1967). The moments of a doubly noncentral t-distribution, *Journal of the American Statistical Association*, **62**, 278–287.

[32] Krishnan, M. (1968). Series representations of the doubly noncentral t-distribution, *Journal of the American Statistical Association*, **63**, 1004–1012.

[33] Kruskal, W. H. (1954). The monotonicity of the ratio of two noncentral t density functions, *Annals of Mathematical Statistics*, **25**, 162–164.

[34] Laubscher, N. F. (1960). Normalizing the noncentral t and F distributions, *Annals of Mathematical Statistics*, **31**, 1105–1112.

[35] Locks, M. O., Alexander, M. J. and Byars, B. J. (1963). *New Tables of the Noncentral t-Distribution*, Report ARL63–19, Wright-Patterson Air Force Base.

[36] Lord, E. (1947). The use of range in place of the standard deviation in the t-test, *Biometrika*, **34**, 41–67, (Correction **39**, 442).

[37] Lord, E. (1950). Power of modified t-test (u-test) based on range, *Biometrika*, **37**, 64–77.

[38] Masuyama, M. (1951). An approximation to the non-central t-distribution with the stochastic paper, *Reports of Statistical Application Research, JUSE*, **1** (3), 28–31.

[39] McKay, A. T. (1932). Distribution of the coefficient of variation and the extended t distribution, *Journal of the Royal Statistical Society*, **95**, 695–698.

[40] Merrington, Maxine and Pearson, E. S. (1958). An approximation to the distribution of noncentral t, *Biometrika*, **45**, 484–491.

[41] Neyman, J., Iwaszkiewicz, K. and Kolodziejczyk, S. (1935). Statistical problems in agricultural experimentation, *Journal of the Royal Statistical Society, Series B*, **2,** 107–180.

[42] Neyman, J. and Tokarska, B. (1936). Errors of the second kind in testing "Student's" hypothesis, *Journal of the American Statistical Association*, **31,** 318–326.

[43] Owen, D. B. (1963). *Factors for one-sided Tolerance Limits and for Variables Sampling Plans*, Sandia Corporation Monograph SCR-607.

[44] Owen, D. B. (1965). The power of Student's t-test, *Journal of the American Statistical Association*, **60,** 320–333.

[45] Owen, D. B. (1968). A survey of properties and applications of the noncentral t-distribution, *Technometrics*, **10,** 445–478.

[46] Owen, D. B. and Amos, D. E. (1963). *Programs for Computing Percentage Points of the Noncentral t-Distributions*, Sandia Corporation Monograph SCR-551.

[47] Park, J. H. (1964). Variations of the noncentral t- and beta-distributions, *Annals of Mathematical Statistics*, **35,** 1583–1593.

[48] Patnaik, P. B. (1955). Hypotheses concerning the means of observations in normal samples, *Sankhyā*, **15,** 343–372.

[49] Pearson, E. S. (1958). Note on Mr. Srivastava's paper on the power function of Student's test, *Biometrika*, **45,** 429–430.

[50] Pearson, E. S. (1963). Some problems arising in approximating to probability distributions, using moments, *Biometrika*, **50,** 95–111. (Appendix p. 112.)

[51] Pearson, E. S. and Adyanthàya, N. K. (1929). The distribution of frequency constants in small samples from non-normal symmetrical and skew populations, *Biometrika*, **21,** 259–286.

[52] Pearson, E. S. and Hartley, H. O. (1954). *Biometrika Tables for Statisticians*, Vol. **I,** London: Cambridge University Press. (2nd edition 1958; 3rd edition 1966)

[53] Resnikoff, G. J. and Lieberman, G. J. (1957). *Tables of the Non-Central t-Distribution*, Stanford: Stanford University Press.

[54] Scheuer, E. M. and Spurgeon, R. A. (1963). Some percentage points of the non-central t-distribution, *Journal of the American Statistical Association*, **58,** 176–182.

[55] Srivastava, A. B. L. (1958). Effect of non-normality on the power function of t-test, *Biometrika*, **45,** 421–429.

[56] Steffens, F. E. (1968). *Probability integrals of doubly noncentral F- and t-distributions with regression applications*, Research Report No. 267, Council for Scientific and Industrial Research, Pretoria, South Africa.

[57] Stone, W. M. and Peura, Bonita (1965). On an asymptotic approximation to the non-central t-distribution, *Annals of Mathematical Statistics*, **36,** 1596 (Abstract).

[58] Wolfowitz, J. (1946). Confidence limits for the fraction of a normal population which lies between two given limits, *Annals of Mathematical Statistics*, **17,** 483–488.

[59] Zaludová, A. H. (1960). The noncentral t-test (q-test) based on range in place of standard deviation, *Acta Technica*, **5,** 143–185.

32

Distributions of Correlation Coefficients

1. Introduction and Genesis

The statistic known as the sample correlation coefficient, based on n pairs of observed values of two characters, represented by random variables $(X_t, Y_t)(t = 1,\ldots,n)$ is

$$(1) \qquad R = \frac{\sum_{t=1}^{n} (X_t - \bar{X})(Y_t - \bar{Y})}{\left[\sum_{t=1}^{n} (X_t - \bar{X})^2 \sum_{t=1}^{n} (Y_t - \bar{Y})^2\right]^{\frac{1}{2}}}$$

where $\bar{X} = n^{-1} \sum_{t=1}^{n} X_t$; $\bar{Y} = n^{-1} \sum_{t=1}^{n} Y_t$. The distribution which is the main topic of this chapter is that of R when

(a) (X_i, Y_i) and (X_j, Y_j) are mutually independent if $i \neq j$, and
(b) the joint distribution of X_t and Y_t has probability density function

(2)

$$p_{X_t,Y_t}(x,y) = \frac{1}{2\pi\sigma_X\sigma_Y\sqrt{1-\rho^2}} \exp\left[-\frac{1}{2(1-\rho^2)}\left\{\left(\frac{x-\xi}{\sigma_X}\right)^2 - 2\rho\left(\frac{x-\xi}{\sigma_X}\right) \times \left(\frac{y-\eta}{\sigma_Y}\right) + \left(\frac{y-\eta}{\sigma_Y}\right)^2\right\}\right]$$

for each $t = 1, 2, \ldots, n(\sigma_X > 0; \sigma_Y > 0; -1 < \rho < 1)$.

Formula (2) is that of the *general bivariate normal distribution*, which will be studied in Chapter 35. We discuss the distribution of R before the parent distribution (2) because R is univariate, (2) is within the class of multivariate distributions, which are studied in volume 3. However, we will use a few of the properties of (2) in order to pursue our analysis of the distribution of R. The first is that ρ is the *population correlation coefficient*, that is

$$\rho = \frac{E[\{X_t - E[X_t]\}\{Y_t - E[Y_t]\}]}{\sqrt{\text{var}(X_t)\,\text{var}(Y_t)}}. \tag{3}$$

The second is that X_t, Y_t each have normal distributions, with $E[X_t] = \xi$; $E[Y_t] = \eta$; $\text{var}(X_t) = \sigma_X^2$; $\text{var}(Y_t) = \sigma_Y^2$. At the times we need to use further properties, they will be stated specifically.

This chapter is primarily concerned with the distribution of R corresponding to (2), but the distributions arising under some other conditions will also be discussed in Section 3. Furthermore, Sections 6–8 are devoted to distributions of serial correlations, and Section 9 to distribution of multiple correlation.

2. Derivation

Since the correlation between the standardized variables $(X_t - \xi)/\sigma_X$ and $(Y_t - \eta)/\sigma_Y$ is the same as that between X_t and Y_t, no generality is lost by taking $\xi = \eta = 0$; $\sigma_X = \sigma_Y = 1$. We now consider the conditional distribution of R for fixed values of $X_1, X_2, \ldots, X_n$. Since the conditional distribution of Y_t, given X_t, is normal, with expected value ρX_t and variance $(1 - \rho^2)$ (remembering that we are taking $\xi = \eta = 0$; $\sigma_X = \sigma_Y = 1$), it follows that $R(1 - R^2)^{-\frac{1}{2}}$ is distributed as $(n - 2)^{-\frac{1}{2}}$ times noncentral t with $(n - 2)$ degrees of freedom and noncentrality parameter

$$\sqrt{\sum_{i=1}^{n} (X_i - \overline{X})^2}\ \frac{\rho}{\sqrt{(1 - \rho^2)}}.$$

To obtain the overall (unconditional) distribution of $R(1 - R^2)^{-\frac{1}{2}}$, we must calculate the expected value of the density function so obtained, over the distribution of $X_1, X_2, \ldots, X_n$. Since the density function depends on the X's only through the function $\sum_{t=1}^{n} (X_t - \overline{X})^2$, we need only use the fact that this function is distributed as χ^2 with $(n - 1)$ degrees of freedom, i.e. as χ^2_{n-1}. Writing $V = R(1 - R^2)^{-\frac{1}{2}}$, the conditional probability density of V is

$$p_V(v \mid S) = \frac{\exp\left[\dfrac{-\rho^2 S}{2(1 - \rho^2)}\right]}{\sqrt{\pi}\,\Gamma(\frac{1}{2}n - 1)} (1 + v^2)^{-\frac{1}{2}(n-1)} \times \sum_{j=0}^{\infty} \frac{\Gamma\left(\dfrac{n - 1 + j}{2}\right)}{j!} \left(\frac{2\rho^2 v^2 S}{(1 - \rho^2)(1 + v^2)}\right)^{\frac{1}{2}j} \tag{4}$$

where $S = \sum_{i=1}^{n} (X_i - \overline{X})^2$.

Since

$$p_S(s) = \left[2^{\frac{1}{2}(n-1)}\Gamma\left(\frac{n-1}{2}\right)\right]^{-1} s^{\frac{1}{2}(n-3)}e^{-\frac{1}{2}s} \qquad (s > 0)$$

and

$$\int_0^\infty s^{\frac{1}{2}j} \exp\left[-\tfrac{1}{2}\rho^2 s(1-\rho^2)^{-1}\right] \cdot p_S(s)\, ds = \frac{2^{\frac{1}{2}j}\Gamma\left(\frac{n-1+j}{2}\right)}{\Gamma\left(\frac{n-1}{2}\right)}(1-\rho^2)^{\frac{1}{2}(n+j-1)},$$

$$p_V(v) = \frac{(1-\rho^2)^{\frac{1}{2}(n-1)}}{\sqrt{\pi}\,\Gamma\left(\frac{n-1}{2}\right)\Gamma(\frac{1}{2}n-1)}(1+v^2)^{-\frac{1}{2}(n-1)} \times \sum_{j=0}^{\infty} \frac{(2\rho)^j\left[\Gamma\left(\frac{n-1+j}{2}\right)\right]^2}{j!}\left(\frac{v^2}{1+v^2}\right)^{\frac{1}{2}j}. \tag{5}$$

Finally, making the transformation $V = R(1 - R^2)^{-\frac{1}{2}}$, we obtain

$$p_R(r) = \frac{(1-\rho^2)^{\frac{1}{2}(n-1)}(1-r^2)^{\frac{1}{2}(n-4)}}{\sqrt{\pi}\,\Gamma\left(\frac{n-1}{2}\right)\Gamma(\frac{1}{2}n-1)} \times \sum_{j=0}^{\infty} \frac{\left[\Gamma\left(\frac{n-1+j}{2}\right)\right]^2}{j!}\cdot(2\rho r)^j \qquad (-1 \le r \le 1). \tag{6.1}$$

(The constant multiplier may be expressed in an alternative form by using the identity $\sqrt{\pi}\,\Gamma\left(\frac{n-1}{2}\right)\Gamma(\frac{1}{2}n-1) = 2^{-(n-3)}\pi\Gamma(n-2)$.)

There are a number of other forms in which the right-hand side of (6.1) may be expressed. These include:

$$p_R(r) = \frac{(n-2)(1-\rho^2)^{\frac{1}{2}(n-1)}(1-r^2)^{\frac{1}{2}(n-4)}}{\pi}\int_0^\infty \frac{dw}{(\cosh w - \rho r)^{n-1}} \tag{6.2}$$

$$p_R(r) = \frac{(n-2)(1-\rho^2)^{\frac{1}{2}(n-1)}(1-r^2)^{\frac{1}{2}(n-4)}}{\pi} \times \int_1^\infty \frac{dw}{(w-\rho r)^{n-1}(w^2-1)^{\frac{1}{2}}} \tag{6.3}$$

$$p_R(r) = \frac{(1-\rho^2)^{\frac{1}{2}(n-1)}(1-r^2)^{\frac{1}{2}(n-4)}}{\pi\Gamma(n-2)}\frac{d^{n-2}}{d(r\rho)^{n-2}}\left\{\frac{\cos^{-1}(-\rho r)}{(1-\rho^2 r^2)^{\frac{1}{2}}}\right\} \tag{6.4}$$

$$(6.5) \qquad p_R(r) = \frac{(n-2)(1-\rho^2)^{\frac{1}{2}(n-1)}(1-r^2)^{\frac{1}{2}(n-4)}}{\sqrt{2}\,(n-1)B(\frac{1}{2},n-\frac{1}{2})(1-\rho r)^{n-\frac{3}{2}}} \times F(\tfrac{1}{2},\tfrac{1}{2};n-\tfrac{1}{2};\tfrac{1}{2}(1+\rho r))$$

$$(6.6) \qquad p_R(r) = \frac{(1-\rho^2)^{\frac{1}{2}(n-1)}(1-r^2)^{\frac{1}{2}(n-4)}}{\pi\Gamma(n-2)}\left(\frac{\partial}{\sin\theta\,\partial\theta}\right)^{n-2}\frac{\theta}{\sin\theta} \quad (\text{with } \theta = \cos^{-1}(\rho r)).$$

In all cases $-1 \leq r \leq 1$. ($F(\cdots)$ is the hypergeometric function.)

Formulas (6.2) and (6.3) are obtainable, one from the other, by simple transformation of the variable in the integral. (6.5) is a direct consequence of (6.1) and, even for moderately large n, the hypergeometric series converges rapidly. Formulas (6.4) and (6.6) are notable in that they express the probability density by a finite number of terms which involve elementary functions only.

In 1915 Fisher [19] obtained the distribution of R in the form (6.6) using a geometrical argument. (Earlier investigations were made by Student [85] and Soper [81].)

This distribution also served Fisher as the initial model for introducing the "fiducial method of inference" (see Chapters 1 and 13, and Chapter 27, Section 7) and has been, in this connection, the subject of numerous discussions in the literature (e.g. Fraser [23]).

For small values of n, simple explicit formulas are available for the cumulative distribution function. A few are given below. They were obtained by Garwood [26]. We use the notation $F_{n,R}(r)$ to represent the value of $\Pr[R \leq r]$ when the sample size is n.

$$(7.1) \qquad \underline{n=3}: F_{3,R}(r) = \pi^{-1}[\cos^{-1}(-r) - \sqrt{(1-r^2)/(1-\rho^2r^2)}\cos^{-1}(-\rho r)]$$

$$(7.2) \qquad \underline{n=4}: F_{4,R}(r) = \rho^{-1}\sqrt{(1-\rho^2)(1-r^2)}\,F_{3,R}(r) - \pi^{-1}[\rho^{-1}\sqrt{1-\rho^2} - \cos^{-1}\rho]$$

$$(7.3) \qquad \underline{n=5}: F_{5,R}(r) = \tfrac{1}{2}\rho^{-1}\sqrt{(1-\rho^2)(1-r^2)}\,F_{4,R}(r) - \tfrac{1}{2}r(1-r^2) \times F_{3,R}(r) - \pi^{-1}\left[\tfrac{1}{2}\rho^{-1}(1+\rho^2)\sqrt{\frac{1-r^2}{1-\rho^2r^2}} \times \cos^{-1}(-\rho r) - \cos^{-1}(-r)\right]$$

$$(7.4) \qquad \underline{n=6}: F_{6,R}(r) = \tfrac{1}{3}\rho^{-1}\sqrt{(1-\rho^2)(1-r^2)}\,F_{5,R}(r) + \tfrac{1}{3}\rho^{-2}(1-\rho^2)rF_{4,R}(r) - \tfrac{1}{3}\rho^{-3}(1-\rho^2)^{\frac{3}{2}}\sqrt{1-r^2}\,F_{3,R}(r) + \pi^{-1}[\tfrac{1}{3}\rho^{-3}(1-4\rho^2)\sqrt{1-\rho^2} - \cos^{-1}\rho].$$

Garwood also obtained the general formula, for $n = 2s+3$,

$$F_R(r) = \pi^{-1}\cos^{-1} r - (1 - r^2)^{\frac{1}{2}}[(2s!)]^{-1}\pi^{-1}(1 - \rho^2)^{s+1} \times \left[\left\{\Delta^s\rho^{2s} - \cdots - \binom{s}{1}\frac{\partial^2}{\partial\rho^2}\Delta^{s-1}\rho^{2s-2} + \binom{s}{2}\frac{\partial^4}{\partial\rho^4}\Delta^{s-2}\rho^{2s-4} + \cdots + (-1)\frac{\partial^{2s}}{\partial\rho^{2s}}\right\} \times \frac{\rho}{1-\rho^2}\frac{\cos^{-1}(-\rho r)}{(1-\rho^2 r^2)^{\frac{1}{2}}}\right] \tag{7.5}$$

which gives the cumulative distribution function explicitly when the sample size is odd.

As n increases, the expressions rapidly become more complicated. However, despite the complexity of the formulas, the density function is represented by a simple curve over the range $-1 \le R \le 1$ with a single mode (antimode if $n < 4$).

We also note that the value of $F_R(0) = \Pr[R \le 0]$ can be evaluated rather simply. Since $(R \le 0)$ is equivalent to

$$\sum_{t=1}^{n} (X_t - \bar{X})Y_t \le 0$$

we need to evaluate the probability of this latter event. For given $X_1, X_2, \ldots, X_n$ the probability is (using results already quoted in this section):

$$\Phi\left(\frac{-\rho}{\sqrt{1-\rho^2}}\sqrt{\sum_{j=1}^{n}(X_j - \bar{X})^2}\right) = \Pr\left[\frac{U}{\sqrt{\left(\sum_{j=1}^{n}(X_j - \bar{X})^2\right)}} \le -\frac{\rho}{\sqrt{1-\rho^2}}\right].$$

Averaging over the distribution of $\sum_{j=1}^{n}(X_j - \bar{X})^2$ (which is chi-squared with $(n - 1)$ degrees of freedom) we see that

$$\Pr[R \le 0] = \Pr\left[t_{n-1}/\sqrt{n-1} \le -\frac{\rho}{\sqrt{1-\rho^2}}\right] = \Pr\left[t_{n-1} \le \frac{-\rho\sqrt{n-1}}{\sqrt{1-\rho^2}}\right]. \tag{8}$$

(This result has been noted by Armsen [3] and Ruben [77].)

The moments of the distribution of R can be expressed in terms of hypergeometric functions (Ghosh [28]):

$$\mu_1' = c_n\rho F(\tfrac{1}{2},\tfrac{1}{2};\tfrac{1}{2}(n+1);\rho^2) \tag{9.1}$$

$$\mu_2' = 1 - \frac{(n-2)(1-\rho^2)}{n-1}F(1,1;\tfrac{1}{2}(n+1);\rho^2) \tag{9.2}$$

$$\mu_3' = c_n[\rho F(\tfrac{1}{2},\tfrac{1}{2};\tfrac{1}{2}(n+1);\rho^2) - \rho^{-1}(n-1)(n-2)] \times \{F(\tfrac{1}{2},\tfrac{1}{2};\tfrac{1}{2}(n-1);\rho^2) - F(\tfrac{1}{2},\tfrac{1}{2};\tfrac{1}{2}(n+1);\rho^2)\} \tag{9.3}$$

$$\mu_4' = 1 + \frac{(n-2)(n-4)(1-\rho^2)}{2(n-1)} F(1,1;\tfrac{1}{2}(n+1);\rho^2) - \frac{n(n-2)(1-\rho^2)}{4\rho^2}[F(1,1;\tfrac{1}{2}(n+1);\rho^2) - 1] \tag{9.4}$$

where

$$c_n = \frac{2}{n-1}\left[\Gamma\left(\frac{n}{2}\right) \Big/ \Gamma\left(\frac{n-1}{2}\right)\right]^2.$$

Ghosh [28] also obtained the following expansions for μ_1' and the first three central moments in inverse powers of $m = (n+6)$:

$$\mu_1' = \rho - \tfrac{1}{2}\rho(1-\rho^2)m^{-1}[1 + \tfrac{9}{4}(3+\rho^2)m^{-1} + \tfrac{3}{8}(121 + 70\rho^2 + 25\rho^4)m^{-2}] + O(m^{-4}) \tag{10.1}$$

$$\mu_2 = \frac{(1-\rho^2)^2}{m}[1 + \tfrac{1}{2}(14 + 11\rho^2)m^{-1} + \tfrac{1}{2}(98 + 130\rho^2 + 75\rho^4)m^{-2}] + O(m^{-4}) \tag{10.2}$$

$$\mu_3 = -\frac{\rho(1-\rho^2)^3}{m^2}[6 + (69 + 88\rho^2)m^{-1} + \tfrac{3}{4}(797 + 1691\rho^2 + 1560\rho^4)m^{-2}] + O(m^{-5}) \tag{10.3}$$

$$\mu_4 = \frac{3(1-\rho^2)^4}{m^2}[1 + (12 + 35\rho^2)m^{-1} + \tfrac{1}{4}(436 + 2028\rho^2 + 3025\rho^4)m^{-2}] + O(m^{-5}). \tag{10.4}$$

Also

$$\beta_1 = \frac{\rho^2}{m}[36 + 6(12 + 77\rho^2)m^{-1} - (162 - 1137\rho^2 - 6844\rho^4)m^{-2}] + O(m^{-4}) \tag{11.1}$$

(Note that the sign of $\sqrt{\beta_1}$ is opposite to that of ρ.)

$$\beta_2 = 3 - \frac{3}{m}[2(1 - 12\rho^2) + (10 + 14\rho^2 - 387\rho^4)m^{-1} + \tfrac{1}{2}(100 + 832\rho^2 + 1503\rho^4 - 14202\rho^6)m^{-2}] + O(m^{-4}). \tag{11.2}$$

For everyday use, one can remember that the bias in R as an estimator of ρ is approximately $-\frac{1}{2}\rho(1-\rho^2)n^{-1}$; and

$$\text{var}(R) \doteqdot (1-\rho^2)^2 n^{-1}. \tag{12}$$

It is interesting to note that

$$E(\sin^{-1} R) = \sin^{-1}\rho.$$

(See Harley [31], [32], Daniels and Kendall [11].)

We conclude this section by giving two relationships satisfied by $p_R(r)$. They are not generally of practical use, but may be helpful in specific problems and have intrinsic interest.

(a) In the equation

$$r\,\frac{\partial p_R(r)}{\partial r} + \frac{(n-3)r^2}{1-r^2}\,p_R(r) = \rho\,\frac{\partial p_R(r)}{\partial \rho} + \frac{n\rho^2}{1-\rho^2}\,p_R(r) \tag{13}$$

(Hotelling [36])

the near-symmetry of the coefficients in R (left-hand) and ρ (right-hand) should be noted.

(b) Introducing the notation

$$p_{n,R}(r) \equiv p_R(r)$$

when the sample size is n we have

$$\begin{aligned}(n-1)(n-2)\{1-(\rho R)^2\}\,p_{n+1,R}(r) \\ = (2n-1)(n-2)\rho\sqrt{1-\rho^2}\,R\sqrt{1-R^2}\,p_{n,R}(r) \\ + (n-1)^2(1-\rho^2)(1-r^2)p_{n-1,R}(r).\end{aligned}$$

(Soper *et al.* [83].)

3. Distribution of *R* in Non-Normal Populations

The distribution of R for samples from non-normal populations has been worked out in detail only for certain special cases. However, for certain bivariate Edgeworth populations, investigations indicate the kinds of variation one might expect with various departures from normality, as measured by the lower moment-ratios. (The assumption that there are n *independent* pairs of observations, each with the *same* joint distribution, has been retained.)

Quensel [74] supposed that cumulants (and mixed cumulants) of order higher than 4 were negligible, and that the population value of the correlation coefficient (ρ) was zero. Gayen [27] extended this work by allowing ρ to be non-zero. He obtained an expansion in the density function in terms of the right-hand side of equations (6.1)–(6.6) (here denoted by $f(r,\rho)$) and its derivatives with respect to ρ. The formula is

$$\begin{aligned}P_R(r) = f(r,\rho) + \frac{n-1}{8n(n+1)}\left\{L_{4,1}\frac{\partial f}{\partial \rho} + L_{4,2}\frac{\partial^2 f}{\partial \rho^2}\right\} \\ + \frac{n-2}{12n(n+1)(n+3)}\left\{L_{6,1}\frac{\partial f}{\partial \rho} + L_{6,2}\frac{\partial^2 f}{\partial \rho^2} + L_{6,3}\frac{\partial^3 f}{\partial \rho^3}\right\}\end{aligned} \tag{14}$$

where the L's are functions of n, ρ and the cumulant ratios $\gamma_{ij} = \kappa_{ij}\kappa_{20}^{-\frac{1}{2}i}\kappa_{02}^{-\frac{1}{2}j}$, namely:

$$L_{4,1} = 3\rho(\gamma_{40} + \gamma_{04}) - 4(\gamma_{31} + \gamma_{13}) + 2\rho\gamma_{22}$$

$$L_{4,2} = \rho^2(\gamma_{40} + \gamma_{04}) - 4\rho(\gamma_{31} + \gamma_{13}) + 2(2 + \rho^2)\gamma_{22}$$

$$\begin{aligned} L_{6,1} = & -15\rho(\gamma_{30}^2 + \gamma_{03}^2) - 9\rho\left(1 + \frac{2}{n-2}\right)(\gamma_{21}^2 + \gamma_{12}^2) - \frac{6}{n-2}\gamma_{30}\gamma_{03} \\ & + 6\left(2 + \frac{1}{n-2}\right)\gamma_{21}\gamma_{12} + 18(\gamma_{30}\gamma_{21} + \gamma_{03}\gamma_{12}) \\ & + \frac{18\rho}{n-2}(\gamma_{30}\gamma_{12} + \gamma_{03}\gamma_{21}) \end{aligned}$$

$$\begin{aligned} L_{6,2} = & -9\rho^2(\gamma_{30}^2 + \gamma_{03}^2) - 3\left(4 + 5\rho^2 - \frac{2(2-5\rho^2)}{n-2}\right)(\gamma_{21}^2 + \gamma_{12}^2) \\ & - \frac{18\rho}{n-2}\gamma_{30}\gamma_{03} + 18\rho\left(2 + \frac{1}{n-2}\right)\gamma_{21}\gamma_{12} + 30\rho(\gamma_{30}\gamma_{21} + \gamma_{03}\gamma_{12}) \\ & - 6\left(2 + \frac{2-5\rho^2}{n-2}\right)(\gamma_{30}\gamma_{12} + \gamma_{03}\gamma_{21}) \end{aligned}$$

$$\begin{aligned} L_{6,3} = & -\rho^3(\gamma_{30}^2 + \gamma_{03}^2) - 3\rho\left(2 + \rho^2 - \frac{2(1-\rho^2)}{n-2}\right)(\gamma_{21}^2 + \gamma_{12}^2) \\ & + 2\left(1 + \frac{3(1-\rho^2)}{n-2}\right)\gamma_{30}\gamma_{03} + 6\left(1 + 2\rho^2 - \frac{1-\rho^2}{n-2}\right)\gamma_{21}\gamma_{12} \\ & + 6\rho^2(\gamma_{30}\gamma_{21} + \gamma_{03}\gamma_{12}) - 6\rho\left(1 + \frac{1-\rho^2}{n-2}\right)(\gamma_{30}\gamma_{12} + \gamma_{03}\gamma_{21}). \end{aligned}$$

Cook [8], [9] obtained expressions (up to and including terms in n^{-2}) of the first four moments of R in terms of the cumulants and cross-cumulants of the parent population (without specifying the exact form of this population distribution).

The second and third terms respectively of (14) can be regarded as corrections to the normal density $f(r,\rho)$, for kurtosis and skewness. Gayen [27] gives values of these corrective terms for certain special cases. When ρ is zero, the terms are small even for n as small as four. However, an example with $\rho = 0.8$ shows that quite substantial corrections can be needed when n is not large.

Gayen further discussed the distribution of $z' = \tanh^{-1} R$ (see next section) obtaining expansions for its expected value, variance, β_1 and β_2. He found that β_1 and β_2 of z' still tend to the normal values of 0 and 3 as n increased, but not so rapidly as when the parent population is normal. Table 1 gives the leading terms in expressions for moments and moment-ratios of R and z'.

Cheriyan [6] has reported results of sampling experiments on distribution of correlation coefficient in random samples from certain bivariate gamma distributions (Chapter 39).

4. Approximations and Tables

From Section 2, it is clear that the distribution of R is so complicated that it is very difficult to use without a practicable approximation or an extensive set of

TABLE 1

	R	z'
Expected Value	$\rho + \frac{1}{n}[-\frac{1}{2}\rho(1-\rho^2) + \frac{1}{8}L_{4,1}]$	$\frac{1}{2}\log\frac{1+\rho}{1-\rho} + \frac{1}{n-1} \times \Big[\frac{1}{2}\rho + \frac{1}{8(1-\rho^2)^2}\{\rho(3-\rho^2) \times (\gamma_{40}+\gamma_{04}) - 4(1+\rho^2) \times (\gamma_{31}+\gamma_{13}) + 2\rho(5+\rho^2)\gamma_{22}\}\Big]$
Variance	$\frac{1}{n}[(1-\rho^2)^2 + \frac{1}{4}L_{4,2}]$	$\frac{1}{n-1}\Big[1 + \frac{1}{4(1-\rho^2)^2} \times \{\rho^2(\gamma_{40}+\gamma_{04}) - 4\rho(\gamma_{31}+\gamma_{13}) + 2(2+\rho^2)\gamma_{22}\}\Big]$

tables. We may note that this distribution is important because it is the distribution of the correlation coefficient in random samples from a bivariate normal population, not because it is suitable for fitting purposes.

David [13] has prepared a useful set of tables. They give values of the probability density $p_R(r)$ to 4 decimal places and of the cumulative distribution function $F_R(r)$ to five decimal places for $n = 3(1)25, 50, 100, 200, 400$. For $n \leq 25$, $r = -1.00(0.05)1.00$ for $\rho = 0.0(0.1)0.4$; $r = -1.00(0.05)0.600(0.025)1.000$ for $\rho = 0.5(0.1)0.9$, with additional values for $r = 0.80(0.01)0.900(0.005)1.000$, for $\rho = 0.9$; for $n > 25$, narrower intervals are used.

The introduction to these tables contains some interesting notes on the distribution of R.

For most practical purposes, approximations to the distributions of R use the transformation

$$z' = \tanh^{-1} R = \tfrac{1}{2}\log\left(\frac{1+R}{1-R}\right).$$

This transformation might be suggested as a variance-equalizing transformation, noting that (from (12))

$$\operatorname{var}(R) \doteqdot (1-\rho^2)^2 n^{-1}$$

and

$$\int (1-\rho^2)^{-1}\,d\rho = \tfrac{1}{2}\log\frac{1+\rho}{1-\rho}.$$

This approach, however, was not explicitly used by Fisher [19] in his original suggestion of this transformation.

Approximate values of moments and moment-ratios of z' are

(15.1) $$\mu_1'(z') \doteqdot \tfrac{1}{2}\log\frac{1+\rho}{1-\rho} + \tfrac{1}{2}\rho(n-1)^{-1}[1+\tfrac{1}{4}(5+\rho^2)(n-1)^{-1}]$$

(15.2) $$\mu_2(z') \doteqdot (n-1)^{-1}[1+\tfrac{1}{2}(4-\rho^2)(n-1)^{-1} + \tfrac{1}{6}(22-6\rho^2-3\rho^4)(n-1)^{-2}]$$

(15.3) $$\mu_3(z') \doteqdot (n-1)^{-3}\rho^3$$

(15.4) $$\mu_4(z') \doteqdot 3(n-1)^{-2}[1+\tfrac{1}{3}(14-3\rho^2)(n-1)^{-1} + \tfrac{1}{12}(184-48\rho^2-21\rho^4)(n-1)^{-2}]$$

(16.1) $$\beta_1(z') \doteqdot (n-1)^{-3}\rho^6$$

(16.2) $$\beta_2(z') \doteqdot 3+2(n-1)^{-1}+(4+2\rho^2-3\rho^4)(n-1)^{-2}.$$

These values were given by Fisher [20] and corrected later by Gayen [27]. (See also Nabeya [62].) Comparing (16.1) and (16.2) with (11.1) and (11.2), it can be seen that the (β_1,β_2) values for z' are much closer to the normal values (0,3) than are the values for R. Also $\operatorname{var}(z')$ does not depend on ρ up to and including terms of order $(n-1)^{-1}$.

The most commonly used approximation is to regard z' as normally distributed with expected value $\tfrac{1}{2}\log\left(\frac{1+\rho}{1-\rho}\right)$ and variance $(n-3)^{-1}$. The latter value is obtained by noting that

$$\begin{aligned}(n-1)^{-1}+\tfrac{1}{2}(4-\rho^2)(n-1)^{-2} &= n^{-1}+n^{-2}+\cdots+\tfrac{1}{2}(4-\rho^2)n^{-2}\ldots\\ &= n^{-1}[1+(3-\tfrac{1}{2}\rho^2)n^{-1}\ldots]\\ &\doteqdot n^{-1}[1-3n^{-1}]^{-1} = (n-3)^{-1}.\end{aligned}$$

To improve the approximation, the expected value may be increased by $(2n-5)^{-1}\rho$.

Another kind of approximation can be based on consideration of the structure of the random variable R. Recalling the argument used, at the beginning of Section 2, in deriving the distribution of R we see that, for a fixed set of values of $X_1, X_2, \ldots, X_n$, $R(1-R^2)^{-\frac{1}{2}}$ is distributed as

$$(n-2)^{-\frac{1}{2}}t'_{n-2}\left(\sqrt{\sum_{j=1}^{n}(X_j-\overline{X})^2}\,\rho\Big/\sqrt{1-\rho^2}\right);$$

that is, as

$$\left(U+\sqrt{\sum_{j=1}^{n}(X_j-\overline{X})^2}\,\rho\Big/\sqrt{1-\rho^2}\right)\Big/\chi_{n-2}$$

where U is a unit normal variable and U and χ_{n-2} are mutually independent. Averaging over the joint distribution of the X's, we see that $R(1-R^2)^{-\frac{1}{2}}$ is distributed as

(17) $$[U+\chi_{n-1}\rho(1-\rho^2)^{-\frac{1}{2}}]/\chi_{n-2}$$

where U, χ_{n-1} and χ_{n-2} are mutually independent. This representation was constructed by Ruben [77], [78], who used it as the basis for the following approximations.

From the representation (17), it follows that

$$\Pr[R \leq r] = \Pr[U + \rho(1-\rho^2)^{-\frac{1}{2}}\chi_{n-2} - r(1-r^2)^{-\frac{1}{2}}\chi_{n-1} \leq 0]. \tag{18}$$

Providing n is not too small, χ_{n-2} and χ_{n-1} may be approximated by normal variates using Fisher's approximation that $\{\sqrt{2\chi_n^2} - \sqrt{2n-1}\}$ is approximately distributed as a unit normal variable. Then

$$U + \rho(1-\rho^2)^{-\frac{1}{2}}\chi_{n-2} - r(1-r^2)^{-\frac{1}{2}}\chi_{n-1}$$

is approximately distributed normally with expected value

$$\rho(1-\rho^2)^{-\frac{1}{2}}(n-\tfrac{5}{2})^{\frac{1}{2}} - r(1-r^2)^{-\frac{1}{2}}(n-\tfrac{3}{2})^{\frac{1}{2}}$$

and with standard deviation

$$[1 + \tfrac{1}{2}\rho^2(1-\rho^2)^{-1} + \tfrac{1}{2}r^2(1-r^2)^{-1}]^{\frac{1}{2}}.$$

Hence, from (18)

$$\Pr[R \leq r] \doteqdot \Phi\left(\frac{r(1-r^2)^{-\frac{1}{2}}(n-\frac{3}{2})^{\frac{1}{2}} - \rho(1-\rho^2)^{-\frac{1}{2}}(n-\frac{5}{2})^{\frac{1}{2}}}{[1+\frac{1}{2}r^2(1-r^2)^{-1} + \frac{1}{2}\rho^2(1-\rho^2)^{-1}]^{\frac{1}{2}}}\right). \tag{19}$$

Stammberger [84] has constructed a nomogram from which it is possible to determine any one of $\Pr[R \leq r]$, r or n, given the other two values.

5. Sample Covariance

The sample covariance can be defined as

$$n^{-1}\sum_{t=1}^{n}(X_t - \bar{X})(Y_t - \bar{Y}).$$

If each of the independent pairs (X_t, Y_t) has a bivariate normal distribution as in (2), this is the maximum likelihood estimator of the population covariance $\rho\sigma_X, \sigma_Y$. We will find it more convenient to consider the distribution of the sums of products of deviations from sample means:

$$C = \sum_{t=1}^{n}(X_t - \bar{X})(Y_t - \bar{Y}). \tag{20}$$

The distribution of C has been derived, using various methods, by Pearson *et al.* [69], Wishart and Bartlett [93], Hirschfeld [34] and Mahalanobis *et al.* [56]. Remembering that the conditional distribution of Y_t given $X_1, X_2, \ldots, X_n$ is normal with expected value $[\eta + (\rho\sigma_Y/\sigma_X)(X_t - \xi)]$ and standard deviation $\sigma_Y\sqrt{1-\rho^2}$, we see that the conditional distribution of C, given $X_1, X_2, \ldots, X_n$ is normal with expected value $(\rho\sigma_Y/\sigma_X)S$ and standard deviation $\sigma_Y\sqrt{(1-\rho^2)S}$ where

$$S = \sum_{t=1}^{n} (X_t - \overline{X})^2.$$

Noting that S is distributed as $\chi^2_{n-1}\sigma^2_X$, it follows that the probability density function of C is

$$(21) \qquad p_C(c) = \frac{1}{\sigma_Y\sqrt{2\pi(1-\rho^2)}} \int_0^\infty \left[\frac{S^{\frac{1}{2}(n-4)} \exp(-\frac{1}{2}S/\sigma^2_X)}{(2\sigma^2_X)^{\frac{1}{2}(n-1)}\Gamma(\frac{1}{2}(n-1))}\right] \times \exp\left[-\frac{(c-\rho\sigma_Y S/\sigma_X)^2}{2\sigma^2_Y(1-\rho^2)S}\right] dS.$$

If $\sigma_X = \sigma_Y = 1$, then

$$(22) \qquad p_C(c) = \frac{e^{c\rho/(1-\rho^2)}}{2^{\frac{1}{2}n}\Gamma(\frac{1}{2}(n-1))\sqrt{\pi(1-\rho^2)}} \int_0^\infty S^{\frac{1}{2}(n-4)} \exp\left[-\frac{S+c^2S^{-1}}{2(1-\rho^2)}\right] dS$$

$$= \frac{|c|^{\frac{1}{2}n-1} e^{c\rho/(1-\rho^2)}}{2^{\frac{1}{2}n-1}\Gamma(\frac{1}{2}(n-1))\sqrt{\pi(1-\rho^2)}} K_{\frac{1}{2}(n-2)}\left(\frac{|c|}{1-\rho^2}\right)$$

where $K_\nu(z)$ denotes the modified Bessel function of the second kind, of order ν. From this formula, distributions of C, with $\sigma_X \neq 1$, $\sigma_Y \neq 1$ are easily derived. Some other, equivalent, forms for the density function have been given by Press [72]. He also pointed out that if n is odd, then it is possible to express the density as a finite series of elementary functions, since

$$K_{\frac{1}{2}n-1}(z) = \sqrt{\frac{\pi}{2z}}\, e^{-z} \sum_{j=0}^{\frac{1}{2}(n-3)} \frac{(\frac{1}{2}(n-3)+j)!}{j!(\frac{1}{2}(n-3)-j)!} \cdot \frac{1}{(2z)^j}.$$

The characteristic function of C is

$$(23) \qquad \{1 - 2it\rho\sigma_X\sigma_Y + t^2(1-\rho^2)\sigma^2_X\sigma^2_Y\}^{-\frac{1}{2}(n-1)}$$

$$= \left[\left\{1 - 2it\frac{\sigma_X\sigma_Y(1+\rho)}{2}\right\}\left\{1 + 2it\frac{\sigma_X\sigma_Y(1-\rho)}{2}\right\}\right]^{-\frac{1}{2}(n-1)}.$$

Hence the distribution of C is also that of $\frac{1}{2}\sigma_X\sigma_Y\{(1+\rho)Z_1 - (1-\rho)Z_2\}$ where Z_1 and Z_2 are independent random variables each distributed as χ^2 with $(n-1)$ degrees of freedom. Note that this distribution is the K-form of Bessel function distribution described in Chapter 12, Section 4.

From the representation

$$(24) \qquad C = \tfrac{1}{2}\sigma_X\sigma_Y\{(1+\rho)Z_1 - (1-\rho)Z_2\}$$

it follows that

$$(25) \qquad \kappa_r(C) = (\tfrac{1}{2}\sigma_X\sigma_Y)^r\{(1+\rho)^r + (-1)^r(1-\rho)^r\}\kappa_r(\chi^2_{n-1}).$$

In particular

$$E(C) = (n-1)\rho\sigma_X\sigma_Y \tag{26.1}$$

$$\text{var}(C) = (n-1)(1+\rho^2)\sigma_X^2\sigma_Y^2 \tag{26.2}$$

$$\alpha_3(C) = \sqrt{\beta_1(C)} = 2\rho(3+\rho^2)(1+\rho^2)^{-\frac{3}{2}}(n-1)^{-\frac{1}{2}} \tag{26.3}$$

$$\alpha_4(C) = \beta_2(C) = 3 + 6(1+6\rho^2+\rho^4)(1+\rho^2)^{-2}(n-1)^{-1}. \tag{26.4}$$

The representation (24) also shows clearly how the distribution of C tends to normality as n tends to infinity.

6. Circular Serial Correlation

In formula (1) for the correlation coefficient, X_t and Y_t are usually thought of as representing observations of *different* characters on the *same* individual. Although this is so in very many applications, it is not formally necessary, and some useful techniques can be based on modifications of this idea. In particular, one can take $Y_t = X_{t+k}$, with appropriate allowance for end-effects. The correlation coefficients so obtained are called *serial correlations*. The absolute value of the difference between t and the subscript of X corresponding to Y_t is called the *lag* of the correlation. (It is clear that similar results are obtained by taking $Y_t = X_{t+k}$, or $Y_t = X_{t-k}$.)

Serial correlations are most commonly employed when the subscript 't' in X_t defines the number of units of time elapsed, from some initial moment, when X_t is observed. However, serial correlations can be, and have been, used when the ordering is non-temporal (e.g. spatial).

It is of interest to note that the mean square successive difference ratio (von Neumann [63])

$$d = \frac{\sum_{t=1}^{n-1}(X_t - X_{t+1})^2}{\sum_{t=1}^{n}(X_t - \bar{X})^2}$$

is closely related to a serial correlation (R_1) of lag 1. In fact

(27)

$$d = \frac{\sum_{t=1}^{n-1}(X_t-\bar{X})^2 - 2\sum_{t=1}^{n-1}(X_t-\bar{X})(X_{t+1}-\bar{X}) + \sum_{t=1}^{n-1}(X_{t+1}-\bar{X})^2}{\sum_{t=1}^{n}(X_t-\bar{X})^2}$$

$$= 2(1-R_1) - \frac{(X_1-\bar{X})^2 + (X_n-\bar{X})^2}{\sum_{t=1}^{n}(X_t-\bar{X})^2}.$$

The simplest way of allowing for end-effects is to regard the data as consisting of $(n - k)$ pairs

$$(X_1,X_{1+k})(X_2,X_{2+k}) \ldots (X_{n-k},X_n).$$

Sometimes, however, mathematical analysis is facilitated by using the *circular serial correlation*, based on the n pairs

$$(X_t,X_{t+k}) \qquad (t = 1,\ldots,n).$$

Putting $X_{n+j} = X_j$ for $j = 1, 2, \ldots, k$. Note that in this case, the denominator of R_k is simply $\sum_{t=1}^{n} (X_t - \overline{X})^2$. If k is small compared with n, it may be hoped that distributions appropriate to circular correlations may be applied to non-circular correlations with relatively small risk of serious error.

In the case of the mean square successive difference ratio, for example, the circular definition would lead to

$$\tilde{d} = \sum_{t=1}^{n} (X_t - X_{t+1})^2 \Big/ \sum_{t=1}^{n} (X_t - \overline{X})^2$$

while the circular serial correlation of lag 1 is

$$\tilde{R}_1 = \sum_{t=1}^{n} (X_t - \overline{X})(X_{t+1} - \overline{X}) \Big/ \sum_{t=1}^{n} (X_t - \overline{X})^2 \tag{28}$$

(with, of course, $X_{n+1} = X_1$). The relationship between $\tilde{d}$ and $\tilde{R}_1$ is

$$\tilde{d} = 2(1 - \tilde{R}_1)$$

which is tidier than that between d and R_1.

We will now consider the distribution of circular serial correlation of lag 1 when the X's are related by equations of form

$$X_t = \rho X_{t-1} + Z_t \qquad (|\rho| < 1; t = 1,2,\ldots,n) \tag{29}$$

where the Z_t's are mutually independent normal random variables each having expected value zero and standard deviation σ, and further Z_t is independent of all X_j for $j < t$. The series is supposed to be started by a random variable X_0, which is distributed normally with expected value zero, and has standard deviation $\sigma(1 - \rho^2)^{-\frac{1}{2}}$. The effect of this rather odd-looking assumption is to ensure that each X_t has the same variance (σ^2) as well as the same expected value (zero).

It is clear that both R_1 and $\tilde{R}_1$ are special cases of ratios of quadratic forms in central normal variables (see also Chapter 29, Section 6). The methods used by von Neumann [63] in deriving the distribution of d (for $\rho = 0$) may also be used in deriving the distributions of serial correlations (for $\rho = 0$ or $\rho \neq 0$). The results possess the characteristic property that the density functions take different forms for different intervals in the range of variation of the variable.

The density function of $\tilde{R}_1$ for the case $\rho = 0$ was obtained by Anderson [2] (see also Koopmans [49]) in the simple form

(30)

$$\Pr[\tilde{R}_1 > r] = \sum_{j=1}^{n} \alpha_j^{-1}(\lambda_j - r)^{\frac{1}{2}(n-3)} \quad (\text{for } \lambda_{m+1} \le r \le \lambda_m, m = 1,2,\ldots,n-1)$$

where $\lambda_j = \cos(2\pi j/n)$;

$$\alpha_j = \begin{cases} \prod_{i\neq j, i=1}^{\frac{1}{2}(n-1)} (\lambda_j - \lambda_i) & (n \text{ odd}) \\ \left[\prod_{i\neq j, i=1}^{\frac{1}{2}(n-2)} (\lambda_j - \lambda_i)\right]\sqrt{1+\lambda_j} & (n \text{ even}). \end{cases}$$

Anderson also obtained formulas for the distribution of $\tilde{R}_l$ (circular serial correlation with lag l). He also considered the distribution of the statistic

(31)
$$\tilde{R}_1' = \frac{\sum_{j=1}^{n} X_j X_{j+1}}{\sum_{j=1}^{n} X_j^2}$$

(where we use our knowledge that $E[X_j] = 0$). Roughly speaking the distribution of $\tilde{R}_1'$ is close to that of $\tilde{R}_1$, with n increased by 1. For the case n odd (and $\rho = 0$), Anderson showed that this is exactly so, provided α_j is multiplied by $\sqrt{1-\lambda_j}$. Madow [55] extended these results to the case $\rho \neq 0$. He used the simple, but effective, device of noting that if $(T_1,T_2,\ldots,T_s) \equiv \mathbf{T}$ is a sufficient set of statistics for parameters θ, then

$$\frac{p_{\mathbf{T}}(\mathbf{t} \mid \rho)}{p_{\mathbf{T}}(\mathbf{t} \mid 0)} = \frac{p_{X_1,\ldots,X_n}(x_1,\ldots,x_n \mid \rho)}{p_{X_1,\ldots,X_n}(x_1,\ldots,x_n \mid 0)}$$

so that the joint distribution of $\mathbf{T}$ for general ρ can easily be derived from that for $\rho = 0$. In this case $\mathbf{T}$ is composed of the numerator and denominator of $\tilde{R}_1'$ (or $\tilde{R}_l, \tilde{R}_1$ etc.). The density function of $\tilde{R}_1$, so obtained is

(32)
$$p_{\tilde{R}_1}(r) = \tfrac{1}{2}(n-3) \times \frac{\prod_{j=1}^{n}(1+\rho^2-\lambda_j\rho)^{\frac{1}{2}}}{(1+\rho^2-2\rho r)^{\frac{1}{2}n-1}} \sum_{j=1}^{m} \alpha_j^{-1}(\lambda_j - r)^{\frac{1}{2}(n-5)}$$

$$(\text{for } \lambda_{m+1} \le r \le \lambda_m, m = 1,\ldots,(n-1)).$$

Daniels [10] obtained the following expansion for the density of $\tilde{R}_1'$

(33)
$$p_{\tilde{R}_1}(r) = \frac{\Gamma(\frac{1}{2}n+1)(1-\rho^n)}{\sqrt{\pi}\,(1+\rho^2-2\rho r)^{\frac{1}{2}n}} \times \left\{\frac{(1-r^2)^{\frac{1}{2}(n-1)}}{\Gamma(\frac{1}{2}n+\frac{1}{2})} - \frac{3}{2^r\Gamma(\frac{3}{2}n+\frac{1}{2})}\frac{d^n}{dr^n}(1-r^2)^{\frac{1}{2}(3n-1)} + \cdots\right\}.$$

(The jth term is

$$(-1)^{j-1}\frac{2j-1}{2^{jn}\Gamma((j-\frac{1}{2})n+\frac{1}{2})}\cdot\frac{d^{jn}}{dr^{jn}}(1-r^2)^{\frac{1}{2}((2j-1)n-1)}.)$$

The *randomization* ($\rho = 0$) *distribution* of $\tilde{R}_1$ (or $\tilde{R}_1'$) is the distribution obtained by considering all possible orderings of the n observed values $X_1, \ldots, X_n$. Since the denominator is unchanged by re-ordering, this is essentially the distribution of the $n!$ possible values of the numerator. Wald and Wolfowitz [86] have studied this distribution.

7. Noncircular Serial Correlation

The distributions of noncircular serial correlations are generally even more complicated than those of the corresponding circular correlations. It is possible, however, to obtain formulas for the moments which are reasonably easy to comprehend.

We will first consider the noncircular serial correlation of lag 1, as

$$\hat{R}_1 = \left(\sum_{j=1}^{n} X_{j-1}X_j\right)\Big/\left(\sum_{j=1}^{n-1} X_j^2\right) \tag{34}$$

with $X_0 = 0$, and first consider its distribution under model (27). The joint density function of $X_1, \ldots, X_n$ is

$$p_{X_1,\ldots,X_n}(x_1,\ldots,x_n) = (2\pi\sigma)^{-\frac{1}{2}n}\exp\left[-\tfrac{1}{2}\sum_{j=1}^{n}(x_j-\rho x_{j-1})^2\right].$$

The distribution of $\hat{R}_1$ clearly does not depend on σ, and we will henceforth take $\sigma = 1$.

The joint moment-generating function of the numerator and denominator of $\hat{R}_1$ is then

$$E\left[\theta_1\sum_{j=1}^{n} X_{j-1}X_j + \theta_2\sum_{j=1}^{n-1} X_j^2\right] = [D_n(\theta_1,\theta_2)]^{-\frac{1}{2}}$$

where (for $n \geq 2$)

$$D_n(\theta_1,\theta_2) = (1+\rho^2-2\theta_2)D_{n-1}(\theta_1,\theta_2) - (\rho+\theta_1)^2 D_{n-2}(\theta_1,\theta_2), \tag{35}$$

with $D_0(\theta_1,\theta_2) = D_1(\theta_1,\theta_2) = 1$.

White [89] evaluated the first and second moments of $\hat{R}_1$ using the formula

$$\mu_m'(\hat{R}_1) = \int_{-\infty}^{0}\int_{-\infty}^{t_m}\cdots\int_{-\infty}^{t_2}\left.\frac{\partial^m\{D_n(\theta_1,t_1)\}^{-\frac{1}{2}}}{\partial\theta_1^m}\right|_{\theta_1=0} dt_1\,dt_2\ldots dt_m. \tag{36}$$

Shenton and Johnson [80] found it more convenient to use the formula for moments about ρ

$$E[(\hat{R}_1-\rho)^m] = \sum_{j=0}^{[\frac{1}{2}m]} a_j^{(m)}\int_0^\infty\left(\frac{d}{d\rho}\right)^{m-2j} H_{m-j}(t)\,dt \tag{37}$$

where

$$H_r(t) = \frac{t^r}{r!}\{D_n(0,t)\}^{-\frac{1}{2}},$$

$$\sum_{j=0}^{[\frac{1}{2}m]} (-1)^j a_j^{(m)} y^{m-2j} = e^{-\frac{1}{2}D^2} y^m \text{ * and } a_0^{(m)} = 1.$$

We now quote some of their results, in the form of asymptotic series

(38.1)

$$E[\hat{R}_1] = \rho - \frac{2(n-2)\rho}{(n+1)(n-1)} - \frac{12\rho^3}{(n+5)(n+3)(n+1)} + \frac{18(n+8)\rho^5}{(n+9)^{\{4\}}} + \frac{24(n+10)(n+12)\rho^7}{(n+13)^{\{5\}}} + \frac{30(n+12)(n+14)(n+16)\rho^9}{(n+17)^{\{6\}}} + \cdots,$$

(38.2)

$$E[(\hat{R}_1 - \rho)^2] = \frac{n^2 - 4n + 7}{(n+1)^{\{3\}}} - \frac{(n^3 - 6n^2 - 25n + 42)\rho^2}{(n+5)^{\{4\}}} + \frac{3(n+19)(n-3)\rho^4}{(n+9)^{\{5\}}} + \frac{3(n^3 + 32n^2 + 111n - 928)\rho^6}{(n+13)^{\{6\}}} + \frac{3(n+12)(n^3 + 42n^2 + 215n - 1950)\rho^8}{(n+17)^{\{7\}}} + \cdots$$

where $n^{\{s\}} = n(n-2)(n-4)\ldots(n - 2\cdot\overline{s-1})$.

For n large the following expansions in powers of n^{-1} can be used

(39.1)
$$E[\hat{R}_1] = \rho - \frac{2\rho}{n} + \frac{4\rho}{n^2} - \frac{2\rho(1 - 8\rho^2 + 4\rho^4)}{n^3(1-\rho^2)^2} + \frac{4\rho(1 - 30\rho^2 + 12\rho^4 - 4\rho^6)}{n^4(1-\rho^2)^3} - \cdots$$

(39.2)
$$E[(\hat{R}_1 - \rho)^2] = \frac{1-\rho^2}{n} - \frac{1 - 14\rho^2}{n^2} + \frac{5 - 78\rho^2 + 76\rho^4}{n^3(1-\rho^2)} + \frac{11 + 316\rho^2 - 692\rho^4 + 344\rho^6}{n^4(1-\rho^2)^2} + \cdots$$

(39.3)
$$\text{var}(\hat{R}_1) = \frac{1-\rho^2}{n} - \frac{1 - 10\rho^2}{n^2} + \frac{5 - 62\rho^2 + 60\rho^4}{n^3(1-\rho^2)} + \frac{11 + 292\rho^2 - 596\rho^4 + 296\rho^6}{n^4(1-\rho^2)^2} + \cdots$$

(In [80] the expansions for $E[\hat{R}_1]$ and $E[(\hat{R}_1 - \rho)^2]$ are given up to and including terms in n^{-6}.)

* D here is the differential operator introduced in Chapter 1.

TABLE 2

Moment Values for the Distribution of $\hat{R}_1$ from [80]

ρ \ n		6	8	10	15	50	100	500
0.0	(b)	0.4254	0.3519	0.3109	0.2530	0.1401	0.0995	0.0447
	(d)	4.7331	3.0022	2.7859	2.7440	2.8919	2.9430	2.9881
0.2	(a)	0.0456	0.0380	0.0323	0.0232	0.0077	0.0039	0.0008
	(b)	0.4244	0.3500	0.3086	0.2502	0.1378	0.0977	0.0438
	(c)	0.0267	0.0418	0.0476	0.0477	0.0241	0.0134	0.0029
	(d)	4.8014	3.0767	2.8625	2.8171	2.9302	2.9648	2.9930
0.4	(a)	0.0902	0.0755	0.0642	0.0463	0.0154	0.0078	0.0016
	(b)	0.4217	0.3445	0.3017	0.2419	0.1304	0.0920	0.0410
	(c)	0.1067	0.1719	0.1999	0.2067	0.1085	0.0610	0.0134
	(d)	5.0031	3.3104	3.1096	3.0613	3.0647	3.0422	3.0103
0.6	(a)	0.1321	0.1114	0.0952	0.0689	0.0230	0.0118	0.0024
	(b)	0.4182	0.3358	0.2901	0.2275	0.1172	0.0815	0.0359
	(c)	0.2392	0.4021	0.4872	0.5372	0.3099	0.1774	0.0394
	(d)	5.3160	3.7301	3.5833	3.5721	3.3839	3.2310	3.0535
0.8	(a)	0.1674	0.1426	0.1228	0.0900	0.0306	0.0157	0.0032
	(b)	0.4155	0.3256	0.2751	0.2067	0.0958	0.0641	0.0272
	(c)	0.4319	0.7466	0.9563	1.1876	0.9171	0.5318	0.1235
	(d)	5.6832	4.3496	4.3732	4.5969	3.8334	3.8032	3.1926
1.0	(a)	0.1893	0.1599	0.1369	0.0998	0.034†	0.019†	0.004†
	(b)	0.4146	0.3167	0.2603	0.1826	0.06†	0.03†	0.006†
	(c)	0.7456	1.2859	1.7182	2.4705	3.89†	5.1†	6.5†
	(d)	6.0955	5.1633	5.5432	6.5897	9.3†	11.3†	13.2†

Key: (a) $E(\hat{R}_1 - \rho)$
(b) standard deviation of $\hat{R}_1$
(c) $\beta_1(\hat{R}_1)$
(d) $\beta_2(\hat{R}_1)$.
$E(\hat{R}_1 - \rho)$ and $\beta_1(\hat{R}_1)$ are zero for $\rho = 0$.

†The values thus marked are tentative due to uncertain round-off error.

Shenton and Johnson also obtained formulas for β_1 and β_2 which they did not publish in [80] though they gave results of calculations using these formulas. These are included in Table 2, reproduced from [80].

The exact distribution of $\hat{R}_1$ has been obtained by Pan [65] for the case when the correlation between X_i and X_j is ρ for $|i - j| = 1$, zero otherwise. In this case $\sum_{j=1}^{n} (X_j - \overline{X})^2$ is distributed as a multiple of χ^2, and it is possible to follow the method of von Neumann [63]. It can be shown that $\hat{R}_1$ is distributed as

$$\left(\sum_{j=1}^{n-1} \lambda_j U_j^2\right) \Big/ \left(\sum_{j=1}^{n-1} U_j^2\right)$$

where $U_1, U_2, \ldots, U_{n-1}$ are independent unit normal variables and

$$\lambda_1 > \lambda_2 > \cdot > \lambda_{n-1},$$

and zero, are the characteristic roots of the matrix **MAM** with

$$\mathbf{M} = \begin{pmatrix} 1-n^{-1} & -n^{-1} & \dots & -n^{-1} \\ n^{-1} & 1-n^{-1} & \dots & -n^{-1} \\ \vdots & \vdots & & \vdots \\ n^{-1} & n^{-1} & \dots & 1-n^{-1} \end{pmatrix};$$

$$\mathbf{A} = \begin{pmatrix} 0 & \frac{1}{2} & 0 & \dots & 0 & 0 \\ 0 & 0 & \frac{1}{2} & \dots & 0 & 0 \\ \vdots & \vdots & \vdots & & \vdots & \vdots \\ 0 & 0 & 0 & \dots & 0 & \frac{1}{2} \\ 0 & 0 & 0 & \dots & 0 & 0 \end{pmatrix}.$$

The values of the λ's are, for odd order λ's,

$$\lambda_{2k-1} = \cos \frac{2k\pi}{n+1} \qquad (k = 1,2,\dots,[\tfrac{1}{2}n])$$

while $\lambda_2, \lambda_4, \dots, \lambda_{2[(n-1)/2]}$ are roots of the equation

$$(1-\lambda)^{-2}\{(-\tfrac{1}{2})^n - \tfrac{1}{2}D_{n-1}(\lambda) - (n+1-n\lambda)D_n(\lambda)\}$$

with

$$D_n(\lambda) = (-\tfrac{1}{2}\lambda)^n \sum_{j=1}^{[\frac{1}{2}n+1]} \binom{n+1}{2j-1} (-1)^{j-1}(\lambda^2-1)^{j-1}.$$

The cumulative distribution function of $\hat{R}_1$ is of different form in each of the intervals $\lambda_{j+1} \le \hat{R}_1 \le \lambda_j$, being

$$F_{\hat{R}_1}(r) = 1 + \pi^{-1} \sum_{i=1}^{[j/2]} (-1)^i \int_{-1}^{1} \frac{(y_{2i}(t) - r)^{\frac{1}{2}(n-3)}}{\prod_{\substack{k=1 \\ k\neq j}}^{n-1} |y_j(t) - \lambda_k|^{\frac{1}{2}}} \frac{dt}{\sqrt{1-t^2}} \tag{40}$$

$$+ \frac{1-(-1)^j}{\pi} (-1)^{\frac{1}{2}(j+1)} \int_{-1}^{1} \frac{(y_j(t) - r)^{\frac{1}{2}n-1}}{\prod_{\substack{k=1 \\ k\neq j}}^{n-1} |y_j(t) - \lambda_k|^{\frac{1}{2}}} \frac{dt}{\sqrt{1-t^2}}$$

for $\lambda_{j+1} \le r \le \lambda_j$.

The integrals in this formula can be evaluated approximately by means of the formula (with N sufficiently large)

$$\int_{-1}^{1} f(y) \frac{dy}{\sqrt{1-y^2}} \doteqdot \pi N^{-1} \sum_{j=1}^{N} f(y_j^{(N)})$$

where

$$y_j^{(N)} = \cos \frac{(2j-1)\pi}{2N}.$$

Note that the variables $U_j^2 \left[\sum_{j=1}^{n-1} U_j^2 \right]^{-1}$ are correlated beta variables with a joint Dirichlet distribution (see Chapter 40) and so $\hat{R}_1$ is a linear function of such variables.

Pan [65] also considered a modified noncircular serial correlation coefficient, obtained, in effect, by dividing a sequence of $2n$ values $X_1, \ldots, X_{2n}$ into two sets — the first n, and the last n values respectively. The coefficient is defined as

$$(41) \quad R_{1,1} = \frac{\sum_{j=1}^{n-1} (X_j - \overline{X}_1)(X_{j+1} - \overline{X}_1) + \sum_{j=n+1}^{2n-1} (X_j - \overline{X}_2)(X_{j+1} - \overline{X}_2)}{\sum_{j=1}^{n} (X_j - \overline{X}_1)^2 + \sum_{j=n+1}^{2n} (X_j - \overline{X}_2)^2}$$

where

$$\overline{X}_1 = n^{-1} \sum_{j=1}^{n} X_j; \ \overline{X}_2 = n^{-1} \sum_{j=n+1}^{2n} X_j.$$

By essentially the same procedure as before it can be shown that $R_{1,1}$ is distributed as $\left(\sum_{j=1}^{n-1} \lambda_j V_j \right) \Big/ \left(\sum_{j=1}^{n-1} V_j \right)$ where the V's are mutually independent variables, each distributed as χ^2 with two degrees of freedom. The λ's have the same values as for $\hat{R}_1$. The cumulative distribution function now takes the much simpler form

(42)

$$\Pr[R_{1,1} \le r] = 1 - \sum_{i=1}^{j} \left[\prod_{\substack{k=1 \\ k \ne j}}^{n-1} (\lambda_j - \lambda_k) \right]^{-1} (\lambda_j - r)^{n-2} \quad \text{for } \lambda_{j+1} \le r \le \lambda_j.$$

Similar investigations of the distribution of $R_{1,1}$ were performed earlier by Watson and Durbin [88] (who also computed tables of 5% significant points for an exact non-circular test of the existence of serial correlation in a series of n observations).

The asymptotic normality of $\hat{R}_l$ for general l has been proved by Pan [65] in a series of papers.

The *randomization* ($\rho = 0$) *distribution* of $\hat{R}_1$ has been studied by David and Fix [14].

8. Leipnik Distribution

The distributions described in Sections 5 and 6 are rather complicated in form. An ingenious method of 'smoothing' the characteristic function (eliminating the discontinuities in the derivative of $p_{\hat{R}_1}'(r)$) proposed by Dixon [15] and Rubin [79] and extended by Leipnik [53] leads to the much simpler (approximate) formula

(43) $$P_{\tilde{R}_1'}(r) = [B(\tfrac{1}{2},\tfrac{1}{2}(n+1))]^{-1}(1-r^2)^{\frac{1}{2}(n-1)}(1+\rho^2-2\rho r)^{-\frac{1}{2}n} \qquad (-1 \le r \le 1).$$

Note that this is simply the first term of the expansion (29), omitting the $-\rho^n$. Daniels [10] has investigated the error of this approximation and obtained an upper bound for it.

Note that if $\rho = 0$ we have

$$p_{\tilde{R}_1'}(r) = [B(\tfrac{1}{2},\tfrac{1}{2}(n+1))]^{-1}(1-r^2)^{\frac{1}{2}(n-1)}$$

which is the distribution of an ordinary (non-serial) correlation for a bivariate normal population with $\rho = 0$, when the sample size is $(n+3)$.

White [89] pointed out that $(1+\rho^2-2\rho r)^{-\frac{1}{2}n}$ can be expressed in terms of Gegenbauer polynomials, $c_j^{(\frac{1}{2}n)}(x)$ as

$$\sum_{j=0}^{\infty} c_j^{(\frac{1}{2}n)}(r)\rho^j$$

where

$$c_j^{(\frac{1}{2}n)}(x) = \sum_{m=0}^{[\frac{1}{2}j]} \frac{(-\frac{1}{2}n)^{(j-m)}(-2x)^{j-2m}}{m!(j-2m)!}.$$

Note that $x^{(r)} = x(x-1)\ldots(x-r+1)$ and so

$$\begin{aligned}(-x)^{(r)} &= (-x)(-x-1)\ldots(-x-r+1)\\ &= (-1)^r x(x+1)\ldots(x+r-1).\end{aligned}$$

The Gegenbauer polynomials are a special kind of Jacobi polynomials, (which were described in Chapter 1). They are orthogonal with respect to the weight function $(1-x^2)^{\frac{1}{2}(n-1)}$ over the interval $-1 \le x \le 1$.

The cumulative distribution function is

(44) $$\Pr[\tilde{R}_1' \le R_0] = I_{(R_0+1)/2}\left(\frac{n+1}{2}, \frac{n+1}{2}\right) - \frac{n}{B\left(\frac{1}{2}, \frac{n+1}{2}\right)} \times (1-R_0^2)^{\frac{1}{2}(n+1)} \sum_{j=1}^{\infty} \frac{\rho^j}{j(n+j)} c_{j-1}^{(\frac{1}{2}n+1)}(R_0).$$

The moment generating function of $\tilde{R}_1'$ is

(45) $$E[e^{it\tilde{R}_1'}] = (2t^{-1})^{\frac{1}{2}n}\Gamma(\tfrac{1}{2}n+1)\sum_{j=0}^{\infty}(i\rho)^j \frac{r(n+j)}{j!\Gamma(n)} J_{\frac{1}{2}n+j}(t)$$

where $J_\nu(t)$ is a Bessel function of order ν.

The rth moment about zero is

$$\mu_r' = \frac{r!}{2^r} \frac{\Gamma\left(\frac{n}{2}+1\right)}{\Gamma(n)} \sum_{j=0}^{[r/2]} \frac{\Gamma(n+r-2j)}{\Gamma(\frac{1}{2}n+r-j+1)} \frac{\rho^{r-2j}}{j!(r-2j)!}. \tag{46}$$

From (46) we find

$$\mu_1' = \frac{n}{n+2}\rho \tag{47.1}$$

$$\mu_2 = \frac{1}{n+2} - \frac{n(n-2)\rho^2}{(n+2)^2(n+4)} \tag{47.2}$$

$$\mu_3 = \frac{1}{(n+2)^2}\left[-\frac{6n\rho}{n+4} + \frac{2n(n-2)(3n-2)\rho^3}{(n+2)(n+4)(n+6)}\right] \tag{47.3}$$

$$\mu_4 = \frac{12}{(n+2)(n+4)}\left[1 - \frac{2n(n^2-8n-4)\rho^3}{(n+2)^2(n+6)} + \frac{n(n^4-16n^3+40n^2-32n+16)\rho^4}{(n+2)^3(n+6)(n+8)}\right] \tag{47.4}$$

(Jenkins [42]).

It might be expected that the variance equalizing transformation $\tanh^{-1}\tilde{R}_1'$ would produce a more nearly normally distributed variable, as is the case for the ordinary product-moment correlation (1). Such a transformation was, indeed, studied by Quenouille [73].

However, from (47.2) it can be seen that

$$\operatorname{var}(\tilde{R}_1') \doteqdot \frac{1-\rho^2}{n} \tag{48}$$

whereas $\operatorname{var}(R) \doteqdot (1-\rho^2)^2/n$. Equation (48) suggests the use of the transformation

$$\tilde{Z} = \sin^{-1}\tilde{R}_1'. \tag{49}$$

Jenkins [40] has shown that (49) is comparable in effectiveness for $\tilde{R}_1'$, as z' is for R. He obtained the formulas

(50.1)
$$E[\tilde{Z}] = \sin^{-1}\rho - \tfrac{3}{2}\rho(1-\rho^2)^{-\frac{1}{2}}n^{-1} + \tfrac{1}{8}\rho(17-2\rho^2)(1-\rho^2)^{-\frac{3}{2}}n^{-2} + O(n^{-3});$$

$$\operatorname{var}(\tilde{Z}) = n^{-1} - \tfrac{1}{2}(2-5\rho^2)(1-\rho^2)^{-1}n^{-2} + O(n^{-3}); \tag{50.2}$$

$$\mu_3(\tilde{Z}) = -3\rho(1-\rho^2)^{-\frac{1}{2}}n^{-2} + O(n^{-3}); \tag{50.3}$$

$$\mu_4(\tilde{Z}) = 3n^{-2} - (8-29\rho^2)(1-\rho^2)^{-1}n^{-3} + O(n^{-4}); \tag{50.4}$$

whence

$$\sqrt{\beta_1} = -3\rho(1-\rho^2)^{-\frac{1}{2}}n^{-\frac{1}{2}}\{1+\tfrac{3}{4}(2-5\rho^2)(1-\rho^2)^{-1}n^{-1}\} + O(n^{-\frac{5}{2}}); \tag{51.1}$$

$$\beta_2 = 3 + 2(7\rho^2-1)(1-\rho^2)^{-1}n^{-1} + O(n^{-2}). \tag{51.2}$$

9. Multiple Correlation Coefficient

The multiple correlation coefficient between a random variable X_0 (the 'dependent' variable) and variables $X_1, X_2, \ldots, X_k$ (the 'independent' variables) with $k \geq 2$, is defined to be the maximum correlation between X_0 and any linear function of the 'independent' variables, that is

$$P_{0.12\ldots k} = \max_{a_1, a_2, \ldots, a_k} \rho\left(X_0, \sum_{j=1}^{k} a_j X_j\right).$$

(Where there is no risk of confusion the subscripts $0.12 \ldots k$ can be omitted.) If the variance-covariance matrix of $X_0, X_1, \ldots, X_k$ is

$$\mathbf{V} = \begin{pmatrix} \text{var}(X_0) & \mathbf{V}_0' \\ \mathbf{V}_0 & \mathbf{V}_{(1)} \end{pmatrix}$$

(where $\mathbf{V}_{(1)}$ is the variance-covariance matrix of $X_1, \ldots, X_k$) then

$$\rho\left(X_0, \sum_{j=1}^{k} a_j X_j\right) = (\mathbf{V}_0'\mathbf{a})/\{(\mathbf{a}'\mathbf{V}_{(1)}\mathbf{a}) \operatorname{var}(X_0)\}^{\frac{1}{2}}. \tag{52}$$

Since it can always be arranged (by appropriate choice of signs of the a_j's) that $\rho(\cdot)$ is not negative, one can choose $\mathbf{a}$ to maximize the square

$$(\mathbf{a}'\mathbf{V}_0\mathbf{V}_0'\mathbf{a})/(\mathbf{a}'\mathbf{V}_{(1)}\mathbf{a}).$$

The maximized value of the square of the correlation coefficient is

$$\mathbf{V}_0'\mathbf{V}^{-1}\mathbf{V}_0/(\operatorname{var}(X_0)),$$

so the multiple correlation coefficient is

$$\sqrt{\mathbf{V}_0'\mathbf{V}^{-1}\mathbf{V}_0/(\operatorname{var}(X_0))}. \tag{53}$$

(Sometimes the *square* is called the multiple correlation coefficient, but we shall not follow this practice.)

Suppose now that $X_0, X_1, \ldots, X_k$ have a joint multinormal distribution and that we have available values of n independent sets of these variables. The quantity $R_{0.12\ldots k}$ obtained by replacing in (53) the elements of $\mathbf{V}$ by their maximum likelihood estimators (i.e. the mean squares and mean products of deviations from sample means) is called the *sample multiple correlation coefficient*. This is, of course, a random variable and has a sampling distribution though the word 'sample' is often omitted from its name when the meaning is otherwise clear. Also, just as $P_{0.12\ldots k}$ is often replaced by P, so $R_{0.12\ldots k}$ is replaced by R, when this is conveniently possible.

Following the method of Ruben [78], already described in Section 2, Hodgson [35] has recently shown that (with $n > k + 1$), $R^2(1 - R^2)^{-1}$ is distributed as

$$[\chi_{k-1}^2 + \{U + P(1 - P^2)^{-\frac{1}{2}}\chi_{n-1}\}^2]/\chi_{n-k-1}^2 \tag{54}$$

where the χ^2's and the unit normal variable U are mutually independent. From the identity (noting that R cannot be negative),

(55)
$$\begin{aligned}\Pr[R \le r] &= \Pr[R^2(1-R^2)^{-1} \le r^2(1-r^2)^{-1}]\\ &= \Pr[\chi^2_{k-1} + \{U + P(1-P^2)^{-\frac{1}{2}}\chi_{n-1}\}^2 - r^2(1-r^2)^{-1}\chi^2_{n-k-1} < 0]\end{aligned}$$

and approximating the distribution of (a) the ratio (54) or (b) the left hand side of the last inequality in (55), Hodgson suggests the following approximations (h being a suitable positive number):

(55)′
$$\frac{(n-k+h-2)^h[R^2(1-R^2)^{-1}]^h - \{k+2h-2+(n+h-2)P^2(1-P^2)^{-1}\}^h}{\sqrt{2}\,h[(n-k-1)^{2h-1}\{R^2(1-R^2)^{-1}\}^{2h} + (2+P^2(1+P^2)^{-1})\{k+(n-1)P^2(1-P^2)^{-1}\}^{2h-1}]^{\frac{1}{2}}}$$

has approximately a unit normal distribution;

(55)″
$$\{R^2(1-R^2)^{-1}\}^h$$

is approximately normally distributed with expected value

$$[(n-k+h-2)^{-1}\{k+2h-2+(n+h-2)[P^2(1-P^2)^{-1}\}]^h$$

and variance

$$2h^2(n-k-1)^{-(2h+1)}\{k+(n-1)P^2(1-P^2)^{-1}\}^{2h-1}(2n-k)(1-P^2)^{-1}.$$

The distribution of R^2 was originally obtained by Fisher [22] in 1928, using a geometrical method (see also Soper [82]). (Special cases had been discussed earlier by Yule [94] and Isserlis [39].) Fisher obtained the formula

(56)
$$p_{R^2}(r^2) = \frac{\Gamma(\frac{1}{2}n)(1-P^2)^{\frac{1}{2}(n-1)}}{\pi\Gamma(\frac{1}{2}(k-1))\Gamma(\frac{1}{2}(n-k))}(r^2)^{\frac{1}{2}k-1}(1-r^2)^{\frac{1}{2}(n-k)-1} \times \int_0^\pi \int_{-\infty}^\infty \frac{\sin^{k-2}\theta}{(\cosh\phi - Pr\cos\theta)^{n-1}}\,d\phi\,d\theta \qquad (r^2 > 0).$$

The integral can be evaluated by expanding the integrand in powers of $\cos\theta$ and integrating term-by-term. The result can be conveniently expressed in terms of a hypergeometric function

(57)
$$p_{R^2}(r^2) = \frac{(1-P^2)^{\frac{1}{2}(n-1)}(r^2)^{\frac{1}{2}k-1}(1-r^2)^{\frac{1}{2}(n-k-1)-1}}{B(\frac{1}{2}k,\frac{1}{2}(n-k-1))} \times F\left(\frac{n-1}{2}, \frac{n-1}{2}; \frac{k}{2}; P^2r^2\right) \qquad (r^2 > 0).$$

It will be noted that, excluding the multiplier $(1-P^2)^{\frac{1}{2}(n-1)}$ and the hypergeometric function we have a standard beta distribution with parameters $\frac{1}{2}k$, $\frac{1}{2}(n-k-1)$. This is, in fact, the distribution of r^2 when $P = 0$. Further, from the representation (57), or by direct expansion, it is possible to express $p_{R^2}(r^2)$ as a mixture of standard beta distributions with parameters $(\frac{1}{2}k + j)$, $\frac{1}{2}(n-k-1)$, with weights the terms in the expansion of the negative binomial

$$\left(\frac{1}{1-P^2} - \frac{P^2}{1-P^2}\right)^{-\frac{1}{2}(n-1)}.$$

That is,

(58)

$$p_{R^2}(r^2) = \sum_{j=0}^{\infty} b_j[B(\tfrac{1}{2}k + j, \tfrac{1}{2}(n-k-1))]^{-1}(r^2)^{\frac{1}{2}k+j-1}(1-r^2)^{\frac{1}{2}(n-k-1)-1} \qquad (r^2 > 0)$$

with

$$b_j = \frac{\Gamma(\frac{1}{2}(n-1)+j)}{j!\,\Gamma(\frac{1}{2}(n-1))} \cdot \frac{(P^2)^j}{(1-P^2)^{\frac{1}{2}(n-1)+j}}.$$

This result was obtained by Gurland [29] using characteristic functions. He also showed that, for $(n-k)$ odd, $R^2(1-P^2)(1-P^2R^2)^{-1}$ is distributed as a mixture of standard beta distributions with parameters $(\frac{1}{2}k + j)$, $\frac{1}{2}(n-k-1)$ and weights given by the terms in the binomial expansion of $\{P^2 + (1-P^2)\}^{\frac{1}{2}(n-k-1)}$ so that

$$\text{(59)} \qquad \Pr[R \le r] = \sum_{j=0}^{\frac{1}{2}(n-k-1)} \binom{\frac{1}{2}(n-k-1)}{j} (P^2)^j \times (1-P^2)^{\frac{1}{2}(n-k-1)-j} I_{\frac{R^2(1-P^2)}{1-P^2R^2}}(\tfrac{1}{2}k + j, \tfrac{1}{2}(n-k-1)).$$

Note that there are only a finite number of terms in this expansion. Detailed numerical investigations of series (58) and (59) are reported in Gurland and Milton [30].

Various methods of derivation of the distribution will be found in Garding [25], Moran [60], Soper [82], and Wilks [91].

The mth moment of R about zero can be expressed in a convenient form, due to Banerjee [5]:

$$\text{(60)} \qquad \mu_m'(R) = \frac{(1-P^2)^{\frac{1}{2}(n-1)}\Gamma(\frac{1}{2}(k+m))}{\Gamma(\frac{1}{2}(n-1+m))} D'^m F(\tfrac{1}{2}(n-1+m), \tfrac{1}{2}n; \tfrac{1}{2}k; P^2)$$

where D' denotes the operator $\frac{1}{2}P^3 \dfrac{\partial}{\partial P}$.

The expected value and variance of R^2 are

(61.1)

$$E[R^2] = 1 - \frac{n-k-1}{n-1}(1-P^2)F(1,1;\tfrac{1}{2}(n+1);P^2)$$

$$= P^2 + \frac{k}{n-1}(1-P^2) - \frac{2(n-k-1)}{n^2-1}P^2(1-P^2) + O(n^{-2})$$

(61.2)

$$\text{var}(R^2) = \frac{(n-k)^2 - 1}{n^2 - 1}(1 - P^2)^2F(2,2;\tfrac{1}{2}(n+3);P^2) - [E[R^2] - 1]^2$$

$$= \begin{cases} \dfrac{4P^2(1-P^2)^2(n-k-1)^2}{(n^2-1)(n+3)} + O(n^{-2}) & \text{for } P \neq 0 \\[2ex] \dfrac{2k(n-k-1)}{(n-1)^2(n+1)} & \text{for } P = 0. \end{cases}$$

(Wishart [92].)

The distribution of R (or R^2) is complicated in form, and considerable attention has been devoted to the construction of useful approximations. It is natural to try Fisher's transformation $z' = \frac{1}{2}\log\dfrac{1+R}{1-R} = \tanh^{-1} R$. However, as for serial correlation, it is clear that this transformation is not very suitable. Gajjar [24] has shown that the limiting distribution of $\sqrt{n-1}\tanh^{-1} R$, as n tends to infinity, is *not* normal but is noncentral χ with k degrees of freedom and noncentrality parameter $(n-1)(\tanh^{-1} P)^2$. Numerical calculations indicate that $\tanh^{-1} R$ will not give generally useful results.

Khatri [48] has proposed two approximations. The first is to regard

$$k^{-1}(n-k-1)(1-P^2)\omega(P)R^2/(1-R^2)$$

with

$$\omega(P) = [k + \{n-k-1+\sqrt{(n-1)(n-k-1)}\}P^2] \times [k + (n-k-1)(2-P^2)P^2]^{-1}$$

as being approximately distributed as noncentral F with k, $(n-k-1)$ degrees of freedom and noncentrality parameter $\frac{1}{2}P^2\omega(P)\sqrt{(n-1)(n-k-1)}$.

The second of Khatri's approximations uses a different multiple of $R^2/(1-R^2)$. The multiplier is $(n-k-1)(1-P^2)\{(n-k-1)P^2+k\}^{-1}$ and the approximate distribution a *central* F distribution with

$$[(n-k-1)P^2+k]^2[(n-k-1)P^2(2-P^2)+k]^{-1},$$

$(n-k-1)$ degrees of freedom.

Khatri suggests that these approximations be used for $n-k-1 \geq 100$ and prefers the second when P^2 is large.

Hodgson [35] has constructed some normal approximations based on the representation (54) discussed above.

(Approximations (55)′ and (55)″ seems to be particularly useful with $h = 1$.)

A simple approximation proposed by Gurland [29] (based on (58)) appears to give excellent results. This is to approximate the distribution of $R^2(1-R^2)^{-1}$ by that of $(n-k-1)^{-1}\{(n-1)\theta + k\}F_{f,n-k}$ with

$$f = (n-1)(\theta+k)^2/\{(n-1)\theta(\theta+2)+k\}$$

where $\theta = P^2(1-P^2)^{-1}$.

It can be shown that $\Pr[R \leq r \mid P]$ is a decreasing function of P. Upper and lower limits for $100(1 - \alpha)\%$ confidence intervals for P can be obtained by solving the equations $\int_{-\infty}^{R} p_R(r \mid P)\, dr = \alpha_1,\ 1 - \alpha_2$ with $(\alpha_1 + \alpha_2) = \alpha$. Kramer [50] gives tables for constructing lower 95% limits (i.e., $\alpha_1 = 0.05$, $\alpha_2 = 0$).

REFERENCES

[1] Allan, F. E. (1930). A percentile table of the relation between the true and the observed correlation coefficient from a sample of 4, *Proceedings of the Cambridge Philosophical Society*, **26,** 536–537.

[2] Anderson, R. L. (1942). Distribution of the serial correlation coefficient, *Annals of Mathematical Statistics*, **13,** 1–13.

[3] Armsen, P. (1956). Note on a sample identity concerned with particular values of the distribution function of correlation coefficients in bivariate normal populations, *Journal of the National Institute of Personal Research*, **6,** 175–176.

[4] Baker, G. A. (1930). The significance of the product-moment coefficient of correlation with special reference to the character of the marginal distributions, *Journal of the American Statistical Association*, **25,** 387–396.

[5] Banerjee, D. P. (1952). On the moments of the multiple correlation coefficient in samples from normal populations, *Journal of the Indian Society of Agricultural Statistics*, **4,** 88–90.

[6] Cheriyan, K. C. (1945). Distributions of certain frequency constants in samples from non-normal populations, *Sankhyā*, **7,** 159–166.

[7] Cheshire, I., Oldis, E. and Pearson, E. S. (1932). Further experiments on the sampling distribution of the correlation coefficient, *Journal of the American Statistical Association*, **27,** 121–128.

[8] Cook, M. B. (1951). Bi-variate k-statistics and cumulants of their joint sampling distribution, *Biometrika*, **38,** 179–195.

[9] Cook, M. B. (1951). Two applications of bi-variate k-statistics, *Biometrika*, **38,** 368–376.

[10] Daniels, H. E. (1956). The approximate distribution of the serial correlation coefficients, *Biometrika*, **43,** 169–185.

[11] Daniels, H. E. and Kendall, M. G. (1958). Short proof of Miss Harley's theorem on the correlation coefficient, *Biometrika*, **45,** 571–572.

[12] David, F. N. (1937). A note on unbiased limits for the correlation coefficient, *Biometrika*, **29,** 157–160.

[13] David, F. N. (1938). *Tables of the Correlation Coefficient*, London: Cambridge University Press.

[14] David, F. N. and Fix, Evelyn (1966). Randomization and the serial correlation coefficient, (*Festschrift for J. Neyman;* (Ed. F. N. David), New York: John Wiley & Sons, Inc., 461–468).

[15] Dixon, W. J. (1944). Further contributions to the problem of serial correlation, *Annals of Mathematical Statistics*, **15,** 119–144.

[16] Durbin, J. (1957). Testing for serial correlation in systems of simultaneous regression equations, *Biometrika*, **44,** 370–377.

[17] Durbin, J. and Watson, G. S. (1950, 1951). Testing for serial correlation in least squares regression, I, II, *Biometrika*, **37,** 409–428, and **38,** 159–178.

[18] Eicker, F. (1960). *On the distribution of a noncircular serial correlation coefficient with lag 1 when the mean of the observations is unknown*, University of North Carolina Mimeo Series, No. 265. (Abstract in *Annals of Mathematical Statistics*, **31,** 234).

[19] Fisher, R. A. (1915). Frequency distribution of the values of the correlation coefficient in samples from an indefinitely large population, *Biometrika*, **10,** 507–521.

[20] Fisher, R. A. (1921). On the "probable error" of a coefficient of correlation deduced from a small sample, *Metron*, **1,** 3–32.

[21] Fisher, R. A. (1924). The influence of rainfall on the yield of wheat at Rothamsted, *Philosophical Transactions of the Royal Society of London, Series B*, **213,** 89–142.

[22] Fisher, R. A. (1928). The general sampling distribution of the multiple correlation coefficient, *Proceedings of the Royal Society of London, Series A*, **121,** 654–673.

[23] Fraser, D. A. S. (1963). On the definition of fiducial probability, *Bulletin of the International Statistical Institute*, **40,** 842–856.

[24] Gajjar, A. V. (1967). Limiting distributions of certain transformations of multiple correlation coefficient, *Metron*, **26,** 189–193.

[25] Garding, L. (1941). The distributions of the first and second order moments, the partial correlation coefficients and the multiple correlation coefficient in samples from a normal multivariate population, *Skandinavisk Aktuarietidskrift*, **24,** 185–202.

[26] Garwood, F. (1933). The probability integral of the correlation coefficient in samples from a normal bi-variate population, *Biometrika*, **25,** 71–78.

[27] Gayen, A. K. (1951). The frequency distribution of the product-moment correlation coefficient in random samples of any size drawn from non-normal universes, *Biometrika*, **38,** 219–247.

[28] Ghosh, B. K. (1966). Asymptotic expansions for the moments of the distribution of correlation coefficient, *Biometrika*, **53,** 258–262.

[29] Gurland, J. (1968). A relatively simple form of the distribution of the multiple correlation coefficient, *Journal of the Royal Statistical Society, Series B*, **30,** 276–283.

[30] Gurland, J. and Milton, R. C. (1970). Further consideration of the distribution of the multiple correlation coefficient, *Journal of the Royal Statistical Society*, **32.**

[31] Harley, Betty I. (1954). A note on the probability integral of the correlation coefficient, *Biometrika*, **41,** 278–280.

[32] Harley, Betty I. (1956). Some properties of an angular transformation for the correlation coefficient, *Biometrika*, **43,** 219–224.

[33] Harley, Betty I. (1957). Relation between the distribution of noncentral t and of a transformed correlation coefficient, *Biometrika*, **44,** 219–224.

[34] Hirschfeld, H. O. (Hartley, H. O.) (1937). The distribution of the ratio of covariance estimates in samples drawn from normal bivariate populations, *Biometrika*, **29,** 65–79.

[35] Hodgson, V. (1967). On the sampling distribution of the multiple correlation coefficient. (Presented at the IMS meetings in Washington, D. C., December, 1967.)

[36] Hotelling, H. (1953). New light on the correlation coefficient and its transforms, *Journal of the Royal Statistical Society, Series B*, **15,** 193–224. (Discussion 225–232).

[37] Hsu, P. L. (1946). On the asymptotic distributions of certain statistics used in testing the independence between successive observations from a normal population, *Annals of Mathematical Statistics*, **17,** 350–354.

[38] Isserlis, L. (1915). On certain probable errors and correlation coefficients of multiple frequency distributions with skew regression, *Biometrika*, **11,** 185–190.

[39] Isserlis, L. (1917). The variation of the multiple correlation coefficient in samples drawn from an infinite population with normal distribution, *Philosophical Magazine*, *6th Series*, **34,** 205–220.

[40] Jenkins, G. M. (1954). An angular transformation for the serial correlation coefficient, *Biometrika*, **41,** 261–265.

[41] Jenkins, G. M. (1954). Tests of hypotheses in the linear autoregressive model, I, *Biometrika*, **41,** 405–419.

[42] Jenkins, G. M. (1956). Tests of hypotheses in the linear autoregressive model, II, *Biometrika*, **43,** 186–199.

[43] Kamat, A. R. and Sathe, Y. S. (1962). Asymptotic power of certain test criteria (based on first and second differences) for serial correlation between successive observations, *Annals of Mathematical Statistics*, **33,** 186–200.

[44] Kendall, M. G. (1949). Rank and product-moment correlation, *Biometrika*, **36,** 177–193.

[45] Kendall, M. G. (1954). Note on bias in the estimation of autocorrelation, *Biometrika*, **41,** 403–404.

[46] Kendall, M. G. (1957). The moments of the Leipnik distribution, *Biometrika*, **44,** 270–272.

[47] Kendall, M. G. (1960). The evergreen correlation coefficient (pp. 274–277 in *Essays in Honor of Harold Hotelling*, [Ed. I. Olkin], Stanford, California: Stanford University Press.)

[48] Khatri, C. G. (1966). A note on a large sample distribution of a transformed multiple correlation coefficient, *Annals of the Institute of Statistical Mathematics, Tokyo*, **18,** 375–380.

[49] Koopmans, T. C. (1942). Serial correlation and quadratic forms in normal variables, *Annals of Mathematical Statistics*, **13,** 14–33.

[50] Kramer, K. H. (1963). Tables for constructing confidence limits on the multiple correlation coefficient, *Journal of the American Statistical Association*, **58,** 1082–1085.

[51] Kullback, S. (1936). A note on the multiple correlation coefficient, *Metron*, **12,** 67–72.

[52] Kuzmin, R. O. (1939). Sur la loi de distribution du coefficient de correlation dans les tirages d'un ensemble normal, *Comptes Rendus* (*Doklady*) *de l'Académie des Sciences, URSS*, **N.S. 22,** 298–301.

[53] Leipnik, R. B. (1947). Distribution of the serial correlation coefficient in a circularly correlated universe, *Annals of Mathematical Statistics*, **18,** 80–87.

[54] Leipnik, R. B. (1958). Note on the characteristic function of a serial-correlation distribution, *Biometrika*, **45,** 559–562.

[55] Madow, W. G. (1945). Note on the distribution of the serial correlation coefficient, *Annals of Mathematical Statistics*, **16,** 308–310.

[56] Mahalanobis, P. C., Bose, R. C. and Roy, S. N. (1937). Normalization of statistical variates and the use of rectangular coordinates in the theory of sampling distributions, *Sankhyā*, **3,** 1–40.

[57] Marriott, F. H. C. and Pope, J. A. (1954). Bias in the estimation of autocorrelations, *Biometrika*, **41,** 390–402.

[58] Mehta, J. S. and Gurland, J. (1969). Some properties and an application of a statistic arising in testing correlation, *Annals of Mathematical Statistics*, **40,** 1736–1745.

[59] Moran, P. A. P. (1948). Rank correlation and product-moment correlation, *Biometrika*, **35,** 203–206.

[60] Moran, P. A. P. (1950). The distribution of the multiple correlation coefficient, *Proceedings of the Cambridge Philosophical Society*, **46,** 521–522.

[61] Murthy, V. K. (1960). On the distribution of the sum of circular serial correlation coefficients and the effect of non-normality on its distribution, *Annals of Mathematical Statistics*, **31,** 239–240, (Abstract).

[62] Nabeya, S. (1951). Note on the moments of the transformed correlation, *Annals of the Institute of Statistical Mathematics, Tokyo*, **3,** 1.

[63] Neumann, J. von (1941). Distribution of the ratio of the mean square successive difference to the variance, *Annals of Mathematical Statistics*, **12,** 367–395.

[64] Olkin, I. and Tate, R. F. (1961). Multivariate correlation models with mixed discrete and continuous variables, *Annals of Mathematical Statistics*, **32,** 448–465.

[65] Pan, Jie-jian (1964). Distribution of the serial correlation coefficients with non-circular statistics (In Chinese) *Shuxue Jinzhan*, **7,** 328–337. (English translation: in *Selected Translations in Mathematical Statistics and Probability*, **7,** (1968), pp. 281–291, American Mathematics Society, Providence, Rhode Island); *Ibid;* (1966), **9,** No. 3, 291–295.

[66] Pearson, E. S. (1929). Some notes on sampling tests with two variables, *Biometrika*, **21,** 337–360.

[67] Pearson, E. S. (1931). The test of significance for the correlation coefficient, *Journal of the American Statistical Association*, **26,** 128–134.

[68] Pearson, K. (1931). Editorial note, *Biometrika*, **22,** 362–367.

[69] Pearson, K., Jeffery, G. B. and Elderton, Ethel M. (1929). On the distribution of the first product moment-coefficient, in samples drawn from an indefinitely large normal population, *Biometrika*, **21,** 164–193; (*Tables* computed by E. M. Elderton, pp. 194–201).

[70] Pepper, J. (1929). Studies in the theory of sampling, *Biometrika*, **21,** 231–258.

[71] Pillai, K. C. S. (1946). Confidence interval for the correlation coefficient, *Sankhyā*, **7,** 415–422.

[72] Press, S. J. (1967). On the sample covariance from a bivariate normal distribution, *Annals of the Institute of Statistical Mathematics, Tokyo*, **19,** 355–361.

[72a] Press, S. J. (1969). On serial correlation, *Annals of Mathematical Statistics*, **40,** 188–196.

[73] Quenouille, M. H. (1948). Some results in the testing of serial correlation coefficients, *Biometrika*, **35,** 261–267.

[74] Quensel, C. E. (1938). The distributions of the second moment and of the correlation coefficient in samples from populations of Type A, *Lunds Universitet, Årsskrift N. F. Afdelning*, (2), **34,** (4), 1–111.

[75] Rider, P. R. (1932). On the distribution of the correlation coefficient in small samples, *Biometrika*, **24,** 382–403.

[76] Romanovsky, V. (1925). On the moments of standard deviations and of correlation coefficient in samples from normal population, *Metron*, **5,** (4), 3–46.

[77] Ruben, H. (1963). Probability of a positive sample correlation (Abstract), *Annals of Mathematical Statistics*, **34,** 694.

[78] Ruben, H. (1966). Some new results on the distribution of the sample correlation coefficient, *Journal of the Royal Statistical Society, Series B*, **28,** 513–525.

[79] Rubin, H. (1945). On the distribution of the serial correlation coefficient, *Annals of Mathematical Statistics*, **16,** 211–215.

[80] Shenton, L. R. and Johnson, W. L. (1965). Moments of a serial correlation coefficient, *Journal of the Royal Statistical Society, Series B*, **27,** 308–320.

[81] Soper, H. E. (1913). On the probable error of the correlation coefficient to a second approximation, *Biometrika*, **9,** 91–115.

[82] Soper, H. E. (1929). The general sampling distribution of the multiple correlation coefficient, *Journal of the Royal Statistical Society, Series A*, **92,** 445–447.

[83] Soper, H. E., Young, A. W., Cave, B. M., Lee, Alice and Pearson, K. (1915–1917). On the distribution of the correlation coefficient in small samples. Appendix II to the papers of "Student" and R. A. Fisher, *Biometrika*, **11,** 328–413.

[84] Stammberger, A. (1968). Ein Nomogramm zur Beurteilung von Korrelationskoeffizienten, *Biometrische Zeitschrift*, **10,** 80–83.

[85] 'Student' (1908). Probable error of a correlation coefficient, *Biometrika*, **6,** 302–310.

[86] Wald, A. and Wolfowitz, J. (1943). An exact test for randomness in the nonparametric case based on serial correlation, *Annals of Mathematical Statistics*, **14,** 378–388.

[87] Watson, G. S. (1956). On the joint distribution of the circular serial correlation coefficients, *Biometrika*, **43,** 161–168.

[88] Watson, G. S. and Durbin, J. (1951). Exact tests of serial correlation using noncircular statistics, *Annals of Mathematical Statistics*, **22,** 446–451.

[89] White, J. S. (1957). Approximate moments for the serial correlation coefficient, *Annals of Mathematical Statistics*, **28,** 798–803.

[90] White, J. S. (1961). Asymptotic expansions for the mean and variance of the serial correlation coefficient, *Biometrika*, **48,** 85–94.

[91] Wilks, S. S. (1932). On the sampling distribution of the multiple correlation coefficient, *Annals of Mathematical Statistics*, **3,** 196–203.

[92] Wishart, J. (1931). The mean and second moment coefficient of the multiple correlation coefficient in samples from a normal population, *Biometrika*, **22,** 353–361.

[93] Wishart, J. and Bartlett, M. S. (1932). The distribution of second order moment statistics in a normal system, *Proceedings of the Cambridge Philosophical Society*, **28,** 455–459.

[94] Yule, G. U. (1907). On the theory of correlation for any number of variables, treated by a new system of notation, *Proceedings of the Royal Society of London, Series A*, **79,** 182–193.

[95] Yule, G. U. (1921). On the time-correlation problem with special reference to the variate-difference correlation method, *Journal of the Royal Statistical Society, Series A*, **84,** 497–526.

33

Miscellaneous

1. Introduction

The number of possible different continuous distributions is limitless. In this chapter we restrict our attention to distributions of some importance in statistical practice or theory. In the final section, we take brief note of a few special distributions of less importance which have been reported in published articles. These are not intended to be exhaustive, but rather to indicate the kinds of special results which have appeared in the literature.

We have taken the opportunity to include (in Section 7) some additional material on approximations to Mills' ratio and on inequalities satisfied by distribution functions, subject to rather general restrictions.

2. Kolmogorov-Smirnov Distributions

This is the name given to a variety of distributions, which are exact or approximate distributions of functions of the *sample cumulative distribution function.* This is defined in the following way. If n sample values $X_1, X_2, \ldots, X_n$ are available the *sample cumulative distribution function,* $\hat{F}_n(x)$, is the proportion of X's which are less than or equal to x. It is also a function of $X_1, X_2, \ldots, X_n$ and so is a random variable if the X_j's can be regarded as random variables.

We shall be concerned exclusively with cases in which the X_j's are mutually independent continuous random variables with common cumulative distribution function (c.d.f.) $F_{X_j}(x) = F(x)$ (for all j). The difference

$$\hat{F}_n(x) - F(x) \tag{1}$$

is a deviation of the sample c.d.f. from the c.d.f. It is clear that the distributions of quantities like $\sup_x \{\hat{F}_n(x) - F(x)\}$, $\inf_x \{\hat{F}_n(x) - F(x)\}$ do not depend on

$F(x)$. (If $Y = f(X)$ be a monotonic function of X then the differences (1) remain the same, but are associated with values $f(x)$, instead of x.)

For this reason, it is possible to choose the function $F(x)$ arbitrarily (provided it is a continuous c.d.f.). A common choice is

$$F(x) = \begin{cases} 0 & (x \leq 0) \\ x & (0 \leq x \leq 1) \\ 1 & (x \leq 1) \end{cases}$$

corresponding to a standard rectangular distribution. However, an exponential has also been used, to take advantage of properties described in Section 6 of Chapter 18.

The exact distributions in this class, with the notable exceptions of (15.1) and (15.2) below, are not usually of simple form (though they can be expressed in terms of elementary functions). Asymptotic distributions (as n tends to infinity) are, on the other hand, often of quite elegant form. Sometimes, it is this latter sub-class which is implied, when the adjective 'Kolmogorov-Smirnov' is used. We shall not restrict the adjective in this way, though we do devote more attention to the asymptotic distributions. An extensive bibliography is contained in Barton and Mallows [14]; interesting remarks on the history and applications of these distributions are included in Steck [151]; Sahler [134] gives a summary of results in unified form.

We will first consider the distribution of the statistic $\sup_x |\hat{F}_n(x) - F(x)|$, proposed by Kolmogorov [82] in 1933. Kolmogorov obtained the following recurrence relation among the quantities R_j:

$$R_{j,k}(y) = n^n e^{-n}(n!)^{-1} \Pr[\sup_x |\hat{F}_n(x) - F(x)| \leq y/n] \qquad (y \text{ an integer})$$

(2)

$$R_{j,k+1}(y) = e^{-1} \sum_{i=0}^{2y-1} [R_{j+1-i,k}(y)/i!] \qquad (|j| \leq y - 1).$$

Together with the values

$$R_{0,0}(y) = 1; \quad R_{j,0}(y) = 0 \text{ for } j \neq 0$$
$$R_{j,k}(y) = 0 \text{ for } |j| \geq y$$

the exact distribution can be evaluated by application of (2). (Massey [101] obtained another system of equations.)

If we put $y = z\sqrt{n}$ and let n tend to infinity we find

$$\text{(3)} \quad \lim_{n\to\infty} \Pr[\sqrt{n} \sup_x |\hat{F}_n(x) - F(x)| \leq z] = 1 - 2 \sum_{j=1}^{\infty} (-1)^{j-1} \exp(-2j^2z^2) = L(z).$$

This result was obtained by Smirnov [149], who introduced the notation $L(z)$.

The probability density function corresponding to the cumulative distribution function is

$$8z \sum_{j=1}^{\infty} (-1)^{j-1} j^2 \exp(-2j^2z^2). \tag{4}$$

The rth moment about zero is

$$\frac{\Gamma(\frac{1}{2}r + 1)}{2^{\frac{1}{2}r-1}} \sum_{j=1}^{\infty} (-1)^{j-1} j^{-r}. \tag{5}$$

In particular, the expected value is $\sqrt{\frac{\pi}{2}} \log_e 2$ ($\doteqdot 0.8687$) and the variance is $\frac{1}{8}\pi^2 - \frac{1}{2}\pi(\log_e 2)^2$ ($\doteqdot 0.6771$).

A short table of the upper percentage points of (3) is given in Table 1 (from Miller [105]).

TABLE 1

$L(z_{1-\alpha}) = 1 - \alpha.$

α	0.20	0.15	0.10	0.05	0.01
$z_{1-\alpha}$	1.07	1.14	1.22	1.36	1.63

It can be seen that for $z > 1$, the first term in the summation in (4) will be much greater than the others and so

$$\Pr[\sqrt{n} \sup_x |\hat{F}_n(x) - F(x)| \leq z] \doteqdot 1 - 2e^{-2z^2}.$$

There are two approximations here — the use of the asymptotic distribution *and* the neglect of all terms after the first in the summation. For smaller z, the expression (equivalent to (3))

$$\lim \Pr[\sqrt{n} \sup_x |\hat{F}_n(x) - F(x)| \leq z] = \sqrt{2\pi}\, z^{-1} \sum_{j=1}^{\infty} \exp[-(2j-1)^2\pi^2]/(8z^2) \tag{6}$$

shows that as an approximation, the formula

$$\sqrt{2\pi}\, z^{-1} \exp[-\tfrac{1}{8}(\pi/z)^2] \tag{7}$$

may be used if z is small.

Massey [101] gave exact values of some upper percentage points for $n = 1(1)20(5)35$. Birnbaum [19] gave exact values of $\Pr[\sup_x |\hat{F}_n(x) - F(x)| < y/n]$ to 5 decimal places for $n = 1(1)100$; $y = 1(1)15$.

We now consider the distribution of $\sup_x (\hat{F}_n(x) - F(x))$. The exact value of $\Pr[\sup_x (\hat{F}_n(x) - F(x)) \leq y/n]$ is

$$(8) \qquad 1 - \frac{n!}{n^n} \cdot y \sum_{j>y}^{n} \frac{(j-y)^n (n-j+y)^{n-j-1}}{j!(n-j)!}$$

(Birnbaum and Tingey [25]). There are tables of upper percentage points of $\sup_x (\hat{F}_n(x) - F(x))$ in [25]. Putting $y = z\sqrt{n}$ and letting n tend to infinity we find

$$(9) \qquad \lim_{n\to\infty} \Pr[\sqrt{n} \sup_x (\hat{F}_n(x) - F(x)) \leq z] = 1 - e^{-2z^2}$$

— an even simpler formula than (3).

Lauwerier [89], extending work of Smirnov [149], obtained the formula

$$(10.1) \qquad \Pr[\sqrt{n} \sup_x (\hat{F}_n(x) - F(x)) \leq z] = 1 - \frac{n!}{n^{n+\frac{1}{2}} e^{-n} \sqrt{2\pi}} e^{-2z^2} \sum_{j=0}^{\infty} f_j(H) n^{-\frac{1}{2}j}.$$

The $f_j(\cdot)$'s are polynomials defined by the identity

$$(10.2) \qquad p^{-1}(e^{2p} - 1)^{-1}(e^{2p} - 1 - 2p) \times \exp\{u^{-2}(1 + \tfrac{1}{2}p^2 - p\coth p - \log[p^{-1}\sinh p]\} \equiv \sum_{j=0}^{\infty} f_j\left(\frac{p}{u}\right) u^j$$

and H^r (in the expressions $f_j(H)$) is to be interpreted as the rth Hermite polynomial with argument $2z$.

From this result one has

$$(11) \qquad \Pr[\sqrt{n} \sup_x (\hat{F}_n(x) - F(x)) \leq z]$$
$$= 1 - e^{-2z^2} \times \left\{1 - \frac{H_1(2\lambda)}{3n^{1/2}} + \frac{3 - H_4(2\lambda)}{36n} - \frac{15H_1(2\lambda) - 12H_3(2\lambda) - 5H_4(2\lambda)}{540n^{3/2}} + \frac{45 - 60H_4(2\lambda) + 32H_6(2\lambda) + 5H_8(2\lambda)}{12960n^2}\right\} + 0(n^{-5/2})$$
$$= 1 - e^{-2z^2}\left\{1 - \frac{2z}{3\sqrt{n}} - \frac{4z^4 - 6z^2}{9n} \cdots\right\} + 0(n^{-3/2}).$$

Birnbaum and Pyke [22] have shown that

$$E[\sup_x (\hat{F}_n(x) - F(x))] = \frac{1}{2} \frac{n!}{n^{n+1}} \sum_{j=0}^{n-1} \frac{n^j}{j!}.$$

Kuiper [86] has suggested the use of the statistic

$$V_n = \sup_x \{\hat{F}_n(x) - F(x)\} - \inf_x \{\hat{F}_n(x) - F(x)\}$$

and has shown that

$$\text{(12)} \quad \Pr[\sqrt{n}\, V_n \leq z] = 1 - 2\sum_{j=1}^{\infty} (4j^2z^2 - 1)e^{-2j^2z^2} - \frac{8z}{3\sqrt{n}} \sum_{j=1}^{\infty} j^2(4j^2z^2 - 3)e^{-2j^2z^2} + 0(n^{-1}).$$

Stephens [154] has obtained the exact distribution of V_n and has given tables of a number of upper and lower percentage points for $n = 2(1)12(2)20(10)80$, 100.

For testing for differences between two cumulative distributions functions, using random samples of sizes n_1, n_2 respectively, the criteria

$$\text{(13.1)} \quad Z_1 = \sqrt{\frac{n_1 n_2}{n_1 + n_2}} \sup_x [\hat{F}_{n_1}(x) - \hat{F}_{n_2}(x)]$$

or

$$\text{(13.2)} \quad Z_2 = \sqrt{\frac{n_1 n_2}{n_1 + n_2}} \sup_x |\hat{F}_{n_1}(x) - \hat{F}_{n_2}(x)|$$

(using an obvious notation) can be used.

The limiting distributions are

$$\text{(14.1)} \quad \lim_{n_1, n_2 \to \infty} \Pr[Z_1 \leq z] = 1 - e^{-2z^2} \qquad (z > 0);$$

$$\text{(14.2)} \quad \lim_{n_1, n_2 \to \infty} \Pr[Z_2 \leq z] = \sum_{j=-\infty}^{\infty} (-1)^j \exp(-2j^2z^2) = L(z) \qquad (z > 0).$$

If more exact values of percentile points of the distribution of Z_1 and Z_2 are required, the tables in Borovkov *et al.* [26] can be used. These give, for $\alpha = 0.95, 0.98, 0.99$ and all integer values of n_1, n_2 with $1 \leq n_1 \leq n_2 \leq 50$, pairs of values $z_i^{(1)}$, $z_i^{(2)}$ (to 4 decimal places) such that

$$\Pr[Z_i < z_i^{(1)}] < \alpha < \Pr[Z_i < z_i^{(2)}],$$

together with values of these probabilities (to 5 decimal places). The distribution is discrete and the tables give the values of the probabilities to the nearest α. The introduction to these tables also gives asymptotic expansions for the values of the probabilities, and there is a table of coefficients to facilitate use of these expansions.

It is often more natural to consider the ratio of the deviation, $(\hat{F}_n(x) - F(x))$, to either $F(x)$, or $(1 - F(x))$, rather than the deviation itself. This becomes apparent when considering the use of a Kolmogorov-Smirnov type statistic in the construction of confidence intervals. Since the statement

$$\sup_x |\hat{F}_n(x) - F(x)| \leq z/\sqrt{n}$$

is equivalent to

$$\hat{F}_n(x) - z/\sqrt{n} \leq F(x) \leq \hat{F}_n(x) + z/\sqrt{n} \text{ for all } x$$

it is clear that if z is so chosen that (exactly or approximately)

$$\Pr[\sqrt{n} \sup_x |\hat{F}_n(x) - F(x)| \leq z] = 1 - \alpha$$

then the band $\hat{F}_n(x) \pm z/\sqrt{n}$ is a confidence band for $F(x)$ with (exact or approximate) confidence coefficient $100(1 - \alpha)\%$. The fact that the band is of constant width $(2z/\sqrt{n})$ may lead to unduly wide limits when $\hat{F}_n(x)$ is near 1 or 0. If it were possible to find z such that, for example

$$\Pr\left[\sup_{0<F(x)} \left\{\frac{\sqrt{n}|\hat{F}_n(x) - F(x)|}{F(x)}\right\} \leq z\right] = 1 - \alpha$$

then the corresponding band would have limits

$$\frac{\hat{F}_n(x)}{1 + z/\sqrt{n}} \quad \text{and} \quad \frac{\hat{F}_n(x)}{1 - z/\sqrt{n}}.$$

Daniels [42] obtained the remarkably simple exact results

$$\Pr\left[\sup_{0<F(x)} \frac{\hat{F}_n(x) - F(x)}{F(x)} < v\right] = \frac{v}{1 + v} \qquad (v > 0) \tag{15.1}$$

and similarly

$$\Pr\left[\sup_{F(x)<1} \frac{F(x) - \hat{F}_n(x)}{1 - F(x)} < v\right] = \frac{v}{1 + v}. \qquad (v > 0) \tag{15.2}$$

Note that these values do not depend on n.

The statistics so far studied in this section reflect maximum deviations (absolute or relative) over the whole range of variation of the observed random variables. Rényi [122] considered the same statistics, but with restricted ranges of the variable. These were either finite — of form $0 < a \leq F(x) \leq b < 1$ — or infinite — of form $a \leq F(x)$ or $F(x) \leq b$. Note that the range is defined by means of restrictions on the values of $F(x)$, rather than of x itself.

Maniya [97] showed that (with $0 < a < b < 1$)

$$\begin{aligned}\lim_{n\to\infty} \Pr[\sqrt{n} \sup_{a\leq F(x)\leq b} \{\hat{F}_n(x) - F(x)\} \leq z] \\ = \Pr[T_1 \leq c_1, T_2 \leq c_2] - e^{-2z^2} \Pr[T_1 \leq c_3, T_2 \geq -c_4],\end{aligned} \tag{16}$$

where T_1, T_2 are unit normal variables with a joint bivariate normal distribution with correlation coefficient $[ab^{-1}(1 - a)^{-1}b]^{\frac{1}{2}}$ and

$$c_1 = z[a(1-a)]^{-1/2};\ c_2 = z[b(1-b)]^{-1/2};$$
$$c_3 = (1-2a)c_1;\ c_4 = [1-2(1-b)]c_2.$$

Rényi [122] showed that (for $0 < a < b < 1, z > 0$)

$$\lim_{n\to\infty} \Pr\left[\sqrt{n} \sup_{a \le F(x)} \left|\frac{\hat{F}_n(x) - F(x)}{F(x)}\right| < z\right] = 4\pi^{-1} \sum_{j=0}^{\infty} (-1)^j (2j+1)^{-1} \exp\left[-\frac{(2j+1)^2\pi^2(1-a)}{8az^2}\right]; \tag{17}$$

$$\lim_{n\to\infty} \Pr\left[\sqrt{n} \sup_{a \le F(x)} \left\{\frac{\hat{F}_n(x) - F(x)}{F(x)}\right\} < z\right] = 2\left[\Phi\left(z\sqrt{\frac{a}{1-a}}\right) - 1\right]; \tag{18}$$

$$\lim_{n\to\infty} \Pr\left[\sqrt{n} \sup_{a \le F(x) \le b} \left\{\frac{\hat{F}_n(x) - F(x)}{F(x)}\right\} < z\right] = \frac{1}{\pi}\int_{-\infty}^{Bz} e^{-\frac{1}{2}u^2}\left(\int_{-\infty}^{A(Bz-u)} e^{-\frac{1}{2}t^2}\, dt\right) du \tag{19}$$

with $B = \sqrt{b/(1-b)}$; $A = \sqrt{a(1-b)/(b-a)}$.

Birnbaum and Lientz [21], give the formula ($b < 1$)

$$\Pr\left[\sup_{F(x)\ge b} \frac{\hat{F}_n(x) - F(x)}{F(x)} \le v\right] = b^n + \frac{v}{1+v}\sum_{k=0}^{n-1}\binom{n}{k} b^k + \sum_{j \le K} (-1)^j \binom{n-k}{j} (b - \phi_{j,k})^j (1-\phi_{j,k})^{n-j-k-1} \quad (-1 \le v \le b^{-1} - 1) \tag{20}$$

where $\phi_{j,k} = \{n(1+c)\}^{-1}(j+k)$ and $K = b(1+c)n - k$.
Replacing b by $(1-a)$ with $a > 0$ in the right-hand side of (20) gives

$$\Pr\left[\sup_{F(x)\le a} \frac{F(x) - \hat{F}_n(x)}{1 - F(x)} \le v\right].$$

The range of variation can be defined in terms of $\hat{F}_n(x)$ instead of $F(x)$. This is a stochastic range, which can vary from sample to sample. Since $\hat{F}_n(x)$ can be equal to 0 or 1, one will usually exclude values of $\hat{F}_n(x)$ which will make the denominator zero, e.g. $\sup\{[\hat{F}_n(x) - F(x)]/[1 - \hat{F}_n(x)]\}$ would have x restricted by $\hat{F}_n(x) \le b < 1$.

The reason for introducing statistics of this type is their use in constructing confidence intervals. These are based (see above and e.g. Wald and Wolfowitz [160]) on the fact that the inequality

$$\sup \frac{|\hat{F}_n(x) - F(x)|}{F(x)} \le z$$

is equivalent to

$$(1+z)^{-1}\hat{F}_n(x) \le F(x) \le (1-z)^{-1}\hat{F}_n(x)$$

for all x in the range of x over which the 'sup' is taken (and similar arguments). If we want a restricted range of x, so as to get a narrower confidence interval, we cannot usually define this in terms of $F(x)$, since $F(x)$ is unknown. However, we can define it in terms of $\hat{F}_n(x)$ which can be calculated from the sample.

As would be expected, the *limiting* distributions (as $n \to \infty$), which have their ranges defined in terms of $\hat{F}_n(x)$, are identical with those for which range is defined, in the same way, in terms of $F(x)$.

The quantity $F(x)$ in the denominator can also be replaced by $\hat{F}_n(x)$. This again, does not affect the limiting distribution. As an example of an exact distribution, we quote Birnbaum and Lientz [21] (correcting Csörgö [39])

$$\text{(21)} \qquad \Pr\left[\sup_{\hat{F}_n(x)\leq b} \frac{F(x) - \hat{F}_n(x)}{1 - \hat{F}_n(x)} < v\right] = 1 - v \sum_{j\leq nb-1} \binom{n}{j} \{1 - [v + (1 - v)j/n]\}^{n-j}\{v + (1 - v)j/n\}^{j-1}.$$

We have not given an exhaustive catalogue of all distributions of this class. Useful papers, giving rather comprehensive accounts of results available at the time they were written, are those of Smirnov [147], Darling [44], Rényi [122], Csörgö [40], Birnbaum and Lientz [21] and Lientz [90].

We now discuss yet further statistics of this kind. The statistics as far discussed share the common feature that they are defined by the values of some multiple of $(\hat{F}_n(x) - F(x))$ at two points at most. There are other statistics, also possessing the property of having distributions not depending on $F(x)$, which are of the nature of averages of the absolute deviation $|\hat{F}_n(x) - F(x)|$ or its square.

One of the best-known is Smirnov's ω_n^2 criterion which we will define as

$$\text{(22)} \qquad \omega_n^2 = n\int_{-\infty}^{\infty} \{\hat{F}_n(x) - F(x)\}^2 \, dF(x).$$

Other statistics of this nature are

$$\text{(23)} \qquad U_n^2 = n\int_{-\infty}^{\infty} \left\{\hat{F}_n(x) - F(x) - \int_{-\infty}^{\infty} [\hat{F}_n(y) - F(y)]\, dF(y)\right\}^2 dF(y)$$

(Stephens [154]), and Sherman's statistic [144], based on the order statistics $X_1' \leq X_2' \leq \cdots \leq X_n'$,

$$\text{(24)} \qquad \frac{1}{2}\sum_{j=1}^{n+1} \left| F(X_j') - F(X_{j-1}') - \frac{1}{n+1} \right|$$

with $F(X_0') = 0$; $F(X_{n+1}') = 1$.

Note that ω_n^2 and U_n^2 can also be expressed in terms of order statistics

$$\sum_{j=1}^{n} \{F(X_j') - (2j - 1)/(2n)\}^2 + \tfrac{1}{12}n^{-1}$$

and

$$\sum_{j=1}^{n} \{F(X_j') - (2j - 1)/(2n) - \bar{F} + \tfrac{1}{2}\}^2$$

respectively, where $\bar{F} = n^{-1} \sum_{j=1}^{n} F(X_j')$.

The distributions of ω_n^2 and U_n^2 are quite complicated. However, Stephens [154] has given exact upper percentile points for $n = 4$, and lower percentile points for $n = 5(1)10$. The moments are, for ω_n^2

$$\mu_1'(\omega_n^2) = \frac{1}{6}; \tag{25.1}$$

$$\mu_2(\omega_n^2) = \frac{4n - 3}{180n}; \tag{25.2}$$

$$\mu_3(\omega_n^2) = \frac{32n^2 - 61n + 30}{3780n^2}; \tag{25.3}$$

$$\mu_4(\omega_n^2) = \frac{496n^3 - 1532n^2 + 1671n - 630}{75600n^3}, \tag{25.4}$$

(from Pearson [113]); and for U_n^2 (from Stephens [153]),

$$\mu_1(U_n^2) = \frac{1}{12}; \tag{26.1}$$

$$\mu_2(U_n^2) = \frac{n - 1}{360n}; \tag{26.2}$$

$$\mu_3(U_n^2) = \frac{(n - 1)(2n - 3)}{7560n^2}; \tag{26.3}$$

$$\mu_4(U_n^2) = \frac{(n - 1)(19n^2 - 51n + 36)}{302400n^3}. \tag{26.4}$$

Pearson and Stephens [114] have found an interesting relation between the limiting distributions (as $n \to \infty$) of ω_n^2 and U_n^2. The characteristic functions are, for ω_∞^2

$$\varphi_{\omega_\infty^2}(t) = \prod_{j=1}^{\infty} \left(1 - \frac{2it}{\pi^2 j^2}\right)^{-\frac{1}{2}}, \tag{27}$$

and for U_∞^2

$$\varphi_{U_\infty^2}(t) = \prod_{j=1}^{n} \left(1 - \frac{it}{2\pi^2 j^2}\right)^{-1}, \tag{28}$$

so that

$$\varphi_{U_\infty^2}(t) = [\varphi_{\omega_\infty^2}(\tfrac{1}{4}t)]^2,$$

whence it follows that

$$\kappa_r(U_\infty^2) = \frac{1}{2^{2r-1}} \kappa_r(\omega_\infty^2). \tag{29}$$

From (28) it can be seen that the limiting distribution (as $n \to \infty$) of U_n^2 is that of $\sum_{j=1}^{\infty} (2\pi^2 j^2)^{-1} W_j$, where the W_j's are independent random variables, each having the standard exponential distribution ($p_{W_j}(w) = e^{-w}, w > 0$). Watson [163] showed that this means that

$$\Pr[U_\infty^2 < v] = 1 - 2 \sum_{j=1}^{\infty} (-1)^{j-1} e^{-2\pi^2 j^2 v}. \tag{30}$$

Comparing this with (3) we see that tables of $L(z)$ can be used to evaluate (30), putting $z = \pi\sqrt{2v}$.

The limiting distribution of ω_n^2 is not of so simple a form. Anderson and Darling [2] give tables of z satisfying $\lim_{n\to\infty} \Pr[\omega_n^2 \leq z] = \alpha$ to 5 decimal places for $\alpha = 0.01(0.01)0.99, 0.999$. They used the formula

$$\lim_{n\to\infty} \Pr[\omega_n^2 \leq z] = \frac{1}{\pi\sqrt{z}} \sum_{j=0}^{\infty} (-1)^j \binom{-\frac{1}{2}}{j} \sqrt{4j+1}\, e^{-z_j'} K_{\frac{1}{4}}(z_j') \tag{31}$$

where $z_j' = (j + \frac{1}{4})^2 z^{-1}$ and $K_{\frac{1}{4}}(\cdot)$ is the modified Bessel function.

Lilliefors [91] has studied the distribution of $\max_x |\hat{F}_n(x) - \hat{F}(x)|$ where $\hat{F}(x)$ is the *estimated* cumulative distribution, assuming a normal form of distribution and using estimated values of expected value and variance.

A similar investigation for ω_n^2 by van Soest (in an unpublished thesis, University of Delft) is mentioned in Barlow [4].

The motivation for using the Sherman statistic (24) is that $F(X_1') \ldots F(X_n')$ will be distributed like order statistics corresponding to n independent standard rectangular variables, and so, in particular $E[F(X_r')] = r/(n+1)$ and $E[F(X_r') - F(X_{r-1}')] = (n+1)^{-1}$. The statistic is thus half the sum of absolute values of deviations of the differences $\{F(X_r') - F(X_{r-1}')\}$ from their common expected value $(n+1)^{-1}$.

Anderson and Darling [2] also obtain the result

$$\lim_{n\to\infty} \Pr\left[n \int_{-\infty}^{\infty} \frac{[\hat{F}_n(x) - F(x)]^2}{F(x)\{1 - F(x)\}}\, dF(x) \leq z\right] \tag{32}$$
$$= \frac{\sqrt{2\pi}}{z} \sum_{j=0}^{\infty} \binom{-\frac{1}{2}}{j} (4j+1) e^{-z_j''} \int_0^{\infty} \exp\left[\frac{z}{8(\omega^2+1)} - z_j''\omega^2\right] d\omega$$

where $z_j'' = \frac{1}{8}(4j+1)^2 \pi^2 z^{-1}$.

3. 'Normal' Distributions

3.1 *Circular Normal Distributions*

The name "circular normal" is sometimes given to a bivariate normal distribution with zero correlation and equal standard deviations. This name is natural, in that the joint probability density function is constant along circular contours. This is a bivariate distribution and, as such, is not discussed in the present chapter, but in Chapter 35.

The *circular normal* distribution which will be discussed here has probability density function

$$(33)\quad p_\theta(t) = [2\pi I_0(k)]^{-1} \exp(k \cos(t - \theta_0)) \quad (0 \le t < 2\pi;\ 0 < \theta_0 < 2\pi)$$

where

$$I_0(k) = \sum_{j=0}^{\infty} (j!)^{-2}(\tfrac{1}{2}k)^{2j}$$

is a Bessel function of zero order of imaginary argument.

The distribution was derived by von Mises [108] in 1918 as a result of requiring the maximum likelihood estimator of θ_0, given independent random variables $\theta_1, \theta_2, \ldots, \theta_n$, each having distribution (33) to satisfy an equation of the form

$$\sum_{j=1}^{n} \sin(\theta_j - \hat{\theta}_0) = 0.$$

Since the maximum likelihood equation is

$$\sum_{j=1}^{n} \left. \frac{\partial \log p_\theta(\theta_j)}{\partial \theta_0} \right|_{\theta_0 = \hat{\theta}_0} = 0$$

it follows that

$$(34)\quad \frac{\partial \log p_\theta(t)}{\partial \theta_0} = k \sin(t - \theta_0) \quad \text{for some } k.$$

We arbitrarily choose to have k positive. The condition $\sum_{j=1}^{n} \sin(\theta_j - \theta_0) = 0$ means that if observed values are represented as points on the circumference of a circle with center at the origin and θ_j is the angle between the horizontal axis and the vector joining the origin to the jth observation point then θ_0 gives an "average" direction. This 'average' is such that the projections of all the observed points onto the vector perpendicular to the average direction sum to zero.

The density function is a maximum at $t = \theta_0$ and a minimum at $t = \theta_0 \pm \pi$ (whichever is in the range 0 to 2π). The maximum and minimum are thus in opposite directions along the same line, called the *axis* of the distribution. The direction θ_0 is called the *pole* of the distribution. The ratio of the maximum to the minimum values of the density function is e^{2k}. If $k = 0$, the distribution is uniform (see Chapter 25). As k increases, the distribution becomes more and more concentrated around the pole.

We have already seen that the maximum likelihood estimator $(\hat{\theta}_0)$ of θ_0 satisfies the equation

$$(35)\quad \sum_{j=1}^{n} \sin(\theta_j - \hat{\theta}_0) = 0.$$

The maximum likelihood estimator $(\hat{k})$ of k satisfies the equation

$$\frac{d \log I_0(\hat{k})}{d\hat{k}} = \frac{1}{n} \sum_{j=1}^{n} \cos(\theta_j - \hat{\theta}_0). \tag{36}$$

Gumbel *et al.* [71] give a table of values of the left-hand side of (36). Once (35) has been solved for $\hat{\theta}_0$, the right-hand side of (36) can be calculated and these tables used in obtaining the value of $\hat{k}$. The tables in [71] can also be used in estimating the noncentrality parameter of a noncentral chi-squared distribution with 2 degrees of freedom (cf. Chapter 28, Section 6).

If the pole (θ_0) is already known, then (36) with $\hat{\theta}_0$ replaced by θ_0 is an equation for the maximum likelihood estimator of k. In this case, the variance of $\hat{k}$ is approximately

$$\left[n \frac{d^2 \log I_0(k)}{dk^2}\right]^{-1}. \tag{37}$$

If only the *axis* is known — i.e. it is known that θ_0 equals either τ or $\tau + \pi$ (but not which) — then the absolute values of angle with the axis can be used.

Gumbel *et al.* [71] give tables of the *square root* of $2\pi p_\theta(t)$ with $\theta_0 = 0$ to three decimal places for $k = 0(0.1)4.0$ and $t = 0°(10°)180°$. The reason for the apparently odd choice of tabled function is that if this is used as the radius vector and t as the angle with the horizontal axis then the area between vectors at angles θ', θ'' with the horizontal is proportional to the probability that $(\theta - \theta_0)$ lies between θ' and θ''. Gumbel *et al.* [71] also give a table of

$$\Pr[-\theta' < \theta - \theta_0 < \theta']$$

to 5 decimal places for $k = 0(0.2)4.0$ and $\theta' = 0°(5°)180°$.

Further tables, in [158], give 5 decimal places, also, for $k = 0(0.05)3.95$, $\theta' = 5°(5°)180°$. These values were obtained by interpolation in the earlier tables. (The earliest tables, by von Mises [108], which appeared in 1918, gave values to 4 decimal places for $k = 0(0.5)3.0$ and $\theta = 0°(15°)(180°.)$ Arguments similar to those of von Mises lead to the following formula for the density function of the *p-dimensional hyperspherical normal distribution* $(p > 2)$:

$$p_\theta(t) = c_p(\sin^{p-1}(t - \theta_0)) \exp(k \cos(t - \theta_0)) \qquad (0 \leq t \leq \pi) \tag{38}$$

where of course

$$c_p = \left[\int_0^\pi (\sin^{p-1}(t - \theta_0)) \exp(k \cos(t - \theta_0))\, dt\right]^{-1}$$
$$= \left[\int_0^\pi (\sin^{p-1} t) \exp(k \cos t)\, dt\right]^{-1}.$$

For the case $p = 3$ (the *spherical normal* distribution),

$$c_3 = k/(2 \sinh k).$$

Generally, for p odd, c_p can be expressed in terms of k, $\sinh k$ and $\cosh k$.

For $p = 3$, the equations for maximum likelihood estimators, $\hat{\theta}_0$ and $\hat{k}$, of θ and k are

$$\sum_{j=1}^{n} \cot (\theta_j - \hat{\theta}_0) = \hat{k} \sum_{j=1}^{n} \sin (\theta_j - \hat{\theta}_0), \tag{39}$$

$$\coth \hat{k} - \hat{k}^{-1} = \sum_{j=1}^{n} \cos (\theta_j - \hat{\theta}_0). \tag{40}$$

Solution of (39) and (40) simultaneously is troublesome. If, however, θ_0 is known, then (39), with $\hat{\theta}_0$ replaced by θ_0, can be used to find the maximum likelihood estimator of k. The variance of this estimator is approximately $(k^{-2} - \operatorname{cosech}^2 k)^{-1}n^{-1}$.

The genesis of the circular ($p = 2$) and spherical ($p = 3$) normal distributions has been described by Langevin [87], Kuhn and Grün [85] and Johnson [78] among others. In fact, the function $\hat{k}$ of $\sum \cos (\theta_j - \hat{\theta}_0)$, defined by (40) is known as the *Langevin function.*

3.2 '*Wrapped-up*' *Normal Distribution*

Imagine a normal distribution with unit standard deviation wrapped around a circle of radius 1, so that the expected value is at a point with radius vector at an angle θ to the horizontal axis. At a point with radius vector at an angle φ $(0 < \varphi < 2\pi)$ to the horizontal axis there will be contributions to the total probability density function of

$$\frac{1}{\sqrt{2\pi}} \exp\left[-\frac{(\varphi \pm 2j\pi - \theta)^2}{2}\right] \qquad (j = 0,1,\ldots).$$

The probability density function of the angle φ $(0 < \varphi < 2\pi)$ is then

$$p_\varphi(t) = \frac{1}{\sqrt{2\pi}} \sum_{j=-\infty}^{\infty} \exp [-\tfrac{1}{2}(t + 2j\pi - \theta)^2] \qquad (0 < t < 2\pi). \tag{41}$$

If $\theta = 0$ we have the standard 'wrapped-up' normal distribution. This distribution, which has some claim to be called a 'circular normal' distribution is, in fact, called a *wrapped-up normal* distribution. It is also called *circular Brownian motion distribution.* A further parameter, σ, can be introduced by supposing the normal distribution which is 'wrapped up' to have standard deviation σ. Then

$$p_\varphi(t) = \frac{1}{\sqrt{2\pi}\sigma} \sum_{j=-\infty}^{\infty} \exp\left[-\frac{1}{2\sigma^2}(t + 2j\pi - \theta)^2\right] \qquad (0 < t < 2\pi). \tag{42}$$

Stephens [153] has shown how, by an appropriate choice of parameters, the 'wrapped up' and 'circular' normal distributions can be brought to close agreement.

4. Distance Distributions

There is a considerable variety of distributions arising from more or less naturally defined geometrical systems of variation. For this section we will describe some distributions of distances. More general discussions can be found in Kendall and Moran [81].

If two points are taken at random within a circle, center at the origin and radius 1, the joint distribution of the coordinates (X_1,Y_1), (X_2,Y_2) is

$$p_{X_1,Y_1;X_2,Y_2}(x_1,y_1;x_2,y_2) = \pi^{-2} \qquad (0 \le x_j^2 + y_j^2 \le 1;\ j = 1,2). \tag{43}$$

The distribution of $R^2 = (X_1 - X_2)^2 + (Y_1 - Y_2)^2$ can be derived by straightforward but rather detailed analysis. The density function of R^2 is

$$p_{R^2}(d) = 1 - \frac{2}{\pi}\sin^{-1}\sqrt{d} - \frac{1}{\pi}\sqrt{d(1 - \tfrac{1}{4}d)} \qquad (0 \le d \le 4). \tag{44}$$

The sth moment of R^2 about zero is

$$\mu_s'(R^2) = \frac{2(2s + 1)!}{(s + 1)!(s + 2)!} \tag{45}$$

whence

$$E[R^2] = 1; \tag{46.1}$$

$$\text{Var}(R^2) = \tfrac{2}{3}; \tag{46.2}$$

$$\alpha_3(R^2) = \sqrt{\beta_1(R^2)} = \tfrac{3}{4}\sqrt{\tfrac{3}{2}} = 0.91856; \tag{46.3}$$

$$\alpha_4(R^2) = \beta_2(R^2) = \tfrac{63}{20} = 3.15 \text{ exactly}. \tag{46.4}$$

The expected value of R is $128/(45\pi) \doteqdot 0.905$ (see ApSimon [3], Watson [162]).

Barton *et al.* [13] obtained a generalization of this distribution by multiplying the density function $p_{R^2}(d)$, given by (44), by $Cd^{\frac{1}{2}a}$ with

$$C = \frac{(a + 2)\pi}{2^{a+4}B(\frac{3}{2},\frac{1}{2}(a + 3))}.$$

This generalization is intended to represent departure from randomness due to attraction ($a < 0$) or repulsion ($a > 0$) between the points. Formulas for the first four moments of the generalized distribution are given in [13], but not reproduced here.

Another generalization has been obtained by Gečiauskas [61] who has studied the distribution of distance between two points independently distributed uniformly over an 'oval' K (a convex region which is symmetrical about a point S the 'center' of the oval). Gečiauskas showed that $\Pr[R \le r]$ can be expressed in the form $4 \times (\text{area of oval})^{-1}[B(r) + C_2]$. In this expression $B(r)$ and C_2

can be determined from the following equations:

$$B(r) + C_2 = \int_0^r y[A(y,1) + C_1(y)]\,dy$$

$$A(r,\lambda) + C_1(r) = \int_0^\lambda t\theta(r,t)\,dt$$

where $\theta(r,\lambda)$ is the mean value of sum of angles of all those sectors of a circle with radius r and center at the point S, which are situated in the oval $K(\lambda)$; and K and $K(\lambda)$ are similar ovals with coefficient of similarity λ. $C_1(r)$ is determined by $A\left(r, \frac{r}{D}\right) + C_1(r) = 0$ and C_2 from the 'boundary condition' $\Pr[R \leq D] = 1$ where D is the largest chord of K.

Among special applications, Gečiauskas showed that if the oval is a square with sides of length a, then

$$\Pr[R \leq r] = \begin{cases} \pi(r/a)^2 - \frac{8}{3}(r/a)^3 + \frac{1}{2}(r/a)^4 & (0 \leq r \leq a) \\ \frac{1}{3} + 4\sqrt{1-(r/a)^2} - [\pi + 2 - 4\sin^{-1}(a/r)](r/a)^2 & \\ + \frac{8}{3}\{1-(r/a)^2\}^{3/2} - \frac{1}{2}(r/a)^4 & (a \leq r \leq a\sqrt{2}). \end{cases} \tag{47}$$

Gečiauskas [62] has obtained similar results for distances between randomly chosen points in ovaloids in 3 dimensions. $4 \times (\text{area})^{-1}$ is replaced by $6 \times (\text{volume})^{-1}$ and $t\theta(\cdot)$ becomes $t^2\theta(\cdot)$ in the formula for $A(r,\lambda) + C_1(r)$.

The rth moment (about zero) of the distance between two independent random points in a p-dimensional sphere of unit radius is

$$\frac{2^r p}{p+r} \cdot \frac{\Gamma(p+1)}{\Gamma(\frac{1}{2}(p+1))} \cdot \frac{\Gamma(\frac{1}{2}(p+1)+\frac{1}{2}r)}{\Gamma(p+1+\frac{1}{2}r)} \qquad \text{(Deltheil [46]).}$$

This shows that the distance is distributed as $2\sqrt{Y}Z$ where Y has a standard beta distribution with parameters $\frac{1}{2}(p+1)$, $\frac{1}{2}(p+1)$ and Z has a truncated power function distribution with density

$$p_Z(z) = pz^{p-1} \qquad (0 \leq z \leq 1).$$

For p large, the distribution is approximately normal with expected value $\sqrt{2}$ and variance $(2p)^{-1}$.

A considerable amount of work has been done on the distribution of distance (X) from a random point to the nearest point of a specified set. For example, Holgate [76] has shown that the distance to the nearest point of a lattice of vertices of equilateral triangles of side a has cumulative distribution function

$$F_X(x) = \begin{cases} \frac{2}{3}\sqrt{3}\,\pi(x/a)^2 & (0 \leq x \leq \frac{1}{2}a) \\ \frac{2}{3}\sqrt{3}\,\pi(x/a)^2 - 4\sqrt{3}\,(x/a)^2\cos^{-1}(\frac{1}{2}(a/x)) & \\ \quad + 2\sqrt{3}\,((x/a)^2 - \frac{1}{4})^{\frac{1}{2}} & (\frac{1}{2}a \leq x \leq a/\sqrt{3}). \end{cases} \tag{48}$$

The expected value of X is $\sqrt{3}(\frac{1}{9} + \frac{1}{12}\log 3)a \doteqdot 0.3510a$.

A paper by Haight [73] contains applications of the kinds of distribution discussed in this section to travel distances between points to a city. Apart from the direct distance, Haight also considers distances along paths consisting of the two radius vectors to the city center, along roads of a rectangular grid, and others.

5. Life Distributions — Transformed Normal Variables

The exponential distribution is used in many situations to represent distribution of 'life-times'. If the departure from this distribution is too pronounced to be ignored, a Weibull distribution may be used. Among others used (Barlow [4]) there is the *linear failure rate* distribution (see also Section 7.2) with density

$$p_X(x) = (1 + \theta x)\exp\{-(x + \tfrac{1}{2}\theta x^2)\} \qquad (x > 0), \tag{49.1}$$

and a distribution with density function

$$p_X(x) = [1 + \theta(1 - e^{-x})]\exp\{-[x + \theta(x + e^{-x} - 1)]\} \qquad (x > 0) \tag{49.2}$$

(These distributions are in standard form. A further parameter can be introduced by considering the distribution of αX, with $\alpha > 0$.)

In this Section we will consider some other distributions which have been suggested to represent 'life-time', based on more or less realistic conjectures about the mechanism leading to conclusion of a 'life'.

We have already discussed a number of simple transformations of normal variables in Chapter 12, from the viewpoint of construction of general systems of distributions. Birnbaum and Saunders [23] were led to a discussion of the distribution of

$$T = \theta[U\sigma + \sqrt{(U\sigma)^2 + 1}]^2 \tag{50}$$

(where θ and σ are positive parameters and U is a unit normal variable) from a model representing time to failure of material subjected to a cyclically repeated stress pattern. They suppose that the jth cycle leads to an increase X_j in length of a 'crack', and that the sum $\sum_{j=1}^{n} X_j$ is approximately normally distributed with expected value $n\mu_0$ and standard deviation $\sigma_0\sqrt{n}$. Then the probability that the crack does not exceed a critical length, ω say, is

$$\Phi\left(\frac{\omega - n\mu_0}{\sigma_0\sqrt{n}}\right) = \Phi\left(\frac{\omega}{\sigma_0\sqrt{n}} - \frac{\mu_0\sqrt{n}}{\sigma_0}\right).$$

It was supposed that 'failure' occurs when the crack length exceeds ω. If T denotes the lifetime (in number of cycles) till failure then

$$\Pr[T \le t] \doteqdot 1 - \Phi\left(\frac{\omega}{\sigma_0\sqrt{t}} - \frac{\mu_0\sqrt{t}}{\sigma_0}\right) = \Phi\left(\frac{\mu_0\sqrt{t}}{\sigma_0} - \frac{\omega}{\sigma_0\sqrt{t}}\right) \tag{51}$$

(it being assumed that probability of negative values of X_j's can be neglected). If (51) be regarded as an exact equation then it follows that

$$U = \frac{\mu_0\sqrt{T}}{\sigma_0} - \frac{\omega}{\sigma_0\sqrt{T}} \tag{52}$$

has a unit normal distribution. Equation (52) can be rewritten

$$T = \frac{\omega}{\mu_0}\left[\frac{U\sigma_0}{2\sqrt{\omega\mu_0}} + \sqrt{\left(\frac{U\sigma_0}{2\sqrt{\omega\mu_0}}\right)^2 + 1}\right]^2$$

which is of the same form as (50) with

$$\theta = \frac{\omega}{\mu_0}; \; \sigma = \frac{\sigma_0}{2\sqrt{\omega\mu_0}}.$$

The parameter θ is simply a multiplier and does not affect the shape of the distribution of T. The moment-ratios of T depend only on σ, and the (β_1,β_2) points lie on a line. (The situation is analogous to that for the lognormal distribution (Chapter 14, Section 3).)

The expected value of $(T/\theta)^r$ is

$$\begin{aligned} E[(T/\theta)^r] &= \sum_{j=0}^{r} \binom{2r}{2j} E[\{(U\sigma)^2 + 1\}^j (U\sigma)^{2(r-j)}] \\ &= \sum_{j=0}^{r} \binom{2r}{2j} \sum_{i=0}^{j} \binom{j}{i} E[U^{2(r-j+i)}]\sigma^{2(r-j+i)} \\ &= \sum_{j=0}^{r} \binom{2r}{2j} \sum_{i=0}^{j} \binom{j}{i} \frac{[2(r-j+i)]!}{4^{r-j+i}(r-j+i)!} \end{aligned} \tag{53}$$

Note that $E[U^s(U^2+1)^t] = 0$ if s is odd.

From (53) we find

$$E[T] = \theta(2\sigma^2 + 1); \tag{54.1}$$

$$\text{Var}(T) = 4\theta^2\sigma^2(5\sigma^2 + 1); \tag{54.2}$$

$$\beta_1(T) = \frac{4\sigma^2(22\sigma^2 + 3)^2}{(5\sigma^2 + 1)^3}; \tag{54.3}$$

$$\beta_2(T) = 3 + \frac{3\sigma^2(372\sigma^2 + 41)}{2(5\sigma^2 + 1)^2}. \tag{54.4}$$

As σ tends to zero, the distribution tends to normality. The ratio

$$(\beta_2(T) - 3)/\beta_1(T)$$

is remarkably stable, varying between 2.88 (as $\sigma \to \infty$) and 3.42 (as $\sigma \to 0$).

T is a monotonically increasing function of U. The median of T corresponds to $U = 0$ and is equal to θ.

Since

$$(U + \sqrt{U^2 + 1})^{-1} = (-U) + \sqrt{(-U)^2 + 1}$$

it follows that the relation

$$T_\alpha T_{1-\alpha} = \theta^2$$

holds between the lower and upper $100\alpha\%$ points of the distribution of T.

Also, the distribution of T^{-1} is the same as that of T, with θ replaced by θ^{-1} (σ remaining unchanged).

Saunders and Birnbaum [136] have proposed the following methods for calculating maximum likelihood estimators of the parameters. The maximum likelihood estimator of θ is the unique positive root of

$$\hat{\theta}^2 - \hat{\theta}(2H + K(\hat{\theta})) + H(\bar{T} + K(\hat{\theta})) = 0 \tag{55}$$

where

$$\bar{T} = n^{-1} \sum_{j=1}^{n} T_j; \; H = \left[n^{-1} \sum_{j=1}^{n} T_j^{-1}\right]^{-1}$$

and

$$K(\hat{\theta}) = \left[n^{-1} \sum_{j=1}^{n} (\hat{\theta} + T_j)^{-1}\right]^{-1}.$$

Having calculated $\hat{\theta}$, the maximum likelihood estimator $\hat{\sigma}$ can be calculated directly from the formula

$$\hat{\sigma} = [\tfrac{1}{2}(\bar{T}\hat{\theta}^{-1} + H^{-1}\hat{\theta}) - 1]^{\frac{1}{2}}.$$

As an initial value to use in the iterative solution of (55), the 'mean mean'

$$(\bar{T}H)^{\frac{1}{2}}$$

is suggested. It is further shown that provided

$$2\bar{T} < 3H + \min(T_1, T_2, \ldots, T_n),$$

the Newton-Raphson iteration method will converge to $\hat{\theta}$ (for *any* initial value between $\bar{T}$ and H). If $2H > \bar{T}$, then iterative calculation of

$$H + \tfrac{1}{2}K(\hat{\theta}) - \sqrt{\tfrac{1}{4}[K(\hat{\theta})]^2 - H(\bar{T} - H)}$$

will converge to $\hat{\theta}$.

The 'mean mean', $(\bar{T}H)^{\frac{1}{2}}$, referred to above has approximate variance (for n large)

$$n^{-1}[\tfrac{1}{2}(1 + \tfrac{3}{2}\sigma^2)(1 + \sigma^2)^{-1}\theta^2\sigma^2].$$

The expected value is approximately

$$\theta[1 + (1 + \tfrac{3}{2}\sigma^2)(1 + \sigma^2)^{-1}\sigma^2 n^{-1}].$$

For $\sigma < 1/\sqrt{2}$ it is suggested that $(\bar{T}H)^{\frac{1}{2}}$ can be used in place of the maximum likelihood estimator of θ.

It might be of interest to consider the distribution of

$$\sqrt{T} = U\sigma + \sqrt{(U\sigma)^2 + 1}.$$

However, calculation of moments is not easy in this case.

We have already encountered a number of other more or less commonly used "life" distributions — e.g. exponential, mixture of exponentials, Weibull. Extensive classes of such distributions have been discussed by Buckland [31].

Ahuja and Nash [1] noted that the cumulative distribution functions

$$F(t) = \exp(-\rho \exp(-t/\sigma)); \tag{56.1}$$

$$F(t) = (1 + \rho \exp(-t/\sigma))^{-\theta}; \tag{56.2}$$

$$F(t) = (1 - \rho \exp(-t/\sigma))^{\theta} \qquad (\sigma \log \rho < t) \tag{56.3}$$

(with $\rho,\sigma,\theta > 0$) which had been used by Gompertz [66] and Verhulst [157] in the first half of the nineteenth century, could be obtained from Pearson Type III, VI and I distributions by simple exponential transformation. Introducing a further parameter $\varphi(> 0)$ they formed distributions with density functions

$$\{\sigma\Gamma(\varphi)\}^{-1}(\rho \exp(-t/\sigma))^{\varphi} \exp(-\rho \exp(-t/\sigma)); \tag{57.1}$$

(Generalized Gompertz)

$$\{\sigma B(\varphi,\theta)\}^{-1}(\rho \exp(-t/\sigma))^{\varphi}\{1 + \rho \exp(-t/\sigma)\}^{-(\varphi+\theta)}; \tag{57.2}$$

(Generalized Logistic)

$$\{\sigma B(\varphi,\theta)\}(\rho \exp(-t/\sigma))^{\varphi}\{1 - \rho \exp(-t/\sigma)\}^{\varphi-1} \qquad (\sigma \log \rho < t) \tag{57.3}$$

(Generalized Exponential).

The moment generating functions are, respectively (with argument u),

$$\rho^{\sigma u}\Gamma(\varphi - \sigma u)/\Gamma(\varphi); \tag{58.1}$$

$$\rho^{\sigma u}\Gamma(\varphi - \sigma u)\Gamma(\theta + \sigma u)/\{\Gamma(\varphi)\Gamma(\theta)\}; \tag{58.2}$$

$$\rho^{\sigma u}\Gamma(\varphi - \sigma u)\Gamma(\varphi + \theta)/\{\Gamma(\varphi)\Gamma(\varphi + \theta - \sigma u)\}. \tag{58.3}$$

An analysis of the ways in which such distributions might arise in population growth problems has been given in [1].

As a further example of the interesting special distributions which can arise from models put together to represent natural situations we will take a result obtained by Saaty [133]. He constructed a model for the results of a "challenge dose" of organisms on a "host". Supposing that

(i) There are originally n organisms in the dose

(ii) Each organism has a probability $\lambda\, dt$ of dividing into two organisms in a (short) period of time dt, and a probability $\mu\, dt$ of dying in this same time interval ($\mu < \lambda$)

(iii) "Response" occurs as soon as the number of organisms reaches N,

Saaty showed that there is a probability $(\mu/\lambda)^n$ that there will *never* be a response. Also, the density function of T, the time before the response, *given* that a response is reached, is approximately (for $n \ll N$)

$$(59)\quad p_T(t) = (\lambda - \mu)\exp[-N(1 - \mu/\lambda)e^{-(\lambda-\mu)t}] \qquad (t > 0)$$
$$\times \sum_{j=0}^{n-1} \frac{1}{j!}\binom{n}{j+1}(\mu/\lambda)^{n-j-1}[(1 - \mu/\lambda)^2 Ne^{-(\lambda-\mu)t}]^{j+1}.$$

Shortley [145] obtained an interesting result by supposing that n has a Poisson distribution with expected value θ. The resulting compound distribution has density function

(60)
$$e^{-\theta}\sum_{i=1}^{\infty} g_n(t)\frac{\theta^i}{i!} = (\lambda - \mu)\sum_{i=1}^{\infty}\frac{(ay)^i}{(i-1)!\,i!} = (\lambda - \mu)\sqrt{ay}\,e^{-(y+a)}I_1(2\sqrt{ay})$$

where $g_n(t)$ equals the expression in Equation (59),

$$y = N(1 - \mu/\lambda)e^{-(\lambda-\mu)t}$$
$$a = \theta(1 - \mu/\lambda)$$

and $I_1(\cdot)$ is a modified Bessel function of first order (Chapter 1, Section 3). Note that this depends on t only through the function y.

Flehinger and Lewis [57] have discussed two "lifetime" distributions constructed from specified hazard rate functions (see Section 7.2). These are shown in the following table.

Hazard Rate	Cumulative Distribution Function
$a + 2b^2t$	$1 - \exp[-at - (bt)^2]$
$a + 3c^3t^2$	$1 - \exp[-at - (ct)^3]$

The parameters a, b, and the argument, t, are all positive. It is, of course, also possible to consider a distribution with hazard rate $(a + 2b^2t + 3c^3t^2)$ and cumulative distribution function

$$1 - \exp[-at - (bt)^2 - (ct)^3]$$

but the extra complexity makes this unattractive.

Flehinger and Lewis also discuss the use of a truncated normal distribution (Chapter 13, Section 7) as a lifetime distribution.

Filadelfina [53] has described a graphical method of estimating parameters of the cumulative distribution function

$$\begin{cases} 1 - e^{-\lambda_0 t} & (0 \le t \le t_1) \\ 1 - c_1e^{-\lambda_1 t} & (t_1 \le t \le t_2) \\ 1 - c_2e^{-\lambda_2 t - \lambda_2' t^2} & (t_2 \le t \le t_3) \\ 1 - c_3t^{-\lambda_3} & (t_3 \le t). \end{cases}$$

The eight parameters (c's and λ's) are linked by three relations so that the function is defined uniquely at $t = t_1, t_2, t_3$.

We finally note a model constructed by Hadwiger [72] which leads to a function of the same mathematical form as the density function of the inverse Gaussian (Wald) distribution, discussed in Chapter 15. Hadwiger sought to obtain a formula for the *reproduction rate function* $\varphi_1(x)$—a function such that the expected number of offspring born to a single individual between ages x_1 and $x_2(> x_1)$ is

$$\int_{x_1}^{x_2} \varphi_1(x)\, dx.$$

Starting from a single individual at time zero, and using $\varphi_n(x)$ to denote the reproduction rate function for the *whole* of the nth generation at time x, we have the relation

$$\varphi_{n+m}(x) = \int_0^x \varphi_n(x - t)\varphi_m(t)\, dt.$$

Making the (arbitrary) assumption that $\varphi_n(x)$ can be expressed in the form of a function $g(x,na)$ where a is a fixed parameter, and passing to a limit, Hadwiger obtained the integral equation

$$g(x,p + q) = \int_0^x g(x - t,p)g(t,q)\, dt$$

(for any $p,q > 0$). This is solved by

$$g(x,a) = \frac{a}{\sqrt{\pi}} x^{-\frac{3}{2}} \exp(Ca - a^2x^{-1} - Ax)$$

(C and A being arbitrary constants) which can be seen to be of the same form as the density function (1) of Chapter 15 (though $g(x,a)$ itself is, of course, not a density function).

6. Mixtures

6.1 *Planck Distributions*

A distribution with probability density function of form

$$p_X(x) = \frac{Kx^3}{e^{\alpha x} - 1} \qquad (x > 0; \alpha > 0) \tag{61}$$

is known in physics as Planck's radiation formula. (The name is also applied to the density function $Kz^{-5}(e^{\alpha/z} - 1)^{-1}$ of $Z = 1/X$.) The family of distributions

$$p_X(x) = \alpha^{f+1}C_f x^f(e^{\alpha x} - 1)^{-1} \qquad (x > 0; \alpha > 0; f > 0)$$

may be called *Planck distributions*. Making the change of variable $Y = \alpha X$ we get the standard form of Planck distribution

(62) $$p_Y(y) = C_f y^f (e^y - 1)^{-1}, \qquad (0 < y)$$

where, of course

$$C_f = \left[\int_0^\infty y^f (e^y - 1)^{-1}\, dy\right]^{-1}.$$

Now

$$\int_0^\infty y^f (e^y - 1)^{-1}\, dy = \sum_{j=1}^\infty \int_0^\infty y^f e^{-jy}\, dy$$

$$= \Gamma(f+1) \sum_{j=1}^\infty j^{-(f+1)}$$

$$= \Gamma(f+1)\zeta(f+1)$$

and so

$$C_f = [\Gamma(f+1)\zeta(f+1)]^{-1}.$$

In particular

$$C_3 = (\tfrac{1}{6})[\zeta(4)]^{-1} = 15\pi^{-4}.$$

The rth moment of Y about zero is

(63)
$$\mu_r'(Y) = C_f/C_{f+r} = (f+1)(f+2)\cdots(f+r)\zeta(f+r+1)/\zeta(f+1).$$

The table below gives values of moments and moment-ratios for $f = 1(1)5$.

f	1	2	3	4	5
Expected value	1.4615	2.7012	3.8322	4.9056	5.9470
Standard deviation	1.3460	1.7479	2.0281	2.2602	2.4670
$\alpha_3^2 = \beta_1$	2.7760	1.3978	0.9731	0.7674	0.6423
$\alpha_4 = \beta_2$	6.9090	5.0120	4.4331	4.1469	3.9672

Since

$$y^f (e^y - 1)^{-1} = y^f \sum_{j=1}^\infty e^{-jy}$$

the distribution of Y can be regarded as a mixture of distributions of $(2j)^{-1}\chi^2_{2(f+1)}$ with weights proportional to $j^{-(f+1)}$ $(j = 1,2,\ldots)$. Equivalently, Y can be regarded as the product of a $\chi^2_{2(f+1)}$ variable and an independent random variable having a zeta distribution (see Chapter 10, Section 3).

Similarly the ratio of two independent variables, each having the same distribution as Y, is distributed as the product of a $F_{2(f+1),2(f+1)}$ variable and an independent random variable Z_1, having the (discrete) distribution

$$\Pr[Z = j_1/j_2] = C_f^2(j_1 j_2)^{-(f+1)} \qquad (j_1,j_2 = 1,2,\ldots).$$

The 'natural' occurrence of this rather remarkable distribution is noteworthy. We noted a distribution similarly taking all positive rational values in Section 1 of Chapter 2.

For values of $f > 1$, the distribution (62) has a single mode at a value y satisfying the equation

$$y(1 - e^{-y})^{-1} = f.$$

For $f = 3$ the mode is at 2.8217. For f large, the mode approximately equals

$$f\{1 - (f+1)e^{-f}\}/(1 - fe^{-f}).$$

The cumulative distribution function of Y is

$$\Pr[Y \le y] = 1 - \frac{1}{\zeta(f+1)} \sum_{r=0}^{f} \frac{y^r}{r!} \sum_{j=1}^{\infty} \frac{(e^{-y})^j}{j^{f-r+1}}. \tag{64}$$

Numerical values can also be obtained using Stegun's tables [152] of Debye functions

$$\int_0^x t^f(e^t - 1)^{-1}\,dt \qquad (\text{for } f = 1,2,3,4).$$

If $X_1, X_2, \ldots, X_n$ are independent random variables, each having distribution (62), then the maximum likelihood estimator, $\hat{\alpha}$, of α satisfies the equation (f being supposed known)

$$n(f+1)\hat{\alpha}^{-1} = \sum_{j=1}^{n} X_j e^{\hat{\alpha}X_j}(e^{\hat{\alpha}X_j} - 1)^{-1}. \tag{65}$$

This equation must be solved numerically.

An unbiased estimator of α can be based on the arithmetic mean

$$\bar{X} = n^{-1}\sum_{j=1}^{n} X_j.$$

Since $E[\bar{X}] = \alpha C_f/C_{f+1}$ it follows that

$$\tilde{\alpha} = (C_{f+1}/C_f)\bar{X} = \{(f+1)\zeta(f+1)/\zeta(f+2)\}\bar{X} \tag{66}$$

is an unbiased estimator of α.

We have, exactly,

$$\begin{aligned} n\,\mathrm{var}(\tilde{\alpha}) &= \alpha^2(C_{f+1}/C_f)^2[(C_f/C_{f+2}) - (C_f/C_{f+1})^2] \\ &= \alpha^2\{C_{f+1}^2/(C_{f+2}C_f) - 1\} \\ &= \alpha^2\left\{\frac{f+2}{f+1}\,\frac{\zeta(f+3)\zeta(f+1)}{[\zeta(f+2)]^2} - 1\right\}. \end{aligned} \tag{67}$$

For large n,

(68) $$n \operatorname{var}(\hat{\alpha}) \doteqdot 2\alpha^2(f+1)^{-1}\left\{\frac{(f+2)\zeta(f+2)}{\zeta(f+1)} - f\right\}^{-1}.$$

For f large, $\zeta(f+1) \doteqdot \zeta(f+2) \doteqdot \zeta(f+3)$ and

(69) $$n \operatorname{var}(\tilde{\alpha}) \doteqdot n \operatorname{var}(\hat{\alpha}) \doteqdot \alpha^2/(f+1).$$

Hence the limiting efficiency (as $n \to \infty$) of $\hat{\alpha}$ is nearly 100% for large f. Since $\tilde{\alpha}$ is much simpler to calculate than $\hat{\alpha}$ and is always unbiased, it seems to be preferable for practical purposes unless f is small. For $f = 3$, the limiting efficiency of $\tilde{\alpha}$ is 99.7%; for $f = 1$, it is 98.9%.

6.2 *Modified (or Modulated) Normal Distributions*

Romanowski [124] has introduced a class of distributions ('modified normal laws of error') obtained by ascribing a truncated (from above) Pareto distribution (see Chapter 19) to the variance of a normal distribution with zero mean. The cumulative distribution function is

$$\Pr[X \le x] = \frac{a+1}{\sigma\sqrt{2\pi}}\int_{-\infty}^{x}\int_{0}^{1} t^{a-\frac{1}{2}} \exp\left(-\tfrac{1}{2}y^2(t\sigma^2)^{-1}\right) dt\, dy.$$

The distribution can also be derived as a limiting distribution of sums of elementary errors taking values $\pm\frac{1}{2}\epsilon$ or zero [124].

The distribution is symmetrical about zero, and has variance

$$\sigma^2 \cdot (a+1)\int_0^1 t \cdot t^a\, dt = \sigma^2(a+1)/(a+2)$$

and

$$\beta_2 = \frac{3(a+2)^2}{(a+1)(a+3)}\cdot$$

(Note that $\beta_2 > 3$, and $\beta_2 \to 3$ as $a \to \infty$.)

Romanowski gave the names '*lineo-normal*' and '*radico-normal*' to the cases $a = 1$ and $a = \frac{1}{2}$ respectively.

Tables of the probability density functions of the *standardized* distribution for values of the variable 0.00(0.01)5.50 and for $a = \frac{1}{2}$, 1 are given to six decimal places in Romanowski and Green [127]. McLane [104] has shown that if $a = p + \frac{1}{2}$, with p an integer, then

$$\Pr[X \le x] = \sum_{j=0}^{p} \frac{(-1)^j(2p+3)(p-j)!}{\sqrt{2\pi}\, 2^{j+1}\sigma^{2j+1}(p+1)!}\int_{-\infty}^{x} y^{2j} \exp\left(-\tfrac{1}{2}y^2\sigma^2\right) dy$$

$$+ (-1)^p \frac{(2p+3)}{\sqrt{2\pi}\,(p+1)!2^{p+2}\sigma^{2p+3}}\int_{-\infty}^{x} y^{2p+2} Ei\left(-\tfrac{1}{2}y^2\sigma^{-2}\right) dy$$

where $Ei[-y] = -\int_y^\infty t^{-1}e^{-t}\, dt$. (The other integrals in the expression can be expressed as incomplete gamma functions).

In Romanowski and Green [128] applications of the distribution to errors of astronomical observations, triangulation errors, and determination of physical standards are described.

A succinct summary of the theory of modified normal distributions is given in Romanowski [125].

In his recent monograph, Romanowski [126] has tabled values of

$$\phi(\lambda) = \frac{a+1}{\sqrt{2\pi}}\sqrt{\frac{a+1}{a+2}}\int_0^1 t^{a-\frac{1}{2}}\exp\left[-\tfrac{1}{2}\lambda^2 t^{-1}(a+1)/(a+2)\right] dt$$

correct to 6 significant figures for $\lambda = 0.00(0.01)5.50$ and

$a = 0$ (equi-normal)
$a = \frac{1}{2}$ (radico-normal)
$a = 1$ (lineo-normal)
$a = 2$ (quadri-normal)
$a = \infty$ (standard normal distribution).

7. Hazard Rates (Mills' Ratio)

If the life-time of an individual be represented by a continuous random variable X with probability density function $p_X(x)$, the probability of surviving for at least a time x is

$$\Pr[X > x] = \int_x^\infty p_X(t)\,dt = 1 - F_X(x).$$

The probability that an individual, having survived to time x, will survive a further time t is

$$\begin{aligned}\Pr[X > x + t \mid X > x] &= \Pr[X > x + t]/\Pr[X > x] \\ &= \frac{F_X(x+t) - F_X(x)}{1 - F_X(x)}.\end{aligned}$$

The "death rate per unit time" between "ages" x and $(x + t)$ is

$$\frac{t^{-1}[F_X(x+t) - F_X(x)]}{1 - F_X(x)}.$$

Taking the limit as $t \to 0$, one obtains the "instantaneous death rate" at age x

$$p_X(x)[1 - F_X(x)]^{-1}. \tag{70}$$

In actuarial work, this is known as the *force of mortality* at age x. The same quantity, for general distributions, has been given the name *hazard rate* (*failure rate* is another term). Generally, in work on hazard rates, it is supposed that the random variable concerned is essentially non-negative (i.e. $F_X(0) = 0$) and we will follow this practice.

Of recent years, a good deal of attention has been given to the classification of distributions by their hazard rate functions. We will briefly summarize some of this work in Section 7.2. In Section 7.1, we will discuss more particularly the evaluation of the reciprocal of the hazard rate for the half-normal (chi with one degree of freedom) distribution.

7.1 *Mills' Ratio*

The hazard rate for the half-normal distribution is

$$\frac{1}{\sqrt{2\pi}} e^{-\frac{1}{2}x^2}\left[\frac{1}{\sqrt{2\pi}}\int_x^\infty e^{-\frac{1}{2}t^2}\,dt\right]^{-1} = Z(x)[1 - \Phi(x)]^{-1} \tag{71}$$

The reciprocal of this quantity,

$$R(x) = [1 - \Phi(x)]/Z(x) \tag{72}$$

("ratio of area to bounding ordinate") is called *Mills' ratio*, on account of its tabulation by Mills [107] in 1926. In this paper tables were presented which have also been included in Pearson [112]. These give values of $R(x)$ to 5 decimal places for $x = 0.00(0.01)4.00(0.05)5.00(0.1)10.0$, and to 10 decimal places for $\Phi(x) = 0.500(0.001)0.999$. More extensive tables by Sheppard [143] were published in 1939. They give $R(x)$ to 12 decimal places for $x = 0.00(0.01)9.50$ and to 24 decimal places for $x = 0.0(0.1)10.0$.

Note that tables of $R(x)$ can be used to calculate the mean and variance,

$$[R(x)]^{-1} \quad \text{and} \quad 1 + x[R(x)]^{-1} - [R(x)]^{-2},$$

respectively, of a unit normal variable truncated from below at x.

Since 1926, there has been a considerable interest in deriving approximate formulas for $R(x)$. Such formulas can be regarded also as approximate formulas for $\Phi(x)$, and, conversely, approximations to $\Phi(x)$ (Chapter 13, Section 5), provide approximations for $R(x)$. Here we concentrate on formulas especially developed as approximations for $R(x)$, with particular attention to formulas of simple structure.

Early in the nineteenth century, Laplace [88] obtained the *asymptotic* expansion

$$R(x) \sim x^{-1} - x^{-3} + 3x^{-5} - 15x^{-7} + \cdots \tag{73}$$

The series is *divergent* for all x, but it possesses the property that $R(x)$ falls between any two partial sums. Thus

$$x^{-1} - x^{-3} < R(x) < x^{-1}$$
$$x^{-1} - x^{-3} < R(x) \leq x^{-1} - x^{-3} + 3x^{-5}$$

and so on. Haldane [74] has pointed out that the terms in (73) can be rearranged as:

(74)

$$R(x) = x^{-1}[1 - (x^2 + 3)^{-1} - 6(x^6 + 13x^4 + 25x^2 + 145)^{-1} + 0(x^{-14})].$$

This is a particular case of the formula

$$\begin{aligned}1 - ht^{-1} + h(h + 1)t^{-2} - h(h + 1)(h + 2)t^{-3} + \cdots \\ = 1 - h(t + h + 1)^{-1} - h(h + 1) \\ \times [t^3 + (3h + 5)t^2 + (h + 1)(3h + 1)t + (h + 2)(h^2 + 7)]^{-1} \\ + O(t^{-7})\end{aligned}$$

with

$$h = \tfrac{1}{2}, \quad t = \tfrac{1}{2}x^2.$$

Closer approximations are obtained at an early stage in the expansion (74) than in (73).

Much better approximations have been developed more recently. In 1941, Gordon [67] showed that

$$\frac{x}{x^2 + 1} < R(x) < \frac{1}{x}$$

and in 1942, Birnbaum [17] improved the lower bound to $2[\sqrt{x^2 + 4} + x]^{-1}$. Further improvements of this kind did not come for some ten years. In 1953, Sampford [135] proved the correctness of the upper bound

$$4[\sqrt{x^2 + 8} + 3x]^{-1} \tag{75}$$

conjectured in 1950 by Birnbaum [18]. (See also Shenton [141].) This is, in fact, a good approximation to $R(x)$ itself, for x greater than 2. In 1959, Boyd [27] pointed out that for large x, (75) exceeds $R(x)$ by only about $2x^{-7}$. (By comparison, $R(x)$ exceeds $2[\sqrt{x^2 + 4} + x]^{-1}$ by about $\tfrac{2}{3}x^{-5}$.)

In the same paper, Boyd also constructed the bounds

$$\begin{aligned}&\frac{\pi}{\sqrt{x^2 + 2\pi} + (\pi - 1)x} \\ &\quad < R(x) < \frac{\pi}{(\pi - 2)\{\sqrt{x^2 + 2\pi(\pi - 2)^{-2}} + 2(\pi - 2)^{-1}x\}}\end{aligned} \tag{76}$$

$$\left(\text{i.e. } \frac{3.141593}{\sqrt{(x^2 + 6.283185)} + 2.141593x} < R(x) < \frac{2.751938}{\sqrt{(x^2 + 4.821227)} + 1.751938x}\right).$$

For large x, the upper bound exceeds $R(x)$ by about $\tfrac{1}{8}x^{-3}$, and $R(x)$ exceeds the lower bound by about $0.43x^{-5}$. An important feature of these bounds is that they are useful for all values of x (> 0). In particular, each tends to $\sqrt{\pi/2} = R(0)$ as x tends to zero. This is not true of the other bounds, so far mentioned, which are useful only for larger values of x (greater than 2, say). The maximum value of the ratio of the upper to the lower bound is less than 1.02.

Boyd's paper may be regarded as the culmination of a series of studies on the construction of bounds and approximations for $R(x)$ of form $\alpha[\sqrt{x^2+\beta}+\gamma]^{-1}$. Among these we note that Komatu [83] obtained the upper bound $2[\sqrt{x^2+2}+x]^{-1}$, which Pollak [116] improved to

$$2[\sqrt{x^2+8\pi^{-1}}+x]^{-1},$$

noting that $8\pi^{-1}$ is the least possible value of b for an upper bound of form $2[\sqrt{x^2+b}+x]^{-1}$. (Note that as $x \to 0$, $2[\sqrt{x^2+8\pi^{-1}}+x]^{-1} \to \sqrt{\pi/2} = R(0)$; for large x, the excess over $R(x)$ is about $0.53x^{-3}$.) We further note that Chu [34] showed that the best possible lower bound for $\Phi(x) - \frac{1}{2}$ (with $x > 0$) of form $\frac{1}{2}[ax^2(1+ax^2)^{-1}]^{\frac{1}{2}}$ is obtained with $a = 2/\pi$.

Boyd also gave the upper bound

$$3[\sqrt{x^2+12}+5x]^{-1}+[\sqrt{x^2+4}+x]^{-1}$$

which is an even better approximation (for large x) than (75), as it exceeds $R(x)$ by only about $4x^{-9}$. It is, however, of more complicated form than (75).

In Table 2 are shown values of

(i) $R(x)$
(ii) lower bound $2[\sqrt{x^2+4}+x]^{-1}$
upper bound $4[\sqrt{x^2+8}+3x]^{-1}$
(iii) the lower and upper bound in (76).

Approximations of different forms have also been developed. The earliest of these, again due to Laplace [88], consist of successive convergents of the

TABLE 2

Approximations to Mills' Ratio $(1-\Phi(x))/Z(x)$

x	(i)	(ii)		(iii)	
	Exact	Lower	Upper	Lower	Upper
0.2	1.075945	—	—	1.067511	1.076992
0.4	0.935667	—	—	0.925365	0.938382
0.6	0.823028	—	—	0.813380	0.827055
0.8	0.731314	—	—	0.723125	0.736112
1.0	0.655680	0.618034	0.666667	0.649045	0.660783
1.2	0.592574	0.566190	0.599479	0.587326	0.597654
1.4	0.539358	0.520656	0.543778	0.535256	0.544206
1.6	0.494040	0.580625	0.496918	0.490851	0.498545
1.8	0.455101	0.445362	0.457006	0.452616	0.459214
2.0	0.421369	0.414214	0.422650	0.419442	0.425080
3.0	0.304590	0.302776	0.304806	0.304001	0.306674
4.0	0.236652	0.236068	0.236701	0.236443	0.237835
5.0	0.192808	0.192582	0.192822	0.192723	0.193518
10.0	0.099029	0.099020	0.099029	0.099025	0.099142

continued fraction

$$R(x) = \cfrac{1}{x + \cfrac{1}{x + \cfrac{2}{x + \cfrac{3}{x + \cdots}}}} \tag{77}$$

This is useful for x moderately large. For small x only, Shenton's [141] continued fraction

$$\frac{\Phi(x) - \frac{1}{2}}{Z(x)} = \cfrac{x}{1 - \cfrac{x^2}{3 + \cfrac{2x^2}{5 - \cfrac{3x^2}{7 + \cfrac{4x^2}{9 - \cdots}}}}} \tag{78}$$

is much better.

Yet further forms of approximation have been constructed by Ruben [130], [131], [132] and by Ray and Pitman [121]. Ruben [130] obtained the asymptotic expansion

$$\begin{aligned} R(x) \sim x^{-1}f_0(x^{-1}) + x^{-3}f_1(2x^{-1}) + 2!x^{-5}f_2(3x^{-1}) + \cdots \\ + x^{-(2n-1)}(n-1)!f_{n-1}(nx^{-1}) + O(x^{-2n-1}) \qquad (\text{as } x \to \infty) \end{aligned} \tag{79}$$

where

$$f_0(t) = \exp(-\tfrac{1}{2}t^2)$$

and

$$f_j(t) = \frac{d}{dt}\frac{f_{j-1}(t) - f_{j-1}(jx^{-1})}{t - jx^{-1}} \qquad (j = 1,2,\ldots).$$

The first three terms of the expansion are (putting $y = \exp(-\frac{1}{2}x^{-2})$):

$$\begin{aligned} yx^{-1} + yx^{-3}\{x^2 - (2 + x^2)y^3\} \\ + 2yx^{-5}\{\tfrac{1}{2}x^4 - (2x^2 + x^4)y^3 + \tfrac{1}{2}(9 + 5x^2 + x^4)y^8\}. \end{aligned}$$

In [131] and [132], Ruben obtained *convergent* expansions of similar form. That derived in [131] is

$$R(x) = x^{-1}\sum_{j=0}^{\infty} j!(\tfrac{1}{2}x^2)^{-2j}f_j(2(j+1)x^{-2}) \tag{80}$$

where

$$f_0(t) = (1 + t)^{-\frac{1}{2}}$$

and

$$f_j(t) = \frac{d}{dt}\frac{f_{j-1}(t) - f_{j-1}(2jx^{-2})}{t - 2jx^{-2}} \qquad (j = 1,2,\ldots).$$

If the series (80) for $R(x)$ is terminated at the Nth term the truncation error is less than

$$\min\left(\frac{(2N)!}{N!}\cdot x^{-1}(2ex^2)^{-N},\ \frac{(4N)!}{(2N)!}\cdot x^{-1}(2\sqrt{2}\,x^2)^{-2N}\right).$$

The results of truncating to the first one, two or three terms are, respectively,

$$\begin{aligned} R(x) &= (x^2+2)^{-\frac{1}{2}} \\ R(x) &= 2(x^2+2)^{-\frac{1}{2}} - (x^2+4)^{\frac{1}{2}} - (x^2+4)^{-\frac{3}{2}} \\ R(x) &= 3(x^2+2)^{-\frac{1}{2}} - 3(x^2+4)^{-\frac{1}{2}} - 3(x^2+4)^{-\frac{3}{2}} \\ &\quad + (x^2+6)^{-\frac{1}{2}} + 2(x^2+6)^{-\frac{3}{2}} + (x^2+6)^{-\frac{5}{2}}. \end{aligned}$$

Although the series is quite rapidly convergent, the accuracy attained with the early terms does not appear to compensate for the greater complexity of the formula.

Shenton [141] has constructed yet another expansion for $R(x)$ in terms of irrational functions, based on the convergents of the Laplace continued fraction (77).

Ray and Pitman [121] obtained further (asymptotic) expansions. These are

$$R(x) \sim x^{-1}\{1 + \tfrac{1}{2}(y^4-1) + \tfrac{1}{12}(y^{16}-4y^4+3) + \tfrac{23}{1440}(y^{36}-6y^{16}+15y^4-10)+\cdots\} \tag{81}$$

and

$$R(x) \sim x^{-1}\{1 + (y-1) + \tfrac{1}{2}(y^4-2y+1) + \tfrac{1}{3}(y^9-3y^4+3y-1)+\cdots\} \tag{82}$$

where, as before, $y = \exp(-\frac{1}{2}x^{-2})$. In (82) the jth term in brackets is

$$c_j \sum_{r=0}^{j-1} \binom{j-1}{r} (-1)^r y^{(j-1-r)^2}$$

where c_j is the coefficient of θ^{j-1} in the expansion of $[1-\log(1+\theta)]^{-1}$.

Ray and Pitman also developed another expansion which gives good results with only the leading terms of the series. They call this a *Laguerre-Gauss* expansion because it is derived from a quadrature formula of this name. The sums of the first one and two terms of this expansion, with upper bounds for the absolute error in $R(x)$, are

Number of terms	Approximation to $R(x)$	Upper bound of error
1	$x^{-1}y$	$\frac{1}{2}x^{-3}$
2	$x^{-1}[0.853553\exp(-0.171573x^{-2}) + 0.146447\exp(-5.828429x^{-2})]$	$\frac{1}{2}x^{-5}$

Raj [120] has obtained bounds for $R(x)$ (i.e. ratio of upper tail area to bounding ordinate) for the standardized Type III density function

$$\left[\Gamma\left(\frac{4}{\alpha_3^2}\right)\right]^{-1}\left(\frac{4}{\alpha_3^2}\right)^{4\alpha_3^{-2}-\frac{1}{2}}\exp\left(-\frac{4}{\alpha_3}\right)\cdot\{1+\tfrac{1}{2}\alpha_3 x\}^{4\alpha_3^{-2}-1}\exp\left(-\frac{2x}{\alpha_3}\right)$$

$$(x > -2/\alpha_3).$$

(Note that $\alpha_3 = \sqrt{\beta_1} > 0$ is the skewness coefficient.) The bounds are

(83)

$$2(1+\tfrac{1}{2}\alpha_3 x)[\sqrt{x^2+4(1+\tfrac{1}{2}\alpha_3 x)}+x]^{-1} \le R(x) \le (1+\tfrac{1}{2}\alpha_3 x)(x+\tfrac{1}{2}\alpha_3)^{-1}.$$

The inequalities are valid for all $x > -2/\alpha_3$, but useful only for x large and positive. Putting $\alpha_3 = 0$, we obtain the limits for the normal distribution Mills' ratio:

$$2[\sqrt{x^2+4}+x]^{-1} \le R(x) \le x^{-1}.$$

The fact that there are considerably better upper bounds than x^{-1} indicates that the upper bound in (83) may be capable of substantial improvement. (Approximate formulas for upper tail areas of Type III (χ^2) distributions (for large x) are given in Wishart [166].)

Shenton and Carpenter [142] have considered a modified Mills' ratio for Type IV distributions. They obtained a continued fraction representation of the ratio of tail area to bounding ordinate for a density function with one parameter decreased by 1. The main use of this representation appears to be in the calculation of tail areas of Type IV distributions.

7.2 *Distributions Classified by Hazard Rate*

Of recent years there have been a number of studies by Barlow and co-workers [5]–[11] on the properties shared by distributions for which $R(x)$ satisfies certain conditions. Among these are distributions with increasing hazard rate (IHR distributions) and with decreasing hazard rate (DHR distributions) — i.e. distributions for which $[R(x)]^{-1}$ is an increasing or decreasing function of x, respectively. The exponential distribution, with a constant hazard rate, is a natural boundary between these two classes. The half-normal (chi with 1 degree of freedom) distribution is a DHR distribution. The Weibull distribution (see (1) of Chapter 20) is IHR if $c > 1$, DHR if $c < 1$. The 'linear failure rate' distribution (Equation (49.1)) has, as its name indicates, a hazard rate which is a linear function of the variable.

Consideration of IHR and DHR classes is usually restricted to continuous non-negative random variables (i.e. with $F_X(0) = 0$) and we will follow this practice.

A broader class than the IHR (DHR) distributions is that of distributions with increasing (decreasing) hazard rate *on the average* (IHRA, DHRA respectively). A distribution belongs to the IHRA (DHRA) class if $x^{-1}\int_0^x [R(t)]^{-1}\,dt$ is an increasing (decreasing) function of x.

It is clear that the classes IHR, DHR do not exhaust all possible distributions, for $R(x)$ need not be a monotone function of x. Nor, of course, are the classes

TABLE 3

Bounds on Distributions Classified by Hazard Rates

Conditions	Values of x	Limits on $1 - F(x)$
IHR	$x \geq \mu_r'^{1/r}$	$1 - F(x) \leq w(x)$ where $\mu_r' = rx^r \int_0^1 t^{r-1}[w(x)]^t\, dt$ $\left(\text{i.e. } \lambda_r = \left(-\frac{x}{\log w(x)}\right)^r \frac{\Gamma_{-\log w(x)}(r)}{\Gamma(r)}\right)$
	$x \leq \mu_r'^{1/r}$	$1 - F(x) \geq \exp(-x/\lambda_r^{1/r})$
DHR	$x \leq r\lambda_r^{1/r}$	$1 - F(x) \geq \exp(-x/\lambda_r^{1/r})$
	$x \geq r\lambda_r^{1/r}$	$1 - F(x) \geq (rx/e)^r\lambda_r$
IHRA	$x < \mu_r'^{1/r}$	$1 - F(x) \geq \exp(-bx)$ where $\mu_r' = x^r(1 - e^{-bx}) + b\int_x^\infty t^r e^{-bt}\, dt$

Notes: $\lambda_r = \mu_r'/\Gamma(r + 1)$

In [7], Barlow and Marshall also give limits for $F(x_2) - F(x_1)$ for any $x_2 > x_1 > 0$ for each of the above cases.

For the case when $r = 1$, and when both the first and second moments are known, better, but considerably more complicated, inequalities are available in Barlow and Marshall [5] and [6].

IHRA, DHRA exhaustive, although these are broader, in that an IHR (DHR) distribution must be IHRA (DHRA), but not necessarily conversely.

Barlow and Marshall [5], [6], [7] give bounds on the distribution of IHR and DHR distributions. Some of these bounds are given in Table 3 above. These bounds can be compared with the well-known Chebyshev-type bounds which are summarized for convenience in Table 4. A few references (Frechet [60], Godwin [65], and Mallows [95], [96]) are included to assist readers wishing for more details.

Note that for $s = 2r$, the Wald inequality becomes one of the Cantelli inequalities. Also for $\mu_r' \leq x^r \leq \mu_{2r}'/\mu_r'$ there is no suitable Cantelli inequality. A detailed account of these inequalities is given in [65] together with further inequalities applicable to sums of independent variables. This latter topic has been further developed by Hoeffding [75] and Bennett [16].

Bennett [16] shows that if $X_1, \ldots, X_n$ are independent with finite expected values $E[X_j]$, variances $[\sigma(X_j)]^2$, and $\Pr[X_j - E[X_j] > M_j] = 0$ for all j then (with $t < 1$)

$$\Pr\left[\sum_{j=1}^n \{X_j - E[X_j]\} \geq t \sum_{j=1}^n M_j\right] \leq [f(t,r)]^B \tag{84}$$

TABLE 4

Bounds on Cumulative Distribution Functions

Names (or References)	Conditions	Values of x	Limits on $1 - F(x)$
Chebyshev	—	$x^r > \mu_r$	$1 - F(x) \leq \mu_r'/x^r$
Cantelli	—	$x^r \geq \mu_{2r}'/\mu_r'$	$1 - F(x) \leq \dfrac{\mu_{2r}' - \mu_r'^2}{(x^r - \mu_r')^2 + \mu_{2r}' - \mu_r'^2}$
		$x^r \leq \mu_r'$	$1 - F(x) \geq 1 - \dfrac{\mu_{2r}' - \mu_r'^2}{(\mu_r' - x^r)^2 + \mu_{2r}' - \mu_r'^2}$
Wald [159]	$r < s$	$1 \geq \mu_r'/x^r \geq \mu_s'/x^s$	$1 - F(x) \leq \dfrac{\mu_s' - \mu_r'\delta^{s-r}}{x^r(x^{s-r} - \delta^r)}$ where δ $(> 0, \neq x)$ satisfies $\mu_r'x^s - \mu_s'x^r + \delta^r(\mu_s' - x^s) + \delta^s(\mu_r' - x^r) = 0$
Gauss–Winkler	$F'(x') \geq F'(x) \geq F'(x'')$ if $x' < x < x''$	$x^r < r^r(r+1)^{-(r-1)}\mu_r'$	$1 - F(x) \leq 1 - x[(r+1)\mu_r']^{-1/r}$
		$x^r > r^r(r+1)^{-(r-1)}\mu_r'$	$1 - F(x) \leq [(1 + r^{-1})x]^{-r}\mu_r'$

where
$$B = \left(\sum_{j=1}^{n} M_j\right) \Big/ (\max_j M_j);\ r = \min_j \{M_j/\sigma(X_j)\}$$

and
$$f(t,r) = (1 + tr^2)^{-1}\{(1 + tr^2)/(1 - t)\}^{r^2(1-t)/(1+r^2)}.$$

In [16] there is a table of $f(t,r)$ to six decimal places for $t = 0.00(0.02)1.00$; $r = 2.0(0.5)5.0$.

We also note a few special results for unimodal distributions. By *unimodal*, we mean that there is a number M such that, for any x_1, x_2, x_3, x_4 with

$$x_1 < x_2 < M < x_3 < x_4,$$

then

$$F(\tfrac{1}{2}(x_1 + x_2)) \leq \tfrac{1}{2}[F(x_1) + F(x_2)]$$

and

$$F(\tfrac{1}{2}(x_3 + x_4)) \geq \tfrac{1}{2}[F(x_3) + F(x_4)].$$

The set of numbers $\mu_1', \mu_2', \ldots, \mu_n'$ can be the first n moments of a distribution if and only if the determinants

$$\begin{vmatrix} 1 & \mu_1' & \cdots & \mu_s' \\ \mu_1' & \mu_2' & \cdots & \mu_{s+1}' \\ \vdots & \vdots & & \vdots \\ \mu_s' & \mu_{s+1}' & \cdots & \mu_{2s}' \end{vmatrix}$$

are non-negative for all s, $2 \leq 2s < n$.

In Johnson and Rogers [79] it is shown that, for unimodal distributions, the same conditions hold with μ_j' replaced by $(j + 1)\mu_j'$. That is, for any unimodal distribution we must have

$$\begin{vmatrix} 1 & 2\mu_1' \\ 2\mu_1' & 3\mu_2' \end{vmatrix} \geq 0; \quad \begin{vmatrix} 1 & 2\mu_1' & 3\mu_2' \\ 2\mu_1' & 3\mu_2' & 4\mu_3' \\ 3\mu_2' & 4\mu_3' & 5\mu_4' \end{vmatrix} \geq 0 \quad \text{and so on.} \tag{85}$$

It can be deduced that for any unimodal distribution with finite values of $\mu_1', \mu_2', \mu_3', \mu_4'$, the value of β_2, for given β_1, must be greater than that given by the parametric equations

$$\beta_1 = \frac{108\theta^4}{(1-\theta)(1+3\theta)^3}; \quad 5\beta_2 - 9 = \frac{72\theta^2(3\theta - 1)}{(1-\theta)(1+3\theta)^2} \qquad (0 \leq \theta < 1).$$

Royden [129] has shown that for unimodal distributions with first and second moments M_1, M_2 respectively about the mode (ϕ):

$$\Pr[|X - \phi| < x] \leq 1 - \frac{(2M_1 - x)^2}{3M_2 - 2xM_1} \qquad (0 \leq x \leq 2M_1); \tag{86.1}$$

$$\Pr[|X - \phi| < x] \geq \frac{x}{2M_1} \qquad (0 \leq x \leq M_1); \tag{86.2}$$

$$\Pr[|X - \phi| < x] \geq 1 - \frac{M_1}{2x} \qquad \left(M_1 \leq x \leq \frac{3M_2}{4M_1}\right); \tag{86.3}$$

$$\Pr[|X - \phi| < x] \geq 1 - \frac{4M_1^2}{3M_2} + \frac{8M_1^3 x}{9M_2^2} \qquad \left(\frac{3M_2}{4M_1} \leq x \leq \frac{M_2}{M_1}\right); \tag{86.4}$$

$$\Pr[|X - \phi| < x] \geq 1 - \frac{\psi - 1}{3t^2 - 4t + \psi} \qquad \left(x \geq \frac{M_2}{M_1}\right) \tag{86.5}$$

where $\psi = 3M_2(4M_1^2)^{-1}$ and

$$\frac{x}{4M_1} = \frac{t^3 - t^2}{3t^2 - 4t + \psi}.$$

Note that from (85), $3M_2 \geq 4M_1^2$ (and so $\psi \geq 1$) for any unimodal distribution.

Another set of inequalities for unimodal distributions with $\Pr[X < 0] = 0$ is

$$\begin{aligned} 1 - F_X(x) &\leq 2\mu_1' x^{-1} - 1 && \text{for } \mu_1' \leq x \leq 3\mu_1'/2 \\ &\leq \mu_1'(2x)^{-1} && \text{for } x \geq 3\mu_1'/2 \end{aligned} \tag{87}$$

This result is ascribed to Mallows by Barlow and Marshall [6].

Selberg [139] has obtained the inequality, valid for unimodal distributions,

$$\Pr[|X - E[X]| < x] \geq 1 - \tau x^{-2} \operatorname{var}(X) \tag{88}$$

where $\tau \doteqdot 0.5654$ satisfies the equation

$$\tau^3 - 9\tau^2 + 3\tau + 1 = 0.$$

Note that this inequality relates to deviation from the expected value, and not from the modal value. Selberg [140] has also obtained the following inequality for asymmetrical deviations from the expected value, valid for *any* distribution (provided of course, the necessary moments are finite). For $0 < \alpha \leq \beta$

(89.1)

$$\Pr[-\alpha < X - E[X] < \beta] \geq \alpha^2(\alpha^2 + \operatorname{var}(X))^{-1} \quad \text{if } \alpha(\beta - \alpha) \geq 2\operatorname{var}(X)$$

(89.2)

$$\Pr[-\alpha < X - E[X] < \beta] \geq 4(\alpha\beta - \operatorname{var}(X))(\alpha + \beta)^{-2} \quad \text{if } \alpha(\beta - \alpha) \leq 2\operatorname{var}(X).$$

Mallows [95], [96] has obtained inequalities valid under a variety of limitations on the values of the cumulative distribution functions and its derivatives.

Among further properties of IHR and DHR distributions, we note that

(i) $(1 - F(x))^{1/x}$ is a decreasing (increasing) function of x if $F(x)$ is IHR (DHR).

(ii) Pólya frequency functions of type 2* (as described in Chapter 12, Section 4) are IHR, but the converse is not necessarily true.

Some inequalities satisfied by moments of order statistics for IHR and DHR distributions are obtained in Barlow and Proschan [11]. These are mostly relations between moments of IHR (DHR) distributions and exponential distributions. They are of use in forming qualitative assessment of the properties of test and estimation procedures constructed for exponential distributions when applied to IHR or DHR distributions.

Note that, denoting the hazard rate by

$$\mu_X(x) = p_X(x)[1 - F_X(x)]^{-1} = -\frac{d}{dx}[1 - F_X(x)]$$

we have

$$1 - F_X(x) = \exp\left[-\int_{-\infty}^{x} \mu_X(t)\,dt\right].$$

The logarithm of the likelihood function corresponding to n independent random variables $X_1, X_2, \ldots, X_n$ can be written in the form

*For which the dimension of the determinant is 2×2

$$\sum_{j=1}^{n} \log\left[-\frac{d}{dx}(1 - F_X(X_j))\right] = \sum_{j=1}^{n} \log \mu_X(X_j) - \sum_{j=1}^{n} \int_{-\infty}^{X_j} \mu_X(t)\,dt.$$

If it is *known* that the common distribution is IHR (or DHR) a 'maximum likelihood' estimator of $\mu_X(x)$ can be obtained by a method described by Marshall and Proschan [99].

The estimator is a function of the order statistics $X'_1 \leq X'_2 \leq \cdots \leq X'_n$ and is given by

$$\hat{\mu}_X(x) = \begin{cases} 0 & (x < X'_1) \\ \hat{\mu}(X_j) & (X'_j \leq x < X'_{j+1}; j = 1,2,\ldots,n-1) \\ \hat{\mu}(X_n) & (x \geq X'_n) \end{cases}$$

where

$$\hat{\mu}(X_j) = \min_{v \geq j+1} \max_{u \leq j} \left[(v - u)\left\{\sum_{i=u}^{v-1} (n - i)(X'_{i+1} - X'_i)\right\}^{-1}\right].$$

If $X'_1 \leq X'_2 \leq \cdots \leq X'_n$ be the order statistics for n independent random variables each having an exponential distribution, then the minimum variance unbiased estimator of $\Pr[X > t]$, $t > 0$, using only $X'_1, X'_2, \ldots, X'_r$, is

$$\left[1 - t\left\{\sum_{j=1}^{r-1} X'_j + (n - r + 1)X'_r\right\}^{-1}\right]^{r-1} \quad \text{if} \quad t < \sum_{j=1}^{r-1} X'_j + (n - r + 1)X'_r$$

and zero otherwise (Chapter 18, Section 6). If the common distribution is not exponential, but is IHR, then it is possible to obtain an upper bound for the expected value of this estimator. Assuming that the expected value of the common distribution is σ (so that in the exponential case the density function is $\sigma^{-1} \exp(-x/\sigma)$, for $x > 0$, and $\Pr[X > t] = \exp(-t/\sigma)$) and that t is less than σ, the estimator has an expected value which is not greater than

$$[(r-1)!]^{-1} \int_{t/\sigma}^{1} (1 - t(y\sigma)^{-1})^{r-1} y^{r-1} e^{-y}\,dy$$
$$+ (1 - t\sigma^{-1})^{r-1}[(r-1)!] \int_{1}^{\infty} y^{r-1} e^{-y}\,dy.$$

The maximum likelihood estimator

$$\exp\left[-rt\left\{\sum_{j=1}^{r-1} X'_j + (n - r + 1)X'_r\right\}\right]$$

on the other hand has expected value *at least*

$$[(r-1)!]^{-1} \int_{0}^{1} e^{-rt/(y\sigma)} y^{r-1} e^{-y} dy + [(r-1)!]^{-1} e^{-rt/\sigma} \int_{1}^{\infty} y^{r-1} e^{-y}\,dy$$

for IHR distributions.

8. Some Special Distributions

We have already noted, in Volume 1, that useful discrete distributions can be formed by analogy with established continuous distributions. We note, in particular, the Zipf (discrete Pareto) and discrete t distributions described in Chapter 10, Section 3. The converse procedure of forming continuous distributions by analogy with discrete distributions can be equally fruitful.

Fraser ([59], p. 257) describes a *continuous Poisson* distribution (ascribing its derivation to Y. S. Lee), which has cumulative distribution function

$$\Pr[X \leq x] = \frac{1}{\Gamma(x + \frac{1}{2})} \int_{\theta}^{\infty} t^{x-\frac{1}{2}} e^{-t}\, dt \qquad (\theta > 0; x > -\tfrac{1}{2}). \tag{90}$$

This distribution has the property that if x is a positive integer or zero

$$\Pr[x - \tfrac{1}{2} < X < x + \tfrac{1}{2}] = e^{-\theta}\theta^{x}/x!. \tag{91}$$

This distribution may be useful as an approximation to a Poisson distribution in analytical investigations.

Another continuous analogue of the Poisson distribution has been proposed by Castoldi [32]. He suggests using a density function of form

$$p(x) = g(\theta)e^{-\theta}\theta^{x}/\Gamma(x + 1) \qquad (\theta > 0;x > 0) \tag{92}$$

with $g(\theta) = e^{\theta}/\int_0^{\infty} \theta^{x}[\Gamma(x + 1)]^{-1}\, dx$. Alternatively, he suggests a mixed distribution with

$$\Pr[X = 0] = 1 - [g(\theta)]^{-1}$$

and density function as in (92) with $g(\theta)$ replaced by 1.

From time to time it becomes necessary to derive distributions arising, in some more or less "natural" way, from combinations of variables having distributions of the more usual kinds. It is not practicable to give a list of such occasional results, for it is clearly possible to invent further variations at will. We will, however, mention here some distributions which have been derived in published papers.

We first note distributions of the maximum or minimum of independent variables having *different* distributions. Such distributions occur naturally when considering failure of a system arising from failure(s) of components having different distributions of life-time. As an example, suppose $X_1, X_2, \ldots, X_n$ are independent random variables, and

$$p_{X_j}(t) = \theta_j^{-1} \exp(-t\theta_j^{-1}) \qquad (\theta_j > 0;\ t > 0).$$

Then the probability that $L = \min(X_1, X_2, \ldots, X_n)$ exceeds l (> 0) is

$$\prod_{j=1}^{n} \Pr[X_j > l] = \exp\left(-l \sum_{j=1}^{n} \theta_j^{-1}\right).$$

Hence the probability density function of L is

$$(93) \qquad p_L(l) = \left(\sum_{j=1}^{n} \theta_j^{-1}\right) \exp\left(-l \sum_{j=1}^{n} \theta_j^{-1}\right) \qquad (l > 0).$$

In a similar way, the distribution of the greatest value

$$G = \max(X_1, X_2, \ldots, X_n)$$

has density function

$$(94) \qquad p_G(g) = \frac{d}{dg} \prod_{j=1}^{n} [1 - \exp(-g\theta_j^{-1})]$$

$$= \left[\sum_{j=1}^{n} \frac{\theta_j^{-1} \exp(-g\theta_j^{-1})}{1 - \exp(-g\theta_j^{-1})}\right] \prod_{j=1}^{n} [1 - \exp(-g\theta_j^{-1})].$$

Among further special cases, we note that the distribution of the lesser of two independent random variables, one having a normal and one an exponential distribution, has been discussed by Gercbah and Kordonski [64].

The *Fisher-Stevens* distribution (Fisher [54], [55], Stevens [155]) can be regarded as the distribution of the maximum of a number of *dependent* random variables $Y_j = Z_j \left\{\sum_{i=1}^{k} Z_i\right\}^{-1}$ $(j = 1, 2, \ldots, k)$, where $Z_1, Z_2, \ldots, Z_k$ are independent random variables each having the same exponential distribution

$$p_{Z_j}(z) = \theta^{-1} e^{-z/\theta} \qquad (\theta > 0; z > 0).$$

Putting $G = \max(Y_1, Y_2, \ldots, Y_k)$

$$(95.1) \qquad \Pr[G \leq g] = 1 - k(1-g)^{n-1} + \binom{k}{2}(1-2g)^{n-1} - \cdots + (-1)^s \binom{k}{s}(1-sg)^{n-1}$$

where s is the greatest integer less than g^{-1}. More generally, the probability that the ith greatest among $Y_1, Y_2, \ldots, Y_k$ does not exceed g is

$$(95.2) \qquad 1 - \sum_{j=1}^{s} (-1)^{j-i} \frac{k!}{(k-j)!(j-i)!(i-1)!} \cdot \frac{(1-jg)^{n-1}}{j}.$$

Tables of the upper 5 per cent points of the distributions of G and the second largest value, to 5 decimal places, are given in Fisher [55] for

$$k = 3(1)10(5)50.$$

A related distribution — that of the sum of the k largest out of the n intervals into which a line of unit length is divided by $(n - 1)$ points independently and randomly chosen (with uniform distribution) on the line — has been derived by Mauldon [102]. Formally this is the distribution of

$$S_k = (Y'_n + Y'_{n-1} + \cdots + Y'_{n-k+1})$$

where $Y'_1, Y'_2, \ldots, Y'_n$ are the order statistics corresponding to

$$Y_j = X'_j - X'_{j-1} \qquad (j = 1,2,\ldots,n)$$

with $X'_1, X'_2, \ldots, X'_{n-1}$ the order statistics corresponding to $(n - 1)$ independent random variables $X_1, \ldots, X_{n-1}$ each having a standard rectangular (uniform) distribution, and $X'_0 = 0$, $X'_n = 1$. Mauldon showed that

$$\text{(96)} \qquad \Pr[S_k \leq s] = \begin{cases} 0 & (s \leq k/n) \\ \sum^*(-1)^{n-j}j^{-1}(js - k)^{n-1}C_j & (k/n \leq s \leq 1) \\ 1 & (s \geq 1) \end{cases}$$

where $\sum^*$ denotes summation with respect to j for $k/s < j \leq n$ and

$$C_j = \frac{1}{k^{n-k-1}(j - k)^{k-1}} \cdot \frac{n!}{(n - j)!(j - k)!k!}.$$

Moments can be found from the special generating function

$$\text{(97)} \qquad E[(1 - S_k t)^{-n}] = (1 - t)^{-k} \prod_{j=k+1}^{n} (1 - kt/j)^{-1}.$$

In particular

$$\text{(98.1)} \qquad E[S_k] = (k/n)\left(1 + \sum_{j=k+1}^{n} j^{-1}\right)$$

and

$$\text{(98.2)} \qquad \text{var}[S_k] = \frac{1}{n^2(n + 1)}\left[k(n - k) + k^2\left\{n \sum_{j=k+1}^{n} j^{-2} - \left(\sum_{j=k+1}^{n} j^{-1}\right)^2 - 2\sum_{j=k+1}^{n} j^{-1}\right\}\right].$$

Distributions of sums of independent random variables with different forms of distribution occur from time to time. For example, the sum of independent normal and Poisson variables has been discussed by Linnik [92]. The distribution of sums of rectangularly distributed and discrete variables has arisen in studies on applying the probability integral transformation to discrete distributions [45].

Also, the distribution of the weighted sum of independent (central) t variables appears in the "fiducial" (Behrens-Fisher) problem of comparing population means when the population variances are unknown (Chapter 27, Section 7), and the distribution of weighted sums of independent F variables in certain multivariate tests proposed by Geisser [63].

The distribution of products of independent random variables has been studied in a systematic way by Springer and Thompson [150] with particular reference to normal and Cauchy distributed variables. Among others, Lomnicki

[93] has extended their work and applied it to products of independent exponential, Weibull and gamma variables. Kotz and Srinivasan [84] have studied similar problems, in particular for variates having Bessel function distributions (and also quotients of such variables). The distribution of the ratio of two independent F-distributed variables has been studied by Schumann and Bradley [137] in connection with comparison of sensitivities of different experimental designs.

Broadbent [29], [30] has studied the distributions of various products and quotients of independent variables and in [30] has proposed lognormal approximations to certain of these distributions. An account of the historical development up to about 1962, and of engineering applications of distributions of products and quotients, is contained in Donahue [47].

The distributions of products of a number of independent χ^2's, and of independent beta-distributed variables occur in the sampling theory of certain statistics used in multivariate analysis and will be described at the appropriate places in the next volume.

REFERENCES

[1] Ahuja, J. C. and Nash, S. W. (1967). The generalized Gompertz-Verhulst family of distributions, *Sankhyà, Series A*, **29,** 141–156. (Also "Stochastic analogue of the modified exponential law of growth," reported at 1965 summer meeting of Biometric Society (WNAR), Riverside, California.)

[2] Anderson, T. W. and Darling, D. A. (1952). Asymptotic theory of certain goodness of fit criteria based on stochastic processes, *Annals of Mathematical Statistics*, **23,** 193–212.

[3] ApSimon, H. G. (1958). A repeated integral, *Mathematical Gazette*, **42,** 52.

[4] Barlow, R. E. (1968). Some recent developments in reliability theory, *Operations Research Center, University of California*, Berkeley, Report ORC 68–19.

[5] Barlow, R. E. and Marshall, A. W. (1964). Bounds for distributions with monotone hazard rate I, II, *Annals of Mathematical Statistics*, **35,** 1234–1257, 1258–1274.

[6] Barlow, R. E. and Marshall, A. W. (1965). Tables of bounds for distributions with monotone hazard rate, *Journal of the American Statistical Association*, **60,** 872–890.

[7] Barlow, R. E. and Marshall, A. W. (1967). Bounds on interval probabilities for restricted families of distributions, *Proceedings of the 5th Berkeley Symposium on Mathematical Statistics and Probability*, **3,** 229–257.

[8] Barlow, R. E., Marshall, A. W. and Proschan, F. (1963). Properties of probability distributions with monotone hazard rate, *Annals of Mathematical Statistics*, **34,** 375–389.

[9] Barlow, R. E. and Proschan, F. (1967). Exponential life test procedures when the distribution has monotone failure rate, *Journal of the American Statistical Association*, **62,** 548–560.

[10] Barlow, R. E. and Proschan, F. (1965). *Mathematical Theory of Reliability*, New York: John Wiley & Sons, Inc.

[11] Barlow, R. E. and Proschan, F. (1966). Tolerance and confidence limits for classes of distributions based on failure rate, *Annals of Mathematical Statistics*, **37,** 1593–1601.

[12] Barrow, D. F. and Cohen, A. C. (1954). On some functions involving Mills' ratio, *Annals of Mathematical Statistics*, **25,** 405–408.

[13] Barton, D. E., David, F. N. and Fix, E. (1963). Random points in a circle and the analysis of chromosome patterns, *Biometrika*, **50,** 23–29.

[14] Barton, D. E. and Mallows, C. L. (1965). Some aspects of the random sequence, *Annals of Mathematical Statistics*, **36,** 236–260.

[15] Bates, C. E. and Orsulak, Jacqueline R. (1968). *A Computer Program for the Kolmogorov Goodness of Fit Test for Normality*, Technical Memo, K-(2/68), U. S. Naval Weapons Laboratory.

[16] Bennett, G. (1968). A one-sided probability inequality for the sum of independent, bounded random variables, *Biometrika*, **55,** 565–569.

[17] Birnbaum, Z. W. (1942). An inequality for Mills' ratio, *Annals of Mathematical Statistics*, **13,** 245–246.

[18] Birnbaum, Z. W. (1950). Effect of linear truncation on a multinormal population, *Annals of Mathematical Statistics*, **21,** 272–279.

[19] Birnbaum, Z. W. (1952). Numerical tabulation of the distribution of Kolmogorov's statistic for finite sample size, *Journal of the American Statistical Association*, **47,** 425–441.

[20] Birnbaum, Z. W., Esary, J. D. and Marshall, A. W. (1966). A stochastic characteristic of wear-out for components and systems, *Annals of Mathematical Statistics*, **37,** 816–825.

[21] Birnbaum, Z. W. and Lientz, B. P. (1969). Exact distributions for some Rényi-type statistics, *Zastosowania Matematyki*, **10,** 179–192.

[22] Birnbaum, Z. W. and Pyke, R. (1958). On some distributions related to the statistic D_n^+, *Annals of Mathematical Statistics*, **29,** 179–187.

[23] Birnbaum, Z. W. and Saunders, S. C. (1969). A new family of life distributions, *Journal of Applied Probability*, **6,** 319–327.

[24] Birnbaum, Z. W. and Saunders, S. C. (1968). A probabilistic interpretation of Miner's rule, *SIAM Journal of Applied Mathematics*, **16,** 637–652.

[25] Birnbaum, Z. W. and Tingey, F. H. (1951). One-sided confidence contours for probability distribution functions, *Annals of Mathematical Statistics*, **22,** 592–596.

[26] Borovkov, A. A., Markova, N. P. and Siycheva, N. M. (1964). *Tables of N. V. Smirnov's Criteria for Homogeneity with Two Samples*, Akademija Nauk SSSR, Novosibirsk. (In Russian)

[27] Boyd, A. V. (1959). Inequalities for Mills' ratio, *Reports of Statistical Application Research, JUSE*, **6,** 44–46.

[28] Breitenberger, E. (1963). Analogues of the normal distributions on the circle and the sphere, *Biometrika*, **50,** 81–88.

[29] Broadbent, S. R. (1954). The quotient of a rectangular or triangular and a general variate, *Biometrika*, **41,** 330–337.

[30] Broadbent, S. R. (1956). Lognormal approximation to products and quotients, *Biometrika*, **43,** 404–417.

[31] Buckland, W. R. (1964). *Statistical Assessment of the Life Characteristic*, London: Griffin; New York: Hafner.

[32] Castoldi, L. (1963). A continuous analogon of Poisson's distributions, *Rendiconti Seminario della Facoltà di Scienze della Università di Cagliari*, **33,** 245–249.

[33] Chang, L. C. (1955). On the ratio of an empirical distribution to the theoretical distribution function, *Acta Mathematica Sinica*, **5,** 347–368. (English translation in *Selected Translations in Mathematical Statistics and Probability*, **4,** (1963), 17–38.)

[34] Chu, J. T. (1955). On bounds for the normal integral, *Biometrika*, **42,** 263–265.

[35] Chung, K. L. (1949). An estimate concerning the Kolmogorov limit distribution, *Transactions of the American Mathematical Society*, **67,** 36–50.

[36] Cox, D. R. (1966). Notes on the analysis of mixed frequency distributions, *British Journal of Mathematical and Statistical Psychology*, **19,** 39–47.

[37] Cramér, H. (1928). On the composition of elementary errors, I, II. *Skandinavisk Aktuarietidskrift*, **11,** 13–74, 141–180.

[38] Csörgö, M. (1965). Exact probability distribution functions of some Rényi type statistics, *Proceedings of the American Mathematical Society*, **16,** 1158–1167.

[39] Csörgö, M. (1965). Exact and limiting probability distributions of some Smirnov type statistics, *Canadian Mathematical Bulletin*, **8,** 93–103.

[40] Csörgö, M. (1965). Some Smirnov type theorems of probability theory, *Annals of Mathematical Statistics*, **36,** 1113–1119.

[41] Csörgö, M. (1967). A new proof of some results of Rényi, *Canadian Journal of Mathematics*, **19,** 550–558.

[42] Daniels, H. E. (1945). The statistical theory of the strength of bundles of threads, *Proceedings of the Royal Society of London, Series A*, **183,** 405–435.

[43] Darling, D. A. (1955). The Cramér-Smirnov test in the parametric case, *Annals of Mathematical Statistics*, **26,** 1–20.

[44] Darling, D. A. (1957). The Kolmogorov-Smirnov, Cramér-von Mises tests, *Annals of Mathematical Statistics*, **28,** 823–838.

[45] David, F. N. and Johnson, N. L. (1950). The probability integral transformation when the variable is discontinuous, *Biometrika*, **37,** 42–49.

[46] Deltheil, R. (1926). *Probabilités Géométriques* (Vol. II, Fasc. II), Paris: Gauthier Villars.

[47] Donahue, J. D. (1964). *Products and quotients of random variables and their applications*, Report ARL 64-115, Office of Aerospace Research, U.S.A.F.

[48] Donsker, M. D. (1952). Justification and extension of Doob's heuristic approach to Kolmogorov-Smirnov theorems, *Annals of Mathematical Statistics*, **23,** 277–281.

[49] Doob, J. L. (1949). Heuristic approach to Kolmogorov-Smirnov theorems, *Annals of Mathematical Statistics*, **20,** 393–403.

[50] Dugué, D. (1941). Sur un nouveau type de courbe de frèquence, *Comptes Rendus de l'Académie des Sciences, Paris*, **213,** 634–635.

[51] Fairthorne, D. (1964). The distances between random points in two concentric circles, *Biometrika*, **51,** 275–277.

[52] Feller, W. (1948). On the Kolmogorov-Smirnov limit theorems for empirical distributions, *Annals of Mathematical Statistics*, **19,** 177–189.

[53] Filadelfina, N. A. (1967). Experimental determination of parameters of a generalized probability distribution for periods of satisfactory production, *Sbornik Trudov Mekhanicheskogo Instituta, Leningrad*, **62,** 12–19. (In Russian)

[54] Fisher, R. A. (1929). Tests of significance in harmonic analysis, *Proceedings of the Royal Society of London, Series A*, **125,** 54–59.

[55] Fisher, R. A. (1940). On the similarity of the distributions found for the tests of significance in harmonic analysis and in Stevens' problems in geometrical probability, *Annals of Eugenics, London*, **10,** 14–17.

[56] Fisher, R. A. (1953). Dispersion on a sphere, *Proceeding of the Royal Society of London, Series A*, **217,** 295–305.

[57] Flehinger, B. J. and Lewis, P. A. (1959). Two-parameter lifetime distributions for reliability studies of renewal processes, *IBM Journal of Research*, **3,** 58–73.

[58] Floriani, W. de (1960). Distribuzione circolare normale, *Bolletino della Centro per la Ricerca Operativa, Universita Milano*, **4**, (3), 30–35.

[59] Fraser, D. A. S. (1968). *The Structure of Inference*, New York: John Wiley & Sons, Inc.

[60] Fréchet, M. (1950). *Generalités sur les Probabilités. Elements Aléatoires*, Paris: Gauthier Villars, (2nd edition).

[61] Gečiauskas, E. (1966). Distribution function of a distance between two random points in an oval, *Lietuvos Matematikos Rinkinys*, **6**, 245–248 (In Russian).

[62] Gečiauskas, E. (1967). Distribution of a distance in an ovaloid, *Lietuvos Matematikos Rinkinys*, **7**, 33–36. (In Russian)

[63] Geisser, S. (1963). Multivariate analysis of variance for a special covariance case, *Journal of the American Statistical Association*, **58**, 660–669.

[64] Gercbah, I. B. and Kordonski, H. B. (1966). *Reliability Methods*, Moscow: Soviet Radio. (In Russian)

[65] Godwin, H. J. (1964). *Inequalities on Distribution Functions*, London: Griffin; New York: Hafner.

[66] Gompertz, B. (1825). On the nature of the function expressive of the law of human mortality, and on a new mode of determining the value of life contingencies, *Philosophical Transactions of the Royal Society of London*, **115**, 513–585.

[67] Gordon, R. D. (1941). Values of Mills' ratio of area to bounding ordinate and of the normal probability integral for large values of the argument, *Annals of Mathematical Statistics*, **12**, 364–366.

[68] Gray, H. L. and Odell, P. L. (1966). On sums and products of rectangular variates, *Biometrika*, **53**, 615–617

[69] Greenwood, J. A. and Durand, D. (1955). The distribution of length and components of the sum of n random unit vectors, *Annals of Mathematical Statistics*, **26**, 233–246.

[70] Gumbel, E. J. (1954). Applications of the circular normal distribution, *Journal of the American Statistical Association*, **49**, 267–297.

[71] Gumbel, E. J., Greenwood, J. A. and Durand, D. (1953). The circular normal distribution: Theory and tables, *Journal of the American Statistical Association*, **48**, 131–152.

[72] Hadwiger, H. (1940). Eine analytische Reproduktionsfunktion für biologische Gesamtheiten, *Skandinavisk Aktuarietidskrift*, **23**, 101–113.

[73] Haight, F. A. (1964). Some probability distributions associated with commuter travel in a homogeneous circular city, *Operations Research*, **12**, 964–975.

[74] Haldane, J. B. S. (1961). Simple approximations to the probability integral and $P(\chi^2,1)$ when both are small, *Sankhyā, Series A*, **23**, 9–10.

[75] Hoeffding, W. (1963). Probability inequalities for sums of bounded random variables, *Journal of the American Statistical Association*, **58**, 13–30.

[76] Holgate, P. (1965). The distance from a random point to the nearest point of a closely packed lattice, *Biometrika*, **52**, 261–263.

[77] Ishii, G. (1959). On the exact probabilities of Rényi's tests, *Annals of the Institute of Statistical Mathematics, Tokyo*, **11**, 17–24.

[78] Johnson, N. L. (1966). Paths and chains of random straight-line segments, *Technometrics*, **8,** 303–317.

[79] Johnson, N. L. and Rogers, C. A. (1950). The moment problem for unimodal distributions, *Annals of Mathematical Statistics*, **22,** 433–439.

[80] Kac, M. (1949). On deviations between theoretical and empirical distributions, *Proceedings of the National Academy of Science, Washington*, **35,** 252–257.

[81] Kendall, M. G. and Moran, P. A. P. (1963). *Geometrical Probability*, London: Griffin.

[82] Kolmogorov, A. N. (1933). Sulla determinazione empirica di une legge di distribuzione, *Giornale dell'Istituto Italiano degli Attuari*, **4,** 83–91.

[83] Komatu, Y. (1955). Elementary inequalities for Mills' ratio, *Reports of Statistical Application Research, JUSE*, **4,** 69–70.

[84] Kotz, S. and Srinivasan, R. (1969). Distribution of product and quotient of Bessel function variates, *Annals of the Institute of Statistical Mathematics, Tokyo*, **21,** 201–210.

[85] Kuhn, W. and Grün, F. (1942). Beziehungen zwischen elastischen Konstante und Dehnungsdoppelbrechung hochelastischer Stoffe, *Kolloid-Zeitschrift*, **101,** 248–271.

[86] Kuiper, N. H. (1960). Test concerning random points on a circle, *Koninklijke Nederlandse Akademie van Wetenschappen, Amsterdam, Series A*, **63,** 38–47.

[87] Langevin, P. (1905). Magnétisme et théorie des électrons, *Annales de Chémie et de Physique*, **5,** 70–127.

[88] Laplace, P. S. (1812). *Théorie Analytique des Probabilités*, Paris: Gauthier Villars.

[89] Lauwerier, H. A. (1963). The asymptotic expansion of the statistical distribution of N. V. Smirnov, *Zeitschrift für Wahrscheinlichkeitstheorie und Verwandte Gebiete*, **2,** 61–68.

[90] Lientz, B. P. (1968). *Distribution of Rényi and Kac type statistics; Power of corresponding tests under Suzuki-type alternatives*, Technical Report No. 51, Department of Mathematics, University of Washington, Seattle, Washington.

[91] Lilliefors, H. W. (1967). On the Kolmogorov-Smirnov test for normality with mean and variance unknown, *Journal of the American Statistical Association*, **62,** 399–402.

[92] Linnik, Yu. V. (1957). On the decomposition of the convolution of Gaussian and Poissonian laws, *Theory of Probability and its Applications*, **2,** 31–57.

[93] Lomnicki, Z. A. (1967). On the distribution of products of random variables, *Journal of the Royal Statistical Society, Series B*, **29,** 513–524.

[94] Maag, U. R. and Stephens, M. A. (1968). The V_{NM} two-sample test, *Annals of Mathematical Statistics*, **39,** 923–935.

[95] Mallows, C. L. (1956). Generalizations of Tchebycheff's inequalities, *Journal of the Royal Statistical Society, Series B*, **18,** 139–176.

[96] Mallows, C. L. (1963). A generalization of the Chebyshev inequalities, *Proceedings of the London Mathematical Society, 3rd Series*, **13,** 385–412.

[97] Maniya, G. M. (1949). Generalization of the criterion of A. N. Kolmogorov, *Doklady Akademii Nauk SSSR*, **49,** 485–497. (In Russian)

[98] Marshall, A. W. (1958). The small sample distribution of $n\omega_n^2$, *Annals of Mathematical Statistics*, **29**, 307–309.

[99] Marshall, A. W. and Proschan, F. (1965). Maximum likelihood estimation for distributions with monotone failure rate, *Annals of Mathematical Statistics*, **36,** 69–77.

[100] Massey, F. J. (1950). A note on the estimation of a distribution function by confidence limits, *Annals of Mathematical Statistics*, **21,** 116–119.

[101] Massey, F. J. (1951). The Kolmogorov-Smirnov test for goodness of fit, *Journal of the American Statistical Association*, **46,** 68–78.

[102] Mauldon, J. G. (1951). Random division of an interval, *Proceedings of the Cambridge Philosophical Society*, **47,** 331–336.

[103] McLane, P. J. (1965). *A formulation of the modified normal distribution in terms of tabulated functions*, National Research Council of Canada, Report AP-PR-31, NCR 8691.

[104] McLane, P. J. (1967). Mathematical considerations of the modified normal distribution, *Bulletin Géodésique*, **83,** 9–20.

[105] Miller, L. H. (1956). Tables of percentage points of Kolmogorov statistics, *Journal of the American Statistical Association*, **51,** 111–121.

[106] Miller, R. L. and Kahn, J. S. (1962). *Statistical Analysis in the Geological Sciences*, New York: John Wiley & Sons, Inc.

[107] Mills, J. P. (1926). Table of the ratio: Area to bounding ordinate for any portion of normal curve, *Biometrika*, **18,** 395–400.

[108] Mises, R. von (1918). Über die "Ganzzahligkeit" der Atomgewichte und verwandte Fragen, *Physikalische Zeitschrift*, **19,** 490–500.

[109] Murty, V. N. (1952). On a result of Birnbaum regarding the skewness of X in a bivariate normal population, *Journal of the Indian Society of Agricultural Statistics*, **4,** 85–87.

[110] Noé, M. and Vandewiele, G. (1968). The calculation of distributions of Kolmogorov-Smirnov type statistics including a table of significance points for a particular case, *Annals of Mathematical Statistics*, **39**, 233–241.

[111] Owen, D. B. and Monk, D. T. (1957). *Tables of the Normal Probability Integral*, Sandia Corporation Technical Memorandum, 64–57–51.

[112] Pearson, E. S. ("Editorial") (1955). The normal probability function: Tables of certain area-ordinate ratios and of their reciprocals, *Biometrika*, **42,** 217–222.

[113] Pearson, E. S. (1963). Comparison of tests for randomness of points on a line, *Biometrika*, **50,** 315–325.

[114] Pearson, E. S. and Stephens, M. A. (1962). The goodness of fit tests based on W_N^2 and U_N^2, *Biometrika*, **49,** 397–402.

[115] Pearson, K. (1930). *Tables for Statisticians and Biometricians*, **2,** London: Cambridge University Press.

[116] Pollak, H. (1957). A remark on 'Elementary inequalities for Mills' ratio' by Y. Komatu, *Reports of Statistical Application Research, JUSE*, **4,** 110.

[117] Pólya, G. (1919). Zur Statistik der sphärischen Verteilung der Fixsterne, *Astonomische Nachrichten*, **208,** 175–180.

[118] Pólya, G. (1930). Sur quelques points de la théorie des probabilités, *Annales de l'Institut Henri Poincaré*, **1,** 117–161.

[119] Pólya, G. (1935). Zwei Aufgaben aus der Wahrscheinlichkeitsrechnung, *Vierteljahresschrift der Naturforschenden Gesellschaft in Zurich*, **80,** 123–130.

[120] Raj, D. (1953). On Mills' ratio for the Type III population, *Annals of Mathematical Statistics*, **24,** 309–312.

[121] Ray, W. D. and Pitman, A. E. N. T. (1963). Chebyshev polynomial and other new approximations to Mills' ratio, *Annals of Mathematical Statistics*, **34,** 892–902.

[122] Rényi, A. (1953). On the theory of order statistics, *Acta Mathematica, Hungarian Academy of Science*, **4,** 191–231.

[123] Rényi, A. (1963). On the distribution function $L(z)$, *Selected Translations in Mathematical Statistics and Probability*, **4,** 219–225, Providence, R. I.: American Mathematical Society.

[124] Romanowski, M. (1964). On the normal law of error, *Bulletin Géodésique*, **73,** 195–216.

[125] Romanowski, M. (1969). Generalized theory of the modified normal distributions, *Metrologia*, **4,** 2, 84–86.

[126] Romanowski, M. (1969). *Normal and modulated normal distributions*, Report No. APH-1597, National Research Council of Canada.

[127] Romanowski, M. and Green, E. (1965). Tabulation of the modified normal distributions, *Bulletin Géodésique*, **78,** 369–377.

[128] Romanowski, M. and Green, E. (1965). Practical applications of the modified normal distributions, *Bulletin Géodésique*, **76,** 1–20.

[129] Royden, H. L. (1953). Bounds on a distribution function when its first n moments are given, *Annals of Mathematical Statistics*, **24,** 361–376.

[130] Ruben, H. (1962). A new asymptotic expansion for the normal probability integral and Mills' ratio, *Journal of the Royal Statistical Society, Series B*, **24,** 177–179.

[131] Ruben, H. (1963). A convergent asymptotic expansion for Mills' ratio and the normal probability integral in terms of rational functions, *Mathematische Annalen*, **151,** 355–364.

[132] Ruben, H. (1964). Irrational fraction approximations to Mills' ratio, *Biometrika*, **51,** 339–345.

[133] Saaty, T. L. (1961). Some stochastic processes with absorbing barriers, *Journal of the Royal Statistical Society, Series B*, **23,** 319–334.

[134] Sahler, W. (1968). A survey on distribution-free statistics based on distances between distribution functions, *Metrika*, **13,** 149–169.

[135] Sampford, M. R. (1953). Some inequalities on Mills' ratio and related functions, *Annals of Mathematical Statistics*, **24,** 130–132.

[136] Saunders, S. C. and Birnbaum, Z. W. (1969). Estimation for a family of life distributions with applications to fatigue, *Journal of Applied Probability*, **6,** 328–347.

[137] Schumann, D. E. W. and Bradley, R. A. (1959). The comparisons of the sensitivities of similar experiments: Model II of the analysis of variance, *Biometrics*, **15,** 405–416.

[138] Sclove, S. C. and Van Ryzin, J. (1969). Estimating the parameters of a convolution, *Journal of the Royal Statistical Society, Series B*, **31,** 181–191.

[139] Selberg, H. L. (1940). Über eine Ungleichung der mathematischen Statistik, *Skandinavisk Aktuarietidskrift*, **23,** 114–120.

[140] Selberg, H. L. (1940). Zwei Ungleichungen zur Ergänzung des Tchebycheffschen Lemmas, *Skandinavisk Akuarietidskrift*, **23,** 121–125.

[141] Shenton, L. R. (1954). Inequalities for the normal integral including a new continued fraction, *Biometrika*, **41,** 177–189.

[142] Shenton, L. R. and Carpenter, J. A. (1964). The Mills' ratio and the probability integral for a Pearson Type IV distribution, *Biometrika*, **52,** 119–126.

[143] Sheppard, W. F. (1939). *The Probability Integral*, British Association Mathematical Tables, **7,** London: Cambridge University Press.

[144] Sherman, B. (1950). A random variable related to the spacing of sample values, *Annals of Mathematical Statistics*, **21,** 339–361.

[145] Shortley, G. (1965). A stochastic model for distributions of biological response times, *Biometrics*, **21,** 562–582.

[146] Smirnov, N. V. (1936). Sur la distribution de ω^2, *Comptes Rendus de l'Académie des Sciences, Paris*, **202,** 449–452.

[147] Smirnov, N. V. (1939). On the deviations of the empirical distribution curve, *Matematicheskii Sbornik* (*N.S.*), **6(48),** 3–26. (In Russian; French Summary)

[148] Smirnov, N. V. (1948). Table for estimating the goodness of fit of empirical distributions, *Annals of Mathematical Statistics*, **19,** 279–281.

[149] Smirnov, N. V. (1949). On the Cramer-von Mises criterion, *Uspekhi Matematicheskikh Nauk* (N.S.), **4,** 196–197. (In Russian)

[150] Springer, M. D. and Thompson, W. E. (1966). The distribution of products of independent random variables, *SIAM Journal of Applied Mathematics*, **14,** 511–526.

[151] Steck, G. P. (1969). The Smirnov two-sample tests as rank tests, *Annals of Mathematical Statistics*, **40,** 1449–1466.

[152] Stegun, Irene A. (1964). Miscellaneous functions. (In *Handbook of Mathematical Functions* (Ed. M. Abramowitz and I. A. Stegun), National Bureau of Standards, Applied Mathematics, Series, **55,** Washington, D. C.: U. S. Government Printing Office.

[153] Stephens, M. A. (1963). Random walk on a circle, *Biometrika*, **50,** 385–390.

[154] Stephens, M. A. (1963, 1964). The distribution of the goodness-of-fit statistic U_N^2, I, II, *Biometrika*, **50,** 303–313 and **51,** 393–397.

[155] Stevens, W. L. (1939). Solution to a geometrical problem in probability, *Annals of Eugenics*, London, **9,** 315–320.

[156] Tang, S. C. (1962). Some theorems on the ratio of empirical distribution to the theoretical distribution, *Pacific Journal of Mathematics*, **12,** 1107–1114.

[157] Verhulst, P. F. (1845). Recherches mathématiques sur la loi d'accroissement de la population, *Nouvelles Mémoires Académie Royale, Science et Lettres, Bruxelles, Séries* 2, **18,** 1–38.

[158] Vistelius, A. B. (1966). *Structural Diagrams*, Pergamon Press, Oxford (Original Russian edition published by Izvestiya Akademii Nauk SSSR, Moscow).

[159] Wald, A. (1938). Generalization of the inequality of Markoff, *Annals of Mathematical Statistics*, **9,** 244–255.

[160] Wald, A. and Wolfowitz, J. (1939). Confidence limits for continuous distributions functions, *Annals of Mathematical Statistics*, **10,** 105–118.

[161] Wald, A. and Wolfowitz, J. (1946). Tolerance limits for a normal distribution, *Annals of Mathematical Statistics*, **17,** 208–215.

[162] Watson, G. N. (1959). A quadruple integral, *Mathematical Gazette*, **43,** 280–283.

[163] Watson, G. S. (1961). Goodness-of-fit tests on a circle, *Biometrika*, **48,** 109–114.

[164] Watson, G. S. and Williams, E. J. (1956). On the construction of significance tests on the circle and the sphere, *Biometrika*, **43,** 344–352.

[165] Wintner, A. (1947). On the shape of the angular case of Cauchy's distribution curves, *Annals of Mathematical Statistics*, **18,** 589–593.

[166] Wishart, J. (1956). χ^2 probabilities for large numbers of degrees of freedom, *Biometrika*, **43,** 92–95.

Acknowledgements

Figures

P. 28 Adapted from Sarhan, A. E. (1954). Estimation of the mean and standard deviation by order statistics, Part I, *Annals of Mathematical Statistics*, **25,** 317–328.

P. 70 & 71 Adapted from Allan, R. R. (1966). Extension of the binomial model of traffic flow to the continuous case, *Proceedings of the Third Conference of the Australian Road Research Board*, **3,** 276–316.

P. 77 Adapted from Hald, A. (1952). *Statistical Tables and Formulas*, New York: John Wiley & Sons, Inc.

P. 85 Adapted from Stammberger, A. (1967). Über einige Nomogramme zür Statistik, *Wissenschaftliche Zeitschrift der Humboldt-Universität Berlin, Mathematisch-Naturwissenschaftliche Reihe*, **16,** 86–93.

P. 101 Adapted from James-Levy, G. E. (1956). A nomogram for the integral law of Student's distribution, *Teoriya Veroyatnostei i ee Primeneniya*, **1,** 271–274. (In Russian) (pp. 246–248 in English translation.)

P. 102 Adapted from Stammberger, A. (1967). Über einige Nomogramme zür Statistik, *Wissenschaftliche Zeitschrift der Humboldt-Universität Berlin, Mathematisch-Naturwissenschaftliche Reihe*, **16,** 86–93.

P. 177 Adapted from Hultquist, R. (1966). Diagramming theorems relative to quadratic forms, *American Statistician*, **20,** (5), 31.

Tables

P. 12 Adapted from Gupta, S. S. and Gnanadesikan, M. N. (1966). *Journal of the American Statistical Association*, **61,** 329–337.

P. 27 Adapted from Govindarajulu, Z. (1966). Best linear estimates under symmetric censoring of the parameters of a double exponential population, *Journal of the American Statistical Association*, **61,** 248–258.

P. 29 Adapted from Sarhan, A. E. (1954). Estimation of the mean and standard deviation by order statistics, Part I, *Annals of Mathematical Statistics*, **26,** 576–592.

P. 61 Adapted from Johnson, N. L. (1950). On the comparison of estimators, *Biometrika*, **37,** 281–287.

P. 103 Adapted from Dickey, J. M. (1967). Expansions of t densities and related complete integrals, *Annals of Mathematical Statistics*, **38,** 503–510.

P. 104 Adapted from Hendricks, W. A. (1936). An Approximation to "Student's" distribution, *Annals of Mathematical Statistics*, **7,** 210–221.

P. 105 Adapted from Hotelling, H. and Frankel, L. R. (1938). The transformation of statistics to simplify their distribution, *Annals of Mathematical Statistics*, **9,** 87–96.

P. 106 Adapted from Peiser, A. M. (1943). Asymptotic formulas for significance levels of certain distributions, *Annals of Mathematical Statistics*, **14,** 56–62.

P. 107 Adapted from Goldberg, H. and Levine, H. (1946). Approximate formulas for the percentage points and normalization of t and χ^2, *Annals of Mathematical Statistics*, **17,** 216–225.

P. 109 Adapted from Wallace, D. L. (1959). Bounds on normal approximations to Student's and the Chi-squared distributions, *Annals of Mathematical Statistics*, **30,** 1121–1130.

P. 110 Adapted from Pardinor, D. A. and Bombay, B. F. (1965). *Technometrics*, **1,** 71–72.

P. 111 Adapted from Moran, P. A. P. (1966). Accurate approximations for t tests. (*In Research Papers in Statistics* (*Festschrift for J. Neyman*)(Ed. F. N. David), pp. 225–230.)

P. 113 Adapted from Pinkham, R. S. and Wilk, M. B. (1951). *Annals of Mathematical Statistics*, **22,** 469–472.

P. 122 Adapted from Gayen, A. K. (1949). The distribution of Student's t in random samples of any size drawn from non-normal universes, *Biometrika*, **36,** 353–369.

P. 142 Adapted from Sankaran, M. (1963). Approximations to the non-central chi-square distribution, *Biometrika*, **50,** 199–204; Fisher, R. A. (1928). The general sampling distribution of the multiple correlation coefficient, *Proceedings of the Royal Society of London*, **121A,** 654–673; Abdel-Aty, S. H. (1954). Approximate formulae for the percentage points and the probability integral of the non-central χ^2-distribution, *Biometrika*, **41,** 538–540.

P. 196 Adapted from Barton, D. E., David, F. N. and O'Neill, A. F. (1960). Some properties of the distribution of the logarithm of non-central F, *Biometrika*, **47,** 417–419.

P. 237 Adapted from Shenton, L. R. and Johnson, W. L. (1965). Moments of a serial correlation coefficient, *Journal of the Royal Statistical Society*, *Series B*, **27,** 308–320.

P. 255 Adapted from Miller, L. H. (1956). Tables of percentage points of Kolmogorov statistics, *Journal of the American Statistical* Association, **51,** 111–121.

P. 280 Adapted from Boyd, A. V. (1959). Inequalities for Mill's ratio, *Reports of Statistical Application Research*, JUSE, **6,** 44–46.

Index

This index is intended as an auxiliary to the detailed table of contents. Most references are to distributions (the word "distribution" being omitted), but there are a few other references. Distributions are only included in this index when they occur in places not easily identifiable from the table of contents.

Applied Probability and Statistics (Continued)

HAHN and SHAPIRO · Statistical Models in Engineering
HALD · Statistical Tables and Formulas
HALD · Statistical Theory with Engineering Applications
HARTIGAN · Clustering Algorithms
HILDEBRAND, LAING and ROSENTHAL · Prediction Analysis of Cross Classifications
HOEL · Elementary Statistics, *Fourth Edition*
HOLLANDER and WOLFE · Nonparametric Statistical Methods
HUANG · Regression and Econometric Methods
JAGERS · Branching Processes with Biological Applications
JESSEN · Statistical Survey Techniques
JOHNSON and KOTZ · Distributions in Statistics
Discrete Distributions
Continuous Univariate Distributions-1
Continuous Univariate Distributions-2
Continuous Multivariate Distributions
JOHNSON and KOTZ · Urn Models and Their Application: An Approach to Modern Discrete Probability Theory
JOHNSON and LEONE · Statistics and Experimental Design: In Engineering and the Physical Sciences, Volumes I and II, *Second Edition*
KEENEY and RAIFFA · Decisions with Multiple Objectives
LANCASTER · An Introduction to Medical Statistics
LEAMER · Specification Searches: Ad Hoc Inference with Non-experimental Data
LEWIS · Stochastic Point Processes
McNEIL · Interactive Data Analysis
MANN, SCHAFER and SINGPURWALLA · Methods for Statistical Analysis of Reliability and Life Data
MEYER · Data Analysis for Scientists and Engineers
OTNES and ENOCHSON · Digital Time Series Analysis
PRENTER · Splines and Variational Methods
RAO and MITRA · Generalized Inverse of Matrices and Its Applications
SARD and WEINTRAUB · A Book of Splines
SEARLE · Linear Models
THOMAS · An Introduction to Applied Probability and Random Processes
WHITTLE · Optimization under Constraints
WILLIAMS · A Sampler on Sampling
WONNACOTT and WONNACOTT · Econometrics
WONNACOTT and WONNACOTT · Introductory Statistics, *Third Edition*
WONNACOTT and WONNACOTT · Introductory Statistics for Business and Economics, *Second Edition*
YOUDEN · Statistical Methods for Chemists
ZELLNER · An Introduction to Bayesian Inference in Econometrics

Tracts on Probability and Statistics

BHATTACHARYA and RAO · Normal Approximation and Asymptotic Expansions
BILLINGSLEY · Convergence of Probability Measures
JARDINE and SIBSON · Mathematical Taxonomy
RIORDAN · Combinatorial Identities

Applied Probability and Statistics

BAILEY · The Elements of Stochastic Processes with Applications to the Natural Sciences

BAILEY · Mathematics, Statistics and Systems for Health

BARTHOLOMEW · Stochastic Models for Social Processes, *Second Edition*

BECK and ARNOLD · Parameter Estimation in Engineering and Science

BENNETT and FRANKLIN · Statistical Analysis in Chemistry and the Chemical Industry

BHAT · Elements of Applied Stochastic Processes

BLOOMFIELD · Fourier Analysis of Time Series: An Introduction

BOX · The Life of a Scientist

BOX and DRAPER · Evolutionary Operation: A Statistical Method for Process Improvement

BOX, HUNTER, and HUNTER · Statistics for Experimenters: An Introduction to Design, Data Analysis and Model Building

BROWN and HOLLANDER · Statistics: A Biomedical Introduction

BROWNLEE · Statistical Theory and Methodology in Science and Engineering, *Second Edition*

BURY · Statistical Models in Applied Science

CHAMBERS · Computational Methods for Data Analysis

CHATTERJEE and PRICE · Regression Analysis by Example

CHERNOFF and MOSES · Elementary Decision Theory

CHOW · Analysis and Control of Dynamic Economic Systems

CLELLAND, deCANI, BROWN, BURSK, and MURRAY · Basic Statistics with Business Applications, *Second Edition*

COCHRAN · Sampling Techniques, *Third Edition*

COCHRAN and COX · Experimental Designs, *Second Edition*

COX · Planning of Experiments

COX and MILLER · The Theory of Stochastic Processes, *Second Edition*

DANIEL · Application of Statistics to Industrial Experimentation

DANIEL and WOOD · Fitting Equations to Data

DAVID · Order Statistics

DEMING · Sample Design in Business Research

DODGE and ROMIG · Sampling Inspection Tables. *Second Edition*

DRAPER and SMITH · Applied Regression Analysis

DUNN · Basic Statistics: A Primer for the Biomedical Sciences, Second Edition

DUNN and CLARK · Applied Statistics: Analysis of Variance and Regression

ELANDT-JOHNSON · Probability Models and Statistical Methods in Genetics

FLEISS · Statistical Methods for Rates and Proportions

GIBBONS, OLKIN and SOBEL · Selecting and Ordering Populations: A New Statistical Methodology

GNANADESIKAN · Methods for Statistical Data Analysis of Multivariate Observations

GOLDBERGER · Econometric Theory

GROSS and CLARK · Survival Distributions: Reliability Applications in the Biomedical Sciences

GROSS and HARRIS · Fundamentals of Queueing Theory

GUTTMAN, WILKS and HUNTER · Introductory Engineering Statistics, *Second Edition*

continued on back

NORMAN L. JOHNSON

University of North Carolina, Chapel Hill

SAMUEL KOTZ

Temple University, Philadelphia

continuous univariate